D0945341

PHYSICAL TESTING OF RUBBER

Second Edition

PHYSICAL TESTING OF RUBBER

Second Edition

R. P. BROWN

Technical Manager, Rapra Technology Ltd, Shawbury, Shrewsbury, UK

ELSEVIER APPLIED SCIENCE PUBLISHERS
LONDON and NEW YORK

ELSEVIER APPLIED SCIENCE PUBLISHERS LTD
Crown House, Linton Road, Barking, Essex IG11 8JU, England

Sole Distributor in the USA and Canada
ELSEVIER SCIENCE PUBLISHING CO., INC
52 Vanderbilt Avenue, New York, NY 10017, USA

WITH 76 ILLUSTRATIONS

British Library Cataloguing in Publication Data

Brown, R. P.
Physical testing of rubber.—2nd ed.
1. Rubber—Testing
I. Title
678′.4 TS1892

Library of Congress Cataloging-in-Publication Data

Brown, Roger (Roger P.)
Physical testing of rubber.

Rev. ed. of: Physical testing of rubbers. c1979.
Bibliography: p.
Includes index.
1. Rubber—Testing. I. Brown, Roger (Roger P.).
Physical testing of rubbers. II. Title.
TS1892.B7 1987 678′.4 86-16196

ISBN 1-85166-047-X

Printed in Great Britain by Galliard (Printers) Ltd, Great Yarmouth

Preface

A text devoted to the physical testing of rubber written by Rapra staff first appeared in 1965 with the publication of the work of the late Dr. J. R. Scott. The first edition of my own book came in 1979 and this second edition reflects further developments in the subject. There have been many changes in the methods used and especially in the instrumentation, but, throughout, the aim of this work has been to present a comprehensive up-to-date account of the procedures used by suppliers and users to characterise and investigate the physical properties of rubber materials.

The particular nature of rubber demands that specific test procedures are used and the many national and international standard test methods which have been produced bear witness to the importance of the subject to industry. The present work collates these many standard methods, comments on their virtues and defects and considers procedures needed for both quality control and for the generation of design data. The content owes much to the experience gained due to Rapra's position over many decades as an international centre for rubber research, as a test house with a history of developing test procedures and making a significant contribution to national and international standardisation.

The book is primarily intended as a reference for those directly concerned with testing rubbers for quality control, research or specification and for students of rubber technology but it is hoped that it will also be of value to those indirectly involved with the evaluation of rubbers, such as design engineers.

R. P. Brown

Contents

Acknowledgements

The author wishes to record his debt to the late Dr J. R. Scott for his work which was the inspiration for, and the basis of, this book. He is also most grateful to Rapra Technology for their support of the work and for their permission to use material, particularly many of the illustrations, generated at Shawbury. Thanks are due to many colleagues at Rapra and friends in industry for their help and supply of information, especially Dr M. M. Hall for his support of the work and Mr R. H. Norman for much helpful advice and criticism. A most important thank you is to Mrs Pat Aldhous for typing the second edition. The author's family have suffered somewhat from the many evenings which he has spent locked in his study producing the book and he is very appreciative of their tolerance and encouragement.

Figures 8.1, 13.2, 15.2, 15.3, 17.1 and 17.2 are taken from British Standards publications by permission of the British Standards Institution; Figs 4.1, 7.2, 8.2, 8.11, 12.2 and 15.5 were supplied by H. W. Wallace and Co. Ltd; Fig. 7.1 by Daventest Ltd; and Fig. 16.2 by Hampden Test Equipment Ltd.

Chapter 1

Introduction

Rubbers are complex materials; in general they exhibit a unique combination of physical properties whilst at the same time a virtually infinite number of vulcanised rubber compounds is possible, yielding a very wide range of properties. These three facts provide very briefly the main reasons why the physical properties of rubbers are of especially great interest to designers, processors and users.

Rubber differs very considerably from other engineering materials; for example it is the most highly deformable material, exhibiting virtually complete recovery, and it is virtually incompressible with a bulk modulus some thousand times greater than its shear or Young's modulus. For the design engineer particularly, it is important that such properties are measured and understood. The fact that so many variations in properties are possible simply means that standard grades hardly exist and one must evaluate every rubber compound which is met with. The basic structure of rubbers and their sensitivity to small compounding or processing changes means that they are prone to unintended variations in properties from batch to batch and present the processor with a difficult quality control problem.

Not surprisingly with such unusual and complicated materials, the procedures used for measuring their physical properties often differ markedly from procedures used with other materials, so that there is a particular technology of rubber testing. Over the years an enormous effort has been put into developing satisfactory procedures both for quality control use and for providing design data, but particularly from the design aspect many procedures are still painfully inadequate.

The difficulty of formulating meaningful test procedures for rubbers is due to a number of reasons, some of which are general to testing materials, but most because of the rubber's intrinsic properties. For example, the tensile stress/strain curve for a rubber is not linear up to a yield point, so that the slope does not give a single figure for modulus, and other measures of stiffness are used, notably the stress at a given strain, over a range up to as high as 1000%. Generally the results obtained in laboratory tests are much more dependent on the test conditions than is the case for tests on metals. For example, the force required to reach a particular elongation could vary by a factor of three when the speed of testing is varied from slow to fast; while the strength at 100°C could be as little as 20% of the strength at 23°C. Another peculiarity is that the modulus of a rubber at say 300% elongation will increase as temperature is increased, for a range of temperatures well above the material's glass transition. Furthermore, previous stressing history is important since test results could be substantially different if previous cyclic stressing had taken place.

These examples are not of course exhaustive but serve to illustrate the fact that this unusual class of materials requires its own carefully worked out testing procedures and that methods and philosophies taken from other materials cannot be simply transferred to rubber if meaningful results are to be obtained.

The aim of this book is to present an up to date account of rubber testing procedures. It intends to be comprehensive in covering all the tests in common, and sometimes not so common, use. Inevitably the bulk of methods are the standard ones, often somewhat arbitrary and primarily intended for quality assurance purposes, but in each case the requirements for obtaining meaningful design data are discussed.

Standard test methods have the unfortunate habit of not being standard, in that different countries and different organisations each have their own 'standards'. Fortunately this tendency has very much diminished in recent times and it is perhaps appropriate here to make a plea for the adoption of recognised standards without modification where there is really no strong technical reason for change. It really goes without saying that this makes for efficiency because if we all use the same, well-documented method, silly disputes due to the effects of apparently minor differences will be lessened. The principal standard methods discussed in this book are those of the International Standards Organisation (ISO) and the British Standards Institution (BSI). Other national methods are referred to, most notably the American Society for Testing and Materials (ASTM), but less emphasis is placed on the various national bodies than was the case in earlier works, reflecting the increased importance of ISO or rather the increased tendency for national methods to be aligned with ISO. Since the 1st Edition there has been a further increase, particularly in British standards, of dual numbering so that the national standard is verbatim the same as ISO. It is a great pity that more national standards bodies do not follow this practice. A survey of the use of various national standards in Europe has been undertaken by ASTM.[1,2] Generally, test methods peculiar to particular commercial companies have not been considered at all.

It is inevitable that between writing and publication there will have been new editions of standards produced. To counteract this as far as possible the likely trends in test methods have been estimated from the current draft proposals in circulation and from the known activities of the relevant committees.

The layout of subject matter in a book on testing is inevitably to some extent subjective, as is the dividing line between what to include and what to omit. Firstly, there is bound to be some overlap between rubbers and plastics and indeed in my own laboratory no real distinction is made, both groups of polymers being tested side by side. In this context I think it useful to refer to a book on the testing of plastics[3] which serves something of a complementary function, and perhaps one should be considering both when dealing with the flexible plastics. There is in particular the question of thermoplastic rubbers. Although this class of materials is not specifically mentioned in each chapter it is believed that in general thermoplastic rubbers can be best tested by the methods used for vulcanised rubber. This is in line with the conclusions drawn in a recent paper[4] which considers the particular testing requirements for thermoplastic rubbers and, more significantly, agrees with the action of ISO Committee TC45 SC2 who have decided to add thermoplastic rubbers to the scope of rubber test methods wherever the method is thought suitable.

Cellular rubbers have been deliberately omitted as it is my belief that this very distinct class of materials should be treated separately, both rubbers and plastics being considered together. Similarly, tests on latex have also been omitted. Ebonite has not been included as it has been accepted by ISO TC45 and TC61 that it should be considered as a thermosetting plastic.

Some tests on simple composite materials have been included, e.g. rubber/metal and rubber/fabric, although the majority of tests on coated fabrics have not been considered as, once again, this particular product type can be considered as a special subject in its own right. Comment is made, as appropriate, about testing finished products but a separate section on this has not been written, simply because such procedures are too specialised for general treatment.

I have lost no sleep in debating what is physical—if popular opinion treats tests as part of the physical spectrum (e.g. ageing tests) then they are physical. The intention has been to include every type of physical test and, hopefully, this has been, in the main, achieved. However, three areas immediately come to mind which do not have their own section, acoustic properties, optical properties and non-destructive testing.

There are no test methods specific to rubber for acoustic properties. A section on general optical properties would also be rather thin, rubbers usually being opaque and their reflective properties and colour are very rarely of great consequence. Particular uses of microscopy, for example for dispersion, have been mentioned and it is recognised that the microscopist often has a very important role to play in one of the very important reasons for testing—failure analysis.

It is not so easy to excuse the lack of a chapter on non-destructive testing. The reason is a mixture of the fact that the major NDT techniques are, in the main, only applied to a few particular rubber products and the realisation that to properly describe all methods would require a book, not a chapter. It is, however, worth remembering that it is not only ultrasonics, radiography, holography and so on which are non-destructive. A number of the more traditional rubber tests, for example electrical properties, many dynamic tests, hardness and dimensional measures leave you with the product intact. There does not appear to be a text book devoted to NDT of polymers but recent reviews concerned with application to rubbers are given in references 5 and 6 and the papers given at a European workshop have been published.[7]

The greatest change in test laboratories in recent times, and the rubber laboratory is no exception, is the improvements made to apparatus by the introduction of automation and, in particular, the application of computers both to control tests and to handle the data produced. These developments can and do influence the test techniques which are used, for example by allowing a difficult procedure to become routine and hence increase its field of application, but primarily advances in instrumentation and data handling are a matter of test equipment and laboratory management as opposed to test methods. Hence, whenever appropriate, comment is made on the form of apparatus now available for any particular test but a chapter on instrument hardware and software has not been included. A publication on test equipment[8] can be said to be complementary to the present volume. In addition to papers describing the automation of particular instruments there have been accounts of automation and data handling in the rubber laboratory, e.g. references 9–11. Not surprisingly, most articles are from the viewpoint of the instrument manufacturer who wishes to promote more expensive equipment. However, users generally recognise that the value of advances in instrumentation in cost saving and improved accuracy can be very great although they do not intrinsically improve the relevance of tests to product performance.

REFERENCES

1. Gorton, K., Brown, P. and Cropper, W. (May 1984). *ASTM Standardization News*.
2. Gorton, K., Brown, P. and Cropper, W. (June 1984). *ASTM Standardization News*.
3. Brown, R. P. (Ed.) (1981). *Handbook of Plastics Test Methods*, George Godwin Ltd.
4. Brown, R. P. and Thorn, A. D. (1985). *Polym. Test.*, **5**(3).
5. Trivisonno, N. M. (1985). *Rubb. Chem. Tech.*, **58**(3).
6. Berger, H., ACS Rubber Division Meeting, Minneapolis, 1981, Paper 44.
7. Ashbee, K. H. G. (Ed.) (1985). *Polymer NDT*, Technomic Publishing Co.
8. Brown, R. P. (Ed.) (1979). *RAPRA Guide to Rubber and Plastics Test Equipment*, Rubber and Plastics Research Association.
9. Clamroth, R. and Picht, W. (1981). *Polym. Test.*, **2**(1).
10. Davies, J. R. (1985). PRI Quality Assurance Conference, Birmingham, Paper 11.
11. Pawlowski, H. and Kootz, J. P. (1983). *Rubb. World*, **188**(4).

Chapter 2

Standards and Standards Organisations

2.1 STANDARDS—TEST METHODS AND SPECIFICATIONS

It does not take much imagination to realise that if there were no standardised test methods trade would be severely impaired, progress would be stunted and chaos would ensue. Fortunately, technologists appear to have a strong sense of order and from the early days of the industry have supported the development of standard procedures and the use of these in product specifications; not that the effort put in should make us complacent, as there is still plenty of room for confusion of the unwary. Because of increased demand for product reliability and fears of liability legislation standards have become even more important in recent years. The British Government, amongst others, have expressed their commitment to standards and there has been much discussion of the role of standards in industrial strategy.[1]

To avoid misunderstanding over terminology it is as well to note that the British Standards Institution call all their documents standards and the word specification is reserved for those standards which specify minimum requirements for materials or products. Other types of standard are Methods of Test, Glossaries of Terms and Codes of Practice. It follows that a specification may refer to several methods of test and that a commercially written specification can refer to nationally standardised test methods.

In terms of trade it is ultimately specifications which are important, with test methods acting as building blocks. For this reason it has made chronological sense for the test method standards to be developed first and indeed this has generally been the case in practice. Now that test methods are well developed at the national and international level it can be argued that rather more of the effort available should be put into specifications, especially in the current economic climate where less money than before is available. Certainly there is no case for the development of standard test methods which are of academic interest only and unlikely to be generally used. This does not mean that such methods should not be developed. Tests are needed for a number of purposes and not all justify the standardisation process. In a discussion of the requirements for physical testing of polymers[2] the different needs for test methods and the particular role of standards has been considered. There is no doubt that further improvements in rubber test procedures are needed and it has been suggested[3] that progress has been slow in recent times and barely in the right direction. However, it will be apparent throughout this book that considerable activity in test method standardisation is still taking place.

2.1.1 Test Methods

In this book we are concerned with methods of test and only indirectly with specifications. Leaving aside for the moment the various sources of standard test methods, one can recognise different styles or types of published methods. This is not a matter of accident but rather one of progression; the most obvious yardstick being the number of options left open to the user. In the simplest case a particular apparatus is specified, one set of mandatory test conditions given and no choice allowed as to the parameters to be reported; this is the form in which the specification writer needs a test method. Unfortunately for those who want a quiet life many national and international test methods have become rather more complex. This is partially a result of compromise but more importantly because the measurements being described are not intrinsically simple and the method will be required for a number of different purposes and probably for many different end products. The specification user must therefore select the particular conditions which best suit his individual purposes. In practice he frequently fails to do this either because he omitted to read the standard carefully enough or because his understanding of it was somewhat limited. As more advanced concepts are being introduced into test method standards so there is an increase in the practice of including explanatory notes. I fear that these do not always achieve their desired aim.

We can conveniently distinguish three different circumstances in which a standard method is used: (a) purely for quality control, (b) as a performance requirement, and (c) for development purposes. In the first case the prime consideration is that precisely the same procedure is always used and also that this procedure is relatively simple and rapid. The test conditions may be completely arbitrary but one set of conditions and one set only is required. If the test is intended, apart from a quality control function, to be a measure of the performance of the product then test conditions will be chosen which have some relevance to the product end use. For development work it is highly probable that a series of conditions will be wanted in the hope that data of use in designing future products will be realised. Committees preparing standard test methods have all these possibilities in mind and the penalty for the user of the standard is that he must understand the subject sufficiently well to make an intelligent selection of conditions to suit his particular purpose. The following example may not stand up to too close an inspection but serves I think to illustrate the point. If a test for resistance to liquids is considered, one would expect a quality control procedure to involve one liquid at one temperature for a relatively short time. The liquid might be a standard fuel such as liquid B of ISO 1817 and the test involve 24 h exposure with volume change being measured. A rapid measuring method such as area change (see Section 16.2.1) may be used to further speed up testing. If the testing was intended to have a performance function then the liquid met in service would be used, for example commercial petrol, and testing continued long enough for equilibrium absorption to be reached. Apart from volume change, other relevant physical properties would be measured before and after exposure. For development purposes testing would be further extended to cover a number of fluids each tested at several temperatures. An international or national standard would attempt to cater for these and other possibilities and would hence include a choice of measuring procedure, test temperature, duration of exposure, properties to be monitored and test liquids. Preferred test parameters might be indicated for use when there were no outside factors influencing the choice.

This is not the place to discuss in any detail what should or should not be included in standard test methods or how they should be written. The quality and the style of those in current existence varies very considerably but it is possible to detect certain general trends in recent years. Standards have become more involved as more factors which cause variability are identified and control of these is specified. At the same time some apparatus has been specified in a more general way, stating what its performance must be without restricting its design or construction to any particular form. This can only be done when all the important parameters have been identified. Standards also become more complicated as the underlying principles of the property being measured become better understood and as more meaningful results are demanded by product designers. It is very much a matter of opinion as to whether we at present have taken this progress too far or not far enough.

2.2 ORGANISATIONS PRODUCING STANDARDS

Generally, the sources of standards can be placed into three groups:

International organisations
National organisations
Individual companies

Despite the argument that in terms of trade it is commercial specifications which are most important, I think it proper and logical to discuss these groups in descending order of scope, i.e. from the international downwards. In practice of course a new test method usually proceeds in the opposite direction from humble beginnings in particular laboratories via national recognition to international status, by that time having become much modified.

2.2.1 International Standards

The ultimate state of unity would be for all countries to be using the same standards. This would obviously be of great value in smoothing the course of international trade and make it easier for technologists to exchange technical information. It is also a very ambitious concept that the countries of the world can compromise on their national procedures and overcome the very great difficulties of language in a field where language is the most important tool of trade.

THE INTERNATIONAL STANDARDS ORGANISATION

In most fields, including rubber, the principal body attempting to achieve the ideal of international agreement is the International Standards Organisation (ISO) which is hence our most important organisation in the standards field. The ISO in its present form is not a very old organisation being formed in 1946 after previous attempts at setting up this sort of body had met with little success.

The work of ISO is administered by a permanent central secretariat which has headquarters at 2, rue de Varembe, 1221 Geneva 20 and has as members more than 70 national standards bodies, one body per country. Apart from central committees concerned with planning, certification, etc., the technical work of ISO is carried out by technical committees each relating to a particular area of industry. The secretariat of each technical committee is held by a member country and each member may join any committee either as a participating (P) member or observing (O) member. The P members have voting rights at committee meetings. The choice of P and O member depends on the country's interest and the finance available.

The committee for Rubber and Rubber Products is TC45 with the United Kingdom holding the secretariat and plastics is covered by TC61. TC45 normally meets once per year, member countries acting as hosts in a sort of rotation. The delegates to the technical committee are decided by the national standard body. Interesting background information to ISO and TC45 can be found in an in depth review[4] published in 1971. Details of structure and procedure have of course changed since then and indeed are continually changing.

Many ISO technical committees operate with an infra structure of sub-committees, working groups and task groups. TC45 has only recently started to use sub-committees and a considerable amount of its work is conducted in working groups reporting directly to the main committee. This difference in the way of working is irrelevant as regards the technical work although the procedural implications are important for those actively concerned. It is probable that WG3 and WG11, both of which have several task groups, will be converted to sub-committees.

The present structure of TC45 is listed below, the secretariat of each sub-committee or group being taken by a member country.

SC1 Hose
- SC1 WG1 Industrial, chemical and oil hoses
- SC1 WG2 Automotive hose
- SC1 WG3 Hydraulic hose
- SC1 WG4 Test methods

SC2 Physical and Degradation Tests
- SC2 WG1 Physical tests
- SC2 WG2 Viscoelastic tests
- SC2 WG3 Degradation tests

WG1 Chemical Tests

WG2 Latex

WG3 Raw materials for use in the rubber industry
WG7 Flexible and semi-rigid cellular materials
WG8 Classification of vulcanised rubber
WG10 Terminology
WG11 Miscellaneous products
WG12 Footwear
WG13 Coated fabrics
WG14 Laboratory tests for the action of flame on rubber
WG15 Application of statistical methods

It can be seen that those of most interest to physical testing are SC2 and WGs 14 and 15 but other working groups have of course an interest in tests to be included in specifications, particularly specialised product oriented methods. The amount of work being undertaken is rather daunting and to aid matters some working groups have a number of task groups to take care of the details of various aspects of their subject.

The order of progress towards an International Standard is that after consideration at task and working group level a document is proposed to the plenary session of the technical committee or appropriate sub-committee for circulation to members for postal vote as a draft proposal (DP). If approved, balloting and commenting take place and the votes and comments are considered by the working group at the next meeting of the technical committee. If agreement is reached the revised document is again circulated this time to all ISO member countries as a draft international standard (DIS). The comments and votes on this are considered at a subsequent meeting and if approved at the plenary session the final revision is sent to the central secretariat for approval by the ISO council and publication as an international standard. It is immediately obvious that this process is slow. Although certain short cuts are possible, it is not unusual for a document to pass through a second draft proposal or second draft standard stage if agreement proves difficult.

ISO standards were first published in 1972; before that time ISO recommendations was the title used. It is not obligatory for the ISO standards to be incorporated into a national system but obviously the whole aim is a little defeated if this is not done. The BSI takes a very positive attitude in this direction following the dictum 'Do it once, do it right, do it internationally'. Wherever possible British standards at least agree with ISO standards technically and the aim is to reproduce them verbatim. To become an ISO standard requires that a document receives the approval of 75% of the members casting a vote. This fairly high percentage helps to ensure that unsatisfactory documents do not get through. Normally, if the UK gives a positive vote at ISO that standard is then adopted as regards technical content into the BSI system.

It is difficult enough to reach agreement on a standard procedure within one country and the problems internationally are considerably greater, not being helped by language difficulties—the official languages of ISO are English, French and Russian. Therefore the slow pace of production of an ISO standard is hardly surprising. Differences between national documents and ISO methods are often not a matter of disagreement but simply that the two time scales are different and things have got out of step. However, there are encouraging signs that progress towards more complete rationalisation is being achieved. There are at present some 240 ISO standards published in the rubber field. These are listed in the ISO Catalogue and additions during the year are publicised in the British Standards publication *BSI News*. In the UK ISO standards can be obtained from the BSI and in other countries through the national standards body. Standards for rubber can also be purchased grouped into three volumes as *ISO Handbook 22*.

OTHER INTERNATIONAL STANDARDS

In the electrical field the International Electrotechnical Commission (IEC) performs the same function as ISO. The work of this body is of interest where rubbers are used in electrical insulation, etc. As regards electrical test methods for rubber, both ISO and BSI adopt the basic procedures and principles standardised by IEC.

Britain has particular interest in the more limited scope of European Standardisation. The European Committee for Standardisation (CEN) was founded in 1961 and comprises the national standards bodies of EEC and EFTA countries. CENELEC is the equivalent body in the electrical field. To many people the concept of European standards is an unnecessary complication, it being argued that there is no need for any activity in between ISO and the national bodies. However, the work of CEN is likely to assume importance where documents are drawn up at the specific request of the EEC in their programme for the removal of technical trade barriers. Such standards will be used in EEC directives which, if ratified, are binding on the community members. CEN standards are voluntary, being adopted only by approving countries. There are also fundamental differences in the voting and approval systems in CEN and EEC and it is apparent that those concerned with standardisation will have to watch the European developments carefully if unsatisfactory standards and legislation are not to be forced upon us.

There are many other international organisations concerned with standards and a short guide to these is given in BS0: Part 1,[5] which makes clear the confusing abbreviations of titles in use.

2.2.2 National Standards

Although generally each country has one principal standards organisation which provides the official membership of ISO, other organisations can issue standards at national level. It is usual to include government departments in this category. It is not practical, and indeed not necessary, to consider here the national standards bodies of all countries but a list of ISO members is given in the appendix to this chapter. The operations of the British Standards Institution (BSI) will be described in some detail, which, apart from being of particular interest to those trading in Britain, serves to illustrate how the process of generating standards at a national level can be undertaken. It must not be assumed that other countries operate in even roughly the same format. It should be noted however that the BSI is one of the longest established and most highly rated of national standards bodies.

THE BRITISH STANDARDS INSTITUTION

The BSI was formed in 1901 and has now developed to the point where it covers an astonishing range of subjects from virtually all branches of industry. Apart from its main function of producing standards it also operates a quality assurance division which operates BSI's certification and assessment schemes and a comprehensive test house.

BSI receives government support but raises the majority of its income from membership fees, the sale of standards and fees from certification and testing services.

Membership is open to virtually anyone, various categories of organisations being defined for the purpose of computing membership fees. It is in fact rather difficult to keep up to date on standards matters without being in membership. Details of new standards, amendments and articles on standards matters generally are published monthly in *BSI News* which is circulated to members. There are also annual editions of the BSI Catalogue which lists all the British Standards available and is issued to members and may be purchased by non-members. It is appropriate to mention here BS0, A Standard for Standards[5–8] which gives in considerable detail an account of BSI structure, procedure and editorial practice.

Structure

The permanent staff of BSI, administrative, technical and editorial operate under a Director General, with the overall responsibility for policy being vested in an Executive Board. Matters of policy relating to convenient groups of industries (e.g. building, engineering) are dealt with by Divisional Councils and within each of these councils is a number of Industry Standards Committees. Of particular interest to us is the Rubber Industry Standards Committee RUM/- which operates within the umbrella of the Multitechnics Council M/-.

We now have an introduction to the numbering system used for BSI committees. The RU refers to rubber and the M for multitechnics. RUM/- has responsibility for authorising the initiation of a standards project and for final approval of drafts before publication. The preparation of standards is carried out by technical committees and sub-committees reporting to the industry committee. Hence we have for example RUM/7 Footwear and RUM/9 Hose. The BSI supplies the secretariat for technical committees and the members are nominated by industry, government departments and research associations. The industry representatives are usually nominated by trade associations and not the individual company, although individual experts are sometimes co-opted.

Physical testing of rubbers is the concern of RUM/36 which currently has four active panels, numbers 3, 4, 5 and 6. Panels rather than sub-committees are used purely for administrative convenience. These *ad hoc* panels, despite being almost off the end of the numbering system, are very important and RUM/36 would get very bogged down without them.

Other committees which should be mentioned in the context of testing are the PLM/RUM series which are joint committees between the plastics and rubber industries. PLM/RUM/10 deals with methods of test for cellular materials, PLM/RUM/6 with accuracy of test machines and PLM/RUM/9 with electrical tests. Unfortunately systems are never as simple as we would like; specialised tests may be considered in product committees and not all products containing rubber are covered in RUM committees. Lastly, the sub-committee RUM/-/1, reporting to the industry committee, exists to co-ordinate international work and hence is the prime link with ISO affairs.

Preparation of British Standards

Consideration of an initial draft is carried out in a technical committee, sub-committee or panel as appropriate. This initial draft may have come from one of a number of sources, for example, being based on work carried out by one of the bodies represented on the committee. The draft works upwards to the main committee, being presented there as a private circulation. When agreement is reached the document is circulated as a draft British standard (DC) to industry for comment. These drafts have no official standing and nowadays are always prepared after careful consideration of the position in ISO. The comments are considered by the technical committee and the amended document passed to the industry committee for approval. The BSI editorial department then vets the document for consistency of layout, etc., before it goes for printing. This procedure is generally efficient and quicker than the ISO procedure but it must be admitted that unfortunate delays do occur, often as a result of too great a pressure of work on the BSI resources.

The technical committee also has the responsibility of keeping its published standards up to date, which for a major revision would mean a complete re-issue and would be dealt with in the same manner as a new standard. Relatively minor changes are dealt with by amendment slips the publication of which, as for new standards and revisions, is announced in *BSI News*.

Each standard is given a separate number although there may be more than one part to a standard, each issued separately. The year of publication is added so that different editions of the same standards can be recognised. The BSI Catalogue gives a complete list of standards in numerical order, but unfortunately not grouped into subjects. For a complete picture of the up-to-date situation it is necessary to consult the year book and the subsequent issues of *BSI News*, remembering that the year book might be 18 months out of date. However, BSI now operates an automatic updating system (SOS) to which clients may subscribe.

OTHER BRITISH NATIONAL STANDARDS

Standards or specifications issued by individual companies are not considered to be of national status, however large or multi-national the concern might be. Specifications issued by local authorities and nationalised industries would be in the same bracket. The Rubber and Plastics Research Association, The British Rubber Manufacturers' Association, the Malaysian Rubber Producers' Association and the Plastics and Rubber Institution do not issue standards.

Government departments, although contributing a great deal to the work of BSI, produce a large amount of their own standardisation. The reasons for this are really similar to those which apply to individual companies—they are unable to wait for the BS system or they have specific requirements unique to themselves. This latter reason applies particularly to the armed forces. However a memorandum of understanding between the government and BSI signed in 1982 states that the government will seek to use British standards rather than to develop its own. Unfortunately the various standards issued by the government departments are rather confusing to outsiders, as indeed are the departments themselves, which appear to undergo frequent change of titles or scope.

USA STANDARDS

American standards, particularly those of ASTM, are widely used in many parts of the world and indeed many companies adopt wholesale methods from ASTM under their own name. The national standards system in the USA differs in many respects from the British, in particular the organisation which publishes the standards of most interest, the American Society for Testing and Materials (ASTM), is not the official national standards body having ISO membership. That function is fulfilled by the American National Standards Institute (ANSI).

American National Standards Institute (ANSI)

ANSI is the premier USA standardisation body and is the counterpart of BSI in being the official ISO representative. It was until a few years ago known as the American Standards Association. ANSI does not itself write standards but approves as American standards those produced by ASTM and other similar organisations. Not all ASTM standards are approved in this way and approval may take place years after the introduction of the standard by ASTM.

American Society for Testing and Materials (ASTM)

Apart from work at ISO, it is the ASTM that most people in the polymer industry think of as representing American standards. ASTM has a membership drawn from similar sources to those of BSI and operates through more than 140 technical committees which in turn have a sub-committee structure. D11 is the committee for rubber.

ASTM standards can be obtained individually but are more usually seen as ASTM books, each being a collection of standards covering a particular subject or related group of subjects. The books are revised annually and although some standards remain unchanged for years there is always a significant amount of new or revised matter. Hence it is advisable to use only the current edition. Although this is rather expensive it is easier for the user than keeping British standards up to date by studying *BSI News*. There are currently some 66 volumes of ASTM standards, those concerning rubbers in particular being volumes 09.01 and 09.02. The ASTM is active in the technical field apart from purely producing standards. It organises conferences and publishes numerous books and reports as well as the journal *Standardization News*.

GERMAN STANDARDS

The reason for including a mention of German standards as opposed to those of France, Russia or wherever in an English language book is that it has been the author's experience that DIN standards are those most commonly met with in Britain after BS and ASTM.

The principal German standards organisation Deutsches Institut für Normung was founded in 1917. It operates rather similarly to the BSI in that it has subscription-paying members from industry, trade associations, etc., and is independent of individual pressures. It has standards committees and some 3800 technical committees and study groups, membership of which is honorary with representatives from all interested areas.

The DIN series of standards, of which there are many thousands, are catalogued into subject groups numerically. For example the 672.0 group is testing elastomers and rubbers. The standards are published separately and are all listed in the DIN catalogue. Many DIN standards are also published in English and Spanish and a few in French. There is a separate yearly book which lists those standards which have been translated.

The use of DIN standards is generally voluntary unless they are incorporated into legislation. The DIN mark on a product signifies that it complies with the relevant DIN standard and manufacturers can be forbidden to use the mark if they do not stick to the rules. The mark is hence similar in concept to the BSI Kitemark although the policing system appears to be less rigorous. The DIN Testing and Supervision Mark appears to be more equivalent to the Kitemark.

OTHER NATIONAL STANDARDS

It is not possible to discuss all of the national standards bodies but it is necessary to be able to identify the source of a standard from the abbreviations like BS and DIN which are used. Taking the whole world and including government standards the total becomes enormous and very confusing. The letters used by ISO members are given in the appendix.

The standards mostly used by any laboratory will depend on whom they are trading with. Information can be gained from the national bodies listed in the appendix.

2.2.3 Company Standards

There must be literally millions of company standards in existence. Although they have relatively little significance in a national or international sense, they are the basis of many commercial contracts and hence are perhaps the most important standards of all. Among their number are some of the best examples of standardisation and also some of the worst; sadly the worst appear very frequently.

Using a commercial standard is like using any standard, the user must be careful that he has the latest edition and that he has read it very carefully and missed none of the detail. A common fault in commercial standards is that rather a lot of detail is missing, for example there must be insufficient information in a test method to be sure that you are carrying it out correctly. All one can do is to talk to the originator of the standard.

It would save a great deal of pain and confusion if those writing commercial specifications would wherever possible use published standard test methods, preferably those of ISO. Special tests will often be needed but there is no point in inventing your own procedure for a straightforward test which has been well standardised. Perhaps a lot of the trouble is that in some cases those writing specifications are not well versed in standardisation outside of their own organisation and that many engineers have a poor understanding of rubbers and their properties.

2.3 UNITS

In the 1st Edition it was noted that it should be unnecessary to state that SI units will be used—that will be assumed to be the case. However the Imperial System lingers on in various isolated outposts. The universal adoption of SI units virtually eliminates the need to include a section on units because there is no question of conversions or explanations of obscure systems. However it is appropriate to make reference to relevant information.

The basic reference[9] is to ISO 1000 which details all the units, multiples and sub-multiples to be used. BS 5555[10] is identical. PD 5686[11] gives advice on the use of SI units given in ISO 1000, explanatory information for British readers and details of EEC directives on units of measurement, whilst BS 3763[12] gives basic features of the SI as promulgated by the Bureau Internationale des Poids et Mesures.

Many organisations have produced their own version of how SI units should be presented but these were generally printed when metric units were relatively new in British industry and would no longer seem to be warranted. However, certain special considerations will apply in any particular industry and both ISO committees TC45 and TC61 have actively considered the subject. Their conclusions on units which are normal to their materials and products have been included in their own procedural documents.

Where there is a need to convert to or from SI units reference can be made to the conversion factors found in BS 350,[13] and PD 6203.[14] The detailed conversion tables of BS 350:Part 2 have been withdrawn as obsolete!

REFERENCES

1. *Standards, Quality and International Competitiveness*, Dept. of Trade, 1982.
2. Brown, R. P. (1984). *Polym. Test.*, **4**, 2–4.
3. Brown, R. P. (1984). *Europ. Rubb. J.*, **166**, 7.
4. Moakes, R. C. (1971). In *Progress of Rubber Technology*, Vol. 35, Instn. of Rubber Industry.
5. BS0, A Standard for Standards, Part 1, 1981. General Principles.
6. BS0:Part 2, BSI and its Committee Procedures.
7. BS0:Part 3, Drafting and Presentation of British Standards.
8. BS0:Part 4, Guide to BSI Editorial Practice.
9. ISO 1000, 1981. SI Units and Recommendations for the Use of Their Multiples and of Certain Other Units.
10. BS 5555, 1981. SI Units and Recommendations for the Use of Their Multiples and of Certain Other Units.
11. PD 5686, 1978. The Use of SI units.
12. BS 3763, 1976. The International System of Units.
13. BS 350:Part 1, 1974 (1983). Basis of Tables.
14. PD 6203, 1967 (1982). Additional Tables for SI Conversion.

APPENDIX

NATIONAL STANDARDS BODIES (ISO MEMBERS)

Albania/Albanie (BSA)
Komiteti i Cmimeve dhe Standarteve,
Prane Keshillit te Ministrave,
Tirana.

Algeria/Algérie (INAPI)
Institut algérien de normalisation et de propriété industrielle,
5 rue Abou Hamou Moussa,
BP 1021—Centre de tri,
Alger.

Argentina/Argentine (IRAM)
Instituto Argentino de Racionalización de Materiales,
Chile 1192,
C. Postal 1098, Buenos Aires.

Australia/Australie (SAA)
Standards Association of Australia,
Standards House,
80–86 Arthur Street,
North Sydney,
NSW 2060.

Austria/Autriche (ON)
Österreichisches Normungsinstitut,
Heinestrasse 38,
Postfach 130,
A-1021 Wien.

Bangladesh (BSTI)
Bangladesh Standards and Testing Institution,
116/A Tejgaon Industrial Area,
Dhaka-8.

Belgium/Belgique (IBN)
Institut belge de normalisation,
Av. de la Brabançonne, 29,
B-1040 Bruxelles.

Brazil/Brésil (ABNT)
Associação Brasileira de Normas Técnicas,
Av. 13 de Maio, no. 13–28° andar,
Caixa Postal 1680,
CEP: 20.003—Rio de Janeiro–RJ.

Bulgaria/Bulgarie (BDS)
Comité de la qualité auprès du Conseil des Ministres,
21 rue du 6 Septembre,
1000 Sofia.

Canada (SCC)
Standards Council of Canada International Standardization Branch,
2000 Argentia Road, Suite 2-401,
Mississauga, Ontario L5N 1V8.

Chile/Chili (INN)
Instituto Nacional de Normalización,
Matis Consiño 64—6° piso,
Casilla 995—Correo 1,
Santiago.

China/Chine (CSBS)
China State Bureau of Standards,
PO Box 820,
Beijing.

Colombia/Colombie (ICONTEC)
Instituto Colombiano de Normas Técnicas,
Carrera 37, No. 52–95,
PO Box 14237,
Bogota.

Cuba (NC)
Comité Estatal de Normalización,
Egido 602 entre Gloria y Apodaca,
Zona postal 2,
La Habana.

Cyprus/Chypre (CYS)
Cyprus Organization for Standards and Control of Quality,
Ministry of Commerce and Industry,
Nicosia.

Czechoslovakia/Tchécoslovaquie (CSN)
Urăd pro normalizaci a měřeni,
Václavské náměsti 19,
113 47 Praha 1.

Denmark/Danemark (DS)
Dansk Standardiseringsraad,
Aurehøjvej 12,
Postbox 77,
DK-2900 Hellerup.

Egypt, Arab Rep. of/Égypte, Rép. arabe d' (EOS)
Egyptian Organization for Standardization,
2 Latin America Street,
Garden City,
Cairo—Egypt.

Ethiopia/Éthiopie (ESI)
Ethiopian Standards Institution,
PO Box 2310,
Addis Ababa.

Finland/Finlande (SFS)
Suomen Standardisoimisliitto SFS,
PO Box 205,
SF-00121 Helsinki.

France (AFNOR)
Association française de normalisation,
Tour Europe,
Cedex 7,
92080 Paris La Défense.

Germany, F.R./Allemagne, R.F. (DIN)
DIN Deutsches Institut für Normung,
Burggrafenstrasse 4-10,
Postfach 1107, D-1000 Berlin 30.

Ghana (GSB)
Ghana Standards Board,
PO Box M-245,
Accra.

Greece/Grèce (ELOT)
Hellenic Organization for Standardization,
Didotou 15,
106 80 Athens.

Hungary/Hongrie (MSZH)
Magyar Szabványügyi Hivatal,
1450 Budapest 9,
Pf. 24.

India/Inde (ISI)
Indian Standards Institution,
Manak Bhavan,
9 Bahadur Shah Zafar Marg,
New Delhi 110002.

Indonesia/Indonésie (DSN)
Dewan Standardisasi Nasional-DSN
(Standardization Council of Indonesia),
Gedung PDIN-LIPI,
Jalan Gatot Subroto,
PO Box 3123,
Jakarta 12190.

Iran (ISIRI)
Institute of Standards and Industrial Research of Iran,
Ministry of Industries,
PO Box 11365-7594, Tehran.

Iraq (COSQC)
Central Organization for Standardization and Quality Control,
Planning Board,
PO Box 13032,
Aljadiria,
Baghdad.

Ireland/Irlande (NSAI)
National Standards Authority of Ireland,
Ballymun Road,
Dublin-9.

Israel/Israël (SII)
Standards Institution of Israel,
42 University Street,
Tel Aviv 69977.

Italy/Italie (UNI)
Ente Nazionale Italiano di Unificazione,
Piazza Armando Diaz 2,
1-20123 Milano.

Ivory Coast/Côte d'Ivoire (DINT)
Direction de la normalisation et de la technologie,
Ministère de Plan et de l'Industrie,
BP V65,
Abidjan.

Jamaica/Jamaïque (JBS)
Jamaica Bureau of Standards,
6 Winchester Road,
PO Box 113,
Kingston 10.

Japan/Japon (JISC)
Japanese Industrial Standards Committee,
c/o Standards Department, Agency of Industrial Science and Technology,
Ministry of International Trade and Industry,
1-3-1 Kasumigaseki, Chiyoda-ku,
Tokyo 100.

Kenya (KEBS)
Kenya Bureau of Standards,
Off Mombasa Road,
Behind Belle Vue Cinema,
PO Box 54974,
Nairobi.

Korea, Dem. Rep. of/Corée, Rép. dém. p. de (CSK)
Committee for Standardization of the Democratic People's Republic of Korea,
Taesong guyok Ryongnam dong,
Pyongyang.

Korea, Rep. of/Corée, Rép. de (KBS)
Bureau of Standards, Industrial Advancement Administration,
2 Chungang-dong Kwach'ŏn-myon,
Kyŏnggi-do 171-11.

Libyan Arab Jamahiriya/Jamahiriya Arabe Libyenne (LYSSO)
Libyan Standards and Patent Section,
Industrial Research Centre,
PO Box 3633,
Tripoli.

Malaysia/Malaisie (SIRIM)
Standards and Industrial Research Institute of Malaysia,
PO Box 35, Shah Alam,
Selangor.

Mexico/Mexique (DGN)
Dirección General de Normas,
Calle Puente de Tecamachalco No. 6,
Lomas de Tecamachalco,
Sección Fuentes,
Naucalpan de Juárez,
53 950 Mexico.

Mongolia/Mongolie (MSC)
State Committee for Prices and Standards of the Mongolian People's Republic,
Ulan Bator II.

Netherlands/Pays-Bas (NNI)
Nederlands Normalisatie-instituut,
Kalfjeslaan 2,
PO Box 5059,
2600 GB Delft.

New Zealand/Nouvelle-Zélande (SANZ)
Standards Association of New Zealand,
Private Bag,
Wellington.

Nigeria (SON)
Standards Organisation of Nigeria,
Federal Ministry of Industries,
4 Club Road,
PMB 01323,
Enugu.

Norway/Norvège (NSF)
Norges Standardiseringsforbund,
Postboks 7020 Homansbyen,
N-0306 Oslo 3.

Pakistan (PSI)
Pakistan Standards Institution,
39 Garden Road, Saddar,
Karachi-3.

Papua New Guinea/Papouasie-Nouvelle-Guinée (PNGS)
National Standards Council,
PO Box 3042,
Boroko.

Peru/Pérou (ITINTEC)
Instituto de Investigación Tecnológica,
Industrial y de Normas Técnicas,
Av. Guardia Civil 400,
Distrito San Borja,
Lima 34.

Phillippines (PSA)
Product Standards Agency,
Ministry of Trade and Industry,
361 Sen. Gil J. Puyat Avenue,
Makati,
Metro Manila 3117.

Poland/Pologne (PKNMiJ)
Polish Committee for Standardization, Measures and Quality Control,
Ul. Elektoraina 2,
00-139 Warszawa.

Portugal (DGQ)
Direcçŏ-Geral da Qualidade,
Rua José Estêvão, 83-A,
1199 Lisboa Codex.

Romania/Roumanie (IRS)
Institut roumain de normalisation,
Rue Ilie Pintilie 5,
Bucarest 1.

Saudi Arabia/Arabie Saoudite (SASO)
Saudi Arabian Standards Organization,
PO Box 3437,
Riyadh-11471.

Singapore/Singapour (SISIR)
Singapore Institute of
Standards and Industrial
Research,
Maxwell Road,
PO Box 2611,
Singapore 9046.

South Africa, Rep. of/
Afrique du Sud, Rép. d' (SABS)
South African Bureau of
Standards,
Private Bag X 191,
Pretoria, 0001.

Spain/Espagne (IRANOR)
Instituto Español de
Normalización,
Calle Fernandez, de la Hoz, 52,
28010 Madrid.

Sri Lanka (SLSI)
Sri Lanka Standards Institution,
53 Dharmapala Mawatha,
PO Box 17,
Colombo 3.

Sudan/Soudan (SSD)
Sudanese Standards Department,
Ministry of Industry,
PO Box 2184,
Khartoum.

Sweden/Suède (SIS)
Standardiseringskommissionen
i Sverige,
Tegnérgatan 11,
Box 3 295,
S-103 66 Stockholm.

Switzerland/Suisse (SNV)
Swiss Association for
Standardization,
Kirchenweg 4,
Postfach,
8032 Zurich.

Syria/Syrie (SASMO)
Syrian Arab Organization for
Standardization and Metrology,
PO Box 11836,
Damascus.

Tanzania/Tanzanie (TBS)
Tanzania Bureau of Standards,
PO Box 9524,
Dar es Salaam.

Thailand/Thaïlande (TISI)
Thai Industrial Standards
Institute,
Ministry of Industry,
Rama VI Street,
Bangkok 10400.

Trinidad and Tobago/
Trinité-et-Tobago (TTBS)
Trinidad and Tobago Bureau
of Standards,
Century Drive,
Trincity Industrial Estate,
Tunapuna,
PO Box 467,
Port of Spain.

Tunisia/Tunisie (INNORPI)
Institut national de la
normalisation et de la propriété
industrielle,
BP 23,
1012 Tunis-Belvédère.

Turkey/Turquie (TSE)
Türk Standardlari Enstitüsü,
Necatibey Cad. 112,
Bakanliklar,
Ankara.

United Kingdom/ Royaume-Uni (BSI)
British Standards Institution,
2 Park Street,
London W1A 2BS.

USA (ANSI)
American National Standards Institute,
1430 Broadway,
New York, NY 10018.

USSR/URSS (GOST)
USSR State Committee for Standards,
Leninsky Prospekt 9,
Moskva 117049.

Venezuela (COVENIN)
Comisión Venezolana de Normas Industriales,
Avda. Andrés Bello,
Edf. Torre Fondo Común,
Piso 11,
Caracas 1050.

Viet Nam, Socialist Republic of/ République socialiste du (TCVN)
Direction générale de standardisation, de métrologie et de contrôle de la qualité,
70 rue Tràn Hung Dao,
Box 81,
Hanoi.

Yugoslavia/Yougoslavie (SZS)
Savezni zavod za standardizaciju,
Slobodana Penezića-Krcuna br. 35,
Pošt. Pregr. 933,
11000 Beograd.

Zambia/Zambie (ZABS)
Zambia Bureau of Standards,
National Housing Authority Building,
PO Box 50259,
Lusaka.

Chapter 3

Limitations of Test Results—Statistics

It is tempting to claim that this is the most important chapter in the book. Whatever property we measure, whatever test method we use we end up with results and the question 'What do the figures really mean?' Results are useless unless we know their significance; significance means statistics. The only way to avoid statistics is to bury your head in the sand and this surprisingly enough seems to be a popular activity.

The unpopularity of statistics is due in part to the subject having been severely neglected in schools and universities, and even today this has not totally been put right. The result is that a great many technologists lack even a basic understanding of the subject. Statisticians have not helped the situation by blinding with science anyone who comes near them and being renowned for the ability to present data such as to apparently prove anything or nothing. To be fair, the situation has been improving enormously in very recent times. Perhaps the greatest influence has been the proliferation of micro-computers. Many statistical calculations are very tedious and when a computer is readily to hand, perhaps directly linked to the test machine, and abundant software is easily obtained, there is a considerable incentive to make use of it. In fact statistics, in so far as we need to understand it for our rubber testing, is not beyond comprehension and once the basic principles have been grasped it becomes relatively easy to see through the deliberate or unintentional misrepresentations (not lies) with which we may be confronted. It also enables one to put one's own results in perspective and can prevent a number of costly mistakes.

Although there has been a lack of appreciation of statistics by people in general this cannot be blamed on a lack of published information. In fact there are a number of excellent publications dealing with the various aspects of statistics, not least of these being British and ISO standards.

Despite stressing the importance of the subject this is not a book on statistics and hence no attempt will be made to give full accounts, but reference will be made to textbooks and similar sources of information. Statistics should be learnt in relatively small doses and the use of very complicated experimental designs without practice is likely to lead to confusion. Every test result should be treated with healthy suspicion, as should the conclusions that can be reached with the aid of statistics—the conclusions are only as good as the data.

General statistics textbooks are not written with our particular subject in mind and they vary in the method of presentation used. It is hence sensible to get used to one text and stick to it. Most classic texts are now dated and presumably out of print, but a 15th edition of that by Fisher[1] was published in 1972. Two volumes edited by Davies[2,3] can be recommended and both have been reprinted fairly recently. A single volume which includes experiment design is that by Cooper,[4] and Jardine[5] has written on statistical methods for quality control.

Published standards make excellent reference textbooks for the relatively limited ground they cover. The BS 2846 series, Statistical Interpretation of Data[6-12] covers considerable ground and has ISO equivalents in ISO 2602,[13] 3207,[14] 2854,[15] 3494,[16] 3301,[17] 5479.[18] BS 5532[19] gives definitions of vocabulary and symbols and is equivalent to ISO 3534.[20] BS 5324,[21] The Application of Statistics to Rubber Testing, is of quite specific interest and is intended to be complementary to general statistics standards and textbooks.

3.1 VARIABILITY

All measurements are subject to variability. We need to know the sources of variability and make a reliable estimate of its magnitude. From this information we may then judge the reliability of our results and hence their significance.

The term population is simply the total number of objects in a large group. The population may be a fixed number such as a consignment of 100 000 parts or it may be virtually infinite as in the total number of parts produced in a continually running job. Universe is a term also used to indicate a large population. In testing terms a population may be, for example, the total number of possible tensile strength results which could be obtained on a particular compound if every scrap of the material ever likely to be made, was tested.

A sample is a selected number of, for example, parts or tensile results taken from the population. (To avoid confusion do not use 'sample' to mean 'test piece'.) Note that sample can have two meanings, physically as in taking five parts from a boxful and in the statistical sense of five test results.

If we take a sheet from a batch of rubber and make five tensile strength measurements, they might read: 16.8; 15.4; 16.3; 17.7 and 17.6. The sources of variability will be:

(a) the intrinsic variability of the sheet rubber arising from the fact that it is not perfectly homogeneous;
(b) the variability due to the testing procedure, including test piece preparation, machine accuracy and operator error.

If we test several sheets we will have an additional source of variability due to variations in moulding; and if we mix several batches we add yet two more sources of variation, those from the mixing procedure and any variation in compounding ingredients.

If we give sheets, nominally the same, to a number of operators there will be variability due to the operators; similarly if we use a number of different test apparatuses we introduce variability due to the machines. Taking things further, we may test sheets in different laboratories and introduce between-laboratory variability.

In practice we do our best to minimise the magnitude of variability by carefully controlling the processing operations and the testing apparatus and procedures. We never eliminate it altogether. Leaving aside for one minute the variations due to processing and assuming that our physical sample is representative of the material, we are left with random variations due to testing and, possibly, bias due perhaps to a fault or error in our apparatus. Results with a large bias, however good the variability, are not satisfactory.

It is the tester's job to produce results representative of the true population of results and to this end he must have knowledge of the likely magnitude of random error and bias due to his machines and operators. This includes variability between operators, variability due to testing at different times and any bias due to poor calibration or inherent in the test method. To detect bias it is almost essential to compare results with those obtained in other laboratories.

It is surprising how large the discrepancies, which arise purely from testing sources, can be. For example, in an interlaboratory exercise involving the crescent tear test the average result recorded by the participants ranged from 104 to 158 N (i.e. one laboratory was 50% higher than another) and this was using test pieces mixed, moulded and cut at one central laboratory. This sort of difference is not particularly unusual and even larger interlaboratory differences are found in more complicated or difficult tests.

This leads to the question of distinguishing testing error from real variations in the material. Whatever test is carried out, there will be genuine variation duc to the material and also variation due to uncontrolled testing errors. It is often very difficult to separate the two. Testing errors can arise, for example, from random variations in test piece geometry due to the limitation of cutting precision, variations in the response of the test apparatus and from fluctuations in the operator's performance. These errors may be large or small and of indeterminate direction so that in the long run they tend to cancel out. More serious is systematic error or bias which is unidirectional. For example, the error due to a machine being wrongly calibrated or an operator consistently misreading a scale.

Testing error apart, our sample of results will not be representative of the whole population if our physical sample is not representative. We must expect differences between repeat mixes and between repeat mouldings because of some variation in the quantities and quality of ingredients used, the efficiency of mixing and the time and temperature of curing, etc. If gross errors are made, some very atypical results will be recorded and it is dangerous to rely heavily on one small sample unless you are sure it is representative. Perhaps one of the most common cases is when an alternative ingredient is being evaluated and this is compared to the original standard formula. The mixes are uniform, the tester does a good job and it is concluded using statistical methods that the new ingredient is an improvement. It is easily forgotten that this conclusion assumed that the samples of each compound were truly representative of the population. If the variability which would arise from repeat mixings is rather larger than the testing error, as is often the case, then tests on a series of repeat mixes may show no difference between the ingredients or even that the new ingredient was worse. To test the validity of the conclusions, we need to know about the variation of repeat mixes.

3.2 ACCURACY AND PRECISION

Accuracy is the closeness of the mean result to the true value (should you be so lucky as to know the true result) whilst precision is the closeness of agreement between results from repeat measurements. To keep variability to a minimum we want our test method to be as reproducible as possible, i.e. we want it to have good precision. However, it is not much good having high precision if the test has a large bias and hence poor accuracy. So we want both, and indeed they are related in that poor precision (reproducibility) will contribute to lowering the accuracy.

The most obvious first step is to maintain the calibration of test equipment by frequent routine checks. Generally, the necessity of this action is well appreciated for such apparatus as the force scales of tensile testing machines, and often maintenance and calibration services are offered by the machine manufacturer. There are British and ISO standards[22,23] which deal specifically with the requirements and accuracy of such machines. It is the less obvious sources of machine error which are more frequently neglected, together with the inaccuracy introduced by mal-operation of the apparatus by the tester. It would be tedious to even attempt to mention all the possible accuracy checks, and many specific points will be raised in the following chapters, but it is worth saying, as a general principle, that inaccuracy can arise from the most unlikely sources and the tester must adopt a most suspicious mind not to mention something of a fetish for detail.

In many tests an error in measuring a dimension is directly seen as an error in the final result—when did you last calibrate your dial gauge or was the width of the dumb-bell cutter measured after the last sharpening? Thermometers are taken for granted; have yours been calibrated recently? Is the kink in the stress/strain curve a result of machine friction? How do you know the ozone concentration meter reads correctly, has somebody bent the pointer since you last used the apparatus, or can you honestly say that you conditioned at $23 \pm 2°C$? All these points seem very obvious and simple, but how do you explain the variation found between laboratories if they do not actually happen? The investigation of interlaboratory variability sometimes reveals unusual lapse of care or misunderstanding of the method of operation.

3.2.1 Laboratory Accreditation

To keep apparatus, procedures and people in the best condition to produce reliable results requires systems and control. Almost certainly the best way of achieving this in a testing laboratory is to be subjected to the disciplines of a recognised accreditation scheme. Forms of accreditation have been applied by major purchasers for many years but it is only relatively recently that national schemes have reached prominence.

The British National Laboratory Accreditation Scheme (NATLAS) requires a laboratory to maintain rigorous procedures for anything from the training of staff and the control of test pieces to, most importantly, the calibration of equipment. To maintain their requirements, which are given in deceptively short form in BS 6460,[24,25] is both time consuming and difficult but anything less than these standards is not ensuring the highest possible quality in the output of the laboratory—the results.

Other national schemes have their own specific requirements and ISO Guide 25[26] is the international equivalent of BS 6460. The major forum for accreditation matters is the International Laboratory Accreditation Conference (ILAC) which has produced or is producing reports on many aspects of accreditation including proficiency testing, quality manuals and calibration intervals.

Accreditation authorities such as NATLAS will produce documents giving guidance on how to achieve their requirements. There is also a British Standard, BS 5781,[27] which gives a specification for measurement and calibration systems and is incorporated into the NATLAS requirements. A bibliography on laboratory accreditation has been produced by Bryson *et al.*[28]

3.2.2 Interlaboratory Comparisons and Precision Statements

The ultimate proof of how your results compare to those of others is to take part in interlaboratory testing exercises—preferably before you have a disagreement with a customer. Interlaboratory comparisons are organised for different purposes. An accreditation body may require a laboratory to take part in tests to gain a measure of their proficiency and a standards committee may run trials to aid with the development of a new test method. The plan of a trial for one purpose will probably not be suitable for another purpose. ISO Guide 43[29] gives advice on interlaboratory tests for the purpose of assessing proficiency.

An increasingly important need for interlaboratory testing is to produce precision statements for standard test methods. A precision statement gives the repeatability and reproducibility found for the test method in a properly organised interlaboratory experiment. Repeatability refers to the precision within one laboratory and reproducibility to the precision between different laboratories. The precision figures can be referred to by any laboratory to allow them to estimate what levels of variability they should reasonably expect. A commentary on precision and accuracy in the context of rubber testing has been given by McCormick.[30]

Precision statements have been incorporated into ASTM standards in a systematic way for some time but are not often found in other standards. ISO TC45 has decided that precision statements should be generated for their test methods and for a start have produced a guide[31] to their generation which expands on the international standard on the subject, ISO 5725[32] (British equivalent BS 5497[33]), for the particular case of rubber. The testing work to produce precision statements for all methods will be enormous.

Whether specifically designed to yield precision statements or not, interlaboratory test results contain very valuable information on the variability of tests and perhaps on the parameters which particularly influence variability. A great many have been carried out by standards committees in the course of developing a new method but unfortunately the results are rarely published. Details of some studies in the USA were given at an ASTM symposium[34] and interlaboratory procedures used in the control of standard Malaysian rubber have been reported.[35]

3.3 RELEVANCE AND SIGNIFICANCE

3.3.1 Relevance

If accuracy or repeatability was our only interest we would limit our testing to the most accurate or precise methods. However, we also want our test to be relevant in the sense that the results have a useful meaning in terms of material or product performance. All tests are not equal; some are more meaningful than others and if a test has no relevance in terms of product performance or material consistency, there is not much point in carrying it out. The word significance is sometimes used to mean relevance and applies to the actual test or property measured, but significance is used in this chapter in the statistical sense as in one material being significantly, for example, stronger than another.

This is not the place in the book to consider in detail the relevance of particular parameters in terms of product performance or quality assurance, as the relevance of each test will be discussed in the appropriate section. It is, however, true that many of our test methods use quite arbitrary conditions and procedures which may reduce their relevance when the results are used to predict service performance. If these arbitrary (or not so arbitrary) conditions are changed in any way, the result will also change and hence is very important that exactly the same conditions are used for results intended to be comparable. Failure to do this is equivalent to introducing error.

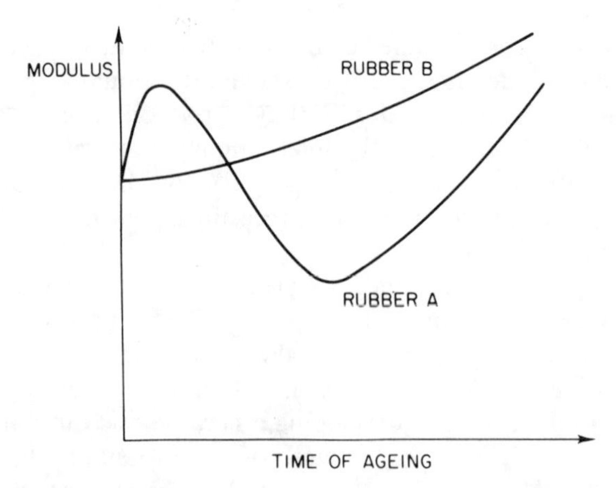

FIG. 3.1. Tensile modulus as a function of time of ageing.

The geometry of test specimens is a notable example of a test method detail because if, as is frequently the case, a fundamental material property cannot be derived from the result independent of test piece geometry, the actual test piece shape and size used are very important. The measurement of electric strength is the sort of example which can be given as a warning. If the result is given in volts per millimetre it is implied that the breakdown voltage is proportional to test piece thickness, which in fact is not generally the case. Unless we know the shape of the breakdown voltage/thickness curve, a test at one thickness is only really relevant for products of that thickness. Note however, that if we were comparing two materials whose breakdown voltage/thickness curves were similar, a test at one thickness would be relevant as regards making a comparison between the two.

In practice it is a common mistake to assume that the relationship between property measured and some parameter such as test piece geometry or time is the same for different materials when in fact it is not. A rather bad example is shown in Fig. 3.1 involving the relationship between tensile modulus and time of accelerated ageing for two rubbers. A comparison between the two at any particular time would be of rather limited use because of the different responses. Furthermore, the tests were made at elevated temperature and the results may or may not be relevant in terms of natural ageing at ambient temperature. This all leads to the conclusion that, in general, tests generating multi-point data, i.e. data as a function of geometry, time, temperature, etc., are likely to yield results of much greater value than a single-point test.

3.3.2 Significance

Significance in the statistical sense is concerned with whether observed differences in results are real or can reasonably be attributed to chance. If the probability of obtaining the observed difference through pure chance is small, say less than 1 in 20, then the difference is said to be significant. This is best illustrated with an example. The set of tensile strength results quoted in Section 3.1 could be compared to other sets obtained on different materials on the same occasion giving say three sets as follows:

Material A	16.8	15.4	16.3	17.7	17.6
Material B	15.67	16.4	14.5	15.8	16.0
Material C	16.4	15.4	14.3	14.7	14.4

The averages of the results for materials A and B are higher than that for C but are they significantly higher? Without the use of statistical tools it is rather difficult to make any sort of judgment. In fact, using a statistical test for significance as is discussed in Section 3.4.2 we can prove that A is significantly greater than C with 95% confidence but that A is not significantly different from B, again with 95% confidence. This is very useful, but the limitations of these findings must be appreciated if one is not to become disenchanted with statistics. The statistical tests proved (with a 1 in 20 chance of being wrong) that results A were significantly greater than C, but not significantly greater than B. They do not prove that compound A is stronger than compound C. We know that results from one sheet of one mix may not be representative of a formulation and would treat these results from a rather inadequate test programme with caution.

In the example given the differences between the average results were relatively small but it is true that tensile strength can be measured accurately with reasonably small variability so that it is not surprising that 10% difference could be proved significant. For other, less reproducible, tests a much greater percentage difference may be needed before the difference can be proved significant. In an electrical resistivity test the mean value for one material may be several times higher than that for a second material but the difference could not be proved significant.

Resistivity tests are generally much more variable than tensile tests and the deduction can be made that significance is not only dependent on the difference between mean values but also on the amount of variability which is inherent in the test.

Circumstances arise where there is an interest in whether there is a significant relationship between two variables, for example natural ageing results and artificial ageing results, i.e. is there a correlation between the two. Rather than make a subjective judgment, statistical tests, as outlined in Section 3.4.4, can be used.

3.3.3 Discrimination

To discriminate between two materials in respect of any property it is necessary to prove a significant difference between two sets of results. We can regard the discriminating power of a test as a measure of the ease with which it will show up real differences between the materials. As was seen in the previous section, the better the reproducibility of the test the smaller the difference between results which can be proved significant, and hence the better the discriminating power. Perhaps more obviously, discriminating power also depends on the magnitude of the difference between results obtained on the two materials. For example, different abrasion machines often show marked differences in the relative performance of materials. If the reproducibility with two machines was similar, the test with the greater discriminating power would be the one that gave the greater relative difference between the materials.

3.4 SAMPLING AND QUALITY CONTROL

The previous section made it clear that the significance of test results depends to a considerable extent on how the physical sample was obtained. The first question, whatever the nature of the work, is whether the samples tested adequately represent the populations being investigated, but in routine quality control there is the added dimension of needing to sample repetitively in time. There is hence a need for a long-term sampling plan and a continuous method for assessing the results.

3.4.1 Sampling

The nature and size of a sample and the frequency of sampling obviously depend on the circumstances. First, the number of test pieces or repeat tests to use per unit item sampled must be decided. Our current standard methods are not consistent, ranging from one to ten or more and it is usually argued, although open to challenge, that the more variable a test the more repeats should be made. There is no doubt that financial considerations have played a large part in the deliberations, witnessed by certain very variable but long-winded methods calling for one test piece only. There is no doubt that to use one test piece only is rarely satisfactory but testing very large numbers will not yield a proportional increase in precision. There is a trend towards five as the preferred number and this level has a lot to recommend it for the more reproducible methods, being just about large enough to make reasonable statistical assessments of variability. An odd number of tests is advantageous if the median (see Section 3.5.1) is to be extracted. In a continuous quality control scheme the number of test pieces used at each point is usually rather less important than the frequency of sampling, i.e. it might be better to use one test piece but check five times as often.

Efficient sampling really boils down to selecting small quantities such that they are truly representative of the much larger whole. The necessity for sheets to be representative of batches and for batches to be representative of the formulation has been mentioned. The direction of test pieces relative to the axes of the sheet and randomisation of their position in the sheet are also important if the sheet cannot be guaranteed homogeneous and isotropic.

When powders are sampled, devices must be used to take representative samples from the sack, drum or other container, bearing in mind that coarse particles tend to separate out.

In the rubber factory, sampling of the product is very much influenced by the fact that rubber production is a batch process and that for moulded products each heat (or lift) constitutes a batch. A common procedure is to sample each batch of compound mixed, but by the time the finished product is rolling off the lines several batches may well be intermixed. The selection of discrete products should preferably be randomised and certainly care must be taken that the sampling procedure is not biased, for example, by sampling at set times which might coincide with a shift change or other external influence. A book of random numbers (a set of tables designed to pick numbers at random without the risk of unconscious bias) is invaluable. Sampling is very much a part of a quality control scheme and information, particularly from a statistical point of view, can be found in the references given in the next section.

3.4.2 Quality Control

There are several versions of the difference between quality control and quality assurance, but whatever the shades of difference between terms we are concerned with maintaining the quality of products to set standards. This embraces the control of incoming materials, the control of compounds produced, and the control of manufacturing processes, and guaranteeing, as far as possible, the quality of the final product. Quality control or assurance schemes utilise physical tests as a most important part of their system. In fact, most of the standardised test methods are principally intended for quality control use and probably, in terms of quantity, the majority of testing carried out is for quality assurance purposes.

Taking quality assurance in a wide sense it is necessary to consider specifications, the relevance of test methods, the accuracy of test methods and the statistically based control schemes which make up the discipline of the quality engineer. We are talking of a specialised subject which happens to involve testing and hence it seems sensible to restrict discussion here to making reference to publications, particularly standards, which are relevant.

A comprehensive tome on quality control is edited by Juran[36] whilst a rather shorter book giving a practical approach to the subject is that by Caplen.[37] There are many others of varying scope. Quality assurance is certainly about people, ultimately it being human fallability that limits the control of quality, and Drury and Fox[38] have considered the subject from this angle. A considerable number of standards have been established in this field and a collection of a number of British standards is given in *BS Handbook 22*.[39] For other standards related to quality control, the ISO and BSI catalogues should be consulted. The level of quality assurance applied in the rubber industry has been investigated by Wain. His reports cover the rubber moulding industry,[40] the polymer and compound supply industry[41] and the views of the customers of the polymer industry.[42] Wain has also produced a guide to writing a quality assurance manual.[43] In recent times there have been a number of conferences and meetings on quality in the rubber industry which is indicative of the general movement towards greater appreciation of the importance of quality in improving profits. However, there appears to be a lack of up to date texts on quality control in our industry.

3.5 TREATMENT OF RESULTS

It may not be entirely logical to consider the treatment of results before experiment design, but one objective of this section is to introduce some of the most useful statistical terms and techniques and it is rather difficult to appreciate experiment design without being familiar with these. The

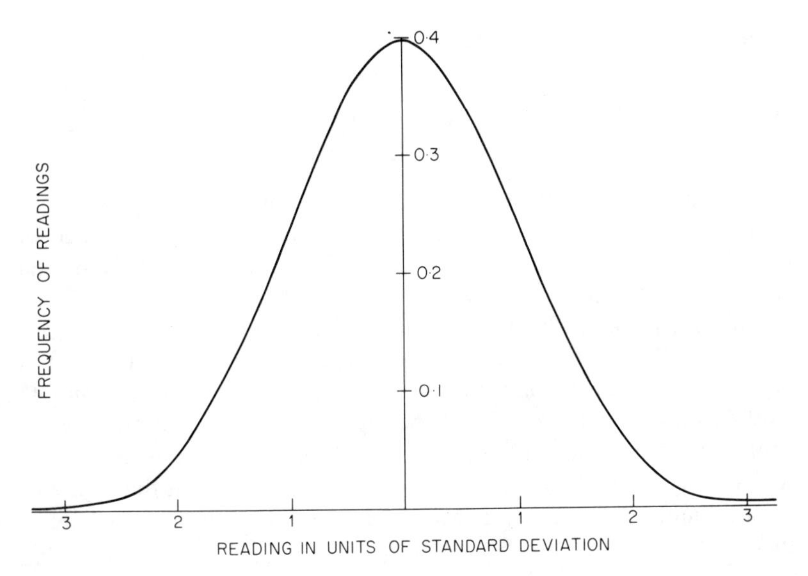

FIG. 3.2. Normal distribution.

form and symbols of BS 2846 'Statistical Interpretation of Data', will, as far as possible, be used. Because of the existence of very clearly written standards and the statistics textbooks referred to at the beginning of the chapter, only brief notes on basic statistical methods will be given here. Incidentally, it is extremely tedious to perform statistical calculations without a programmable calculator or micro-computer.

3.5.1 Distribution of Results

The very fact that a collection of results exhibit variability means that they have a distribution and by far the most common form of distribution is that called normal or Gaussian as shown in Fig. 3.2. A great number of statistical techniques are based on the assumption that the distribution of the data is at least approximate to normal and, for results where this is not the case, it may be convenient to apply simple transformations, such as taking logarithms, to the readings. An important example of a non-Gaussian distribution is the skew distribution known as double exponential which may be exhibited by certain ultimate strength properties, notably tensile strength, and this is shown in Fig. 3.3.

The most common measure of the central tendency of a set of n values is the arithmetic mean defined by

$$\bar{x} = \frac{1}{n}\sum_{i=1}^{n} x_i$$

i.e. the sum of all the values divided by the number of readings.

The median is the middle value when the results are arranged in ascending or descending order. It is less affected by extreme values than the mean but is subject to slightly greater variance. The mean and median are coincident for a normal distribution (Fig. 3.2) but this is not the case for a skew distribution (Fig. 3.3), where the median is in fact a more 'typical' value. The median is calculated from a sample by arranging the results in order and crossing off the highest and lowest values together until only one (or two) are left. If two values are left an estimate of the median may be made by taking their mean. It is apparent that if there is an odd number of results the median can be found simply by inspection and no calculation is involved.

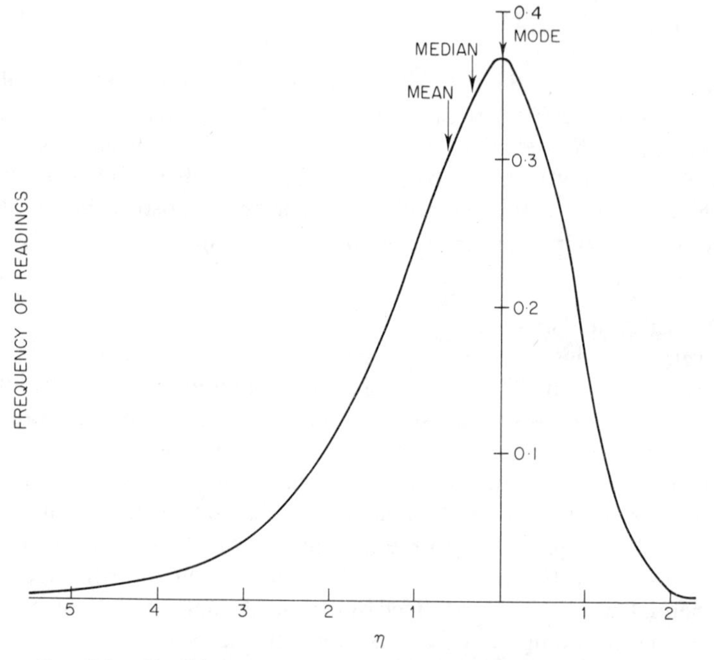

FIG. 3.3. Double exponential distribution, $F(s) = e^{\eta}\exp(-e^{-\eta})$.

The mode is the most frequently occurring value (i.e. the most typical value) and is coincident with both mean and median for a normal distribution. The mode is the best measure of central tendency for a double exponential distribution and hence is in theory the measure which should be calculated for tensile strength results, although in practice it is not widely used. Much of our appreciation of the distribution of tensile strength data results from the work of Kase and Yabuta and a useful account of the earlier work has been given by Heap.[44] The mode is relatively difficult to calculate but estimators are available and a procedure for five results, applicable to a doubly exponential distribution, is as follows:

Arrange the results in decreasing order $x_1 x_2 x_3$, etc., and calculate the mode from:

$$\text{Mode} = 0.327x_1 + 0.269x_2 + 0.217x_3 + 0.153x_4 + 0.034x_5$$

These weighting factors are as given in BS 5324;[21] their derivation is not given and they are not identical to those given by Heap.

Most recently, both the general applicability of a doubly exponential distribution to strength data and the validity of the mode as a measure of central tendency have been questioned by Barker and Smith.[45] They suggest that the median is a better practical parameter to use and can be estimated directly from the results without assumptions about the distribution. If an estimate of the mode was required it can be calculated from:

$$\text{Mode} = x + 0.450\,05\,s$$

They also point to errors in BS 5324.

The most useful measure of dispersion or variability is the standard deviation which is calculated from:

$$s = \sqrt{\frac{\sum_{i=1}^{n} (x_i - \bar{x})^2}{n-1}}$$

It is defined as the square root of the average of the squared deviation of each value from the mean, but $(n-1)$ rather than n is used as the divisor because, especially with a small sample, there is a tendency for the root mean square value to underestimate the standard deviation of the whole population. The standard deviation squared is the variance.

It is often more convenient to use the coefficient of variation which is the standard deviation expressed as a percentage of the mean, i.e.

$$v = \frac{s}{\bar{x}} \times 100\%$$

Then, the relative variability of sets of data having different means can be directly compared.

Perhaps the most obvious measure of dispersion is the range (i.e. the numerical difference between the highest and lowest values) but this is not so useful as the standard deviation as a basis for further statistical calculations. However, for small samples, sound estimates of (or checks on) the standard deviation can be made from the range by multiplying by the appropriate factor as given below:

n	3	4	5	6	7	8	9	10
Factor	0.591	0.486	0.430	0.395	0.370	0.351	0.337	0.325

The frequency of individual readings decreases as they become further from the mean and for the normal distribution approximately 5% may be expected to lie beyond ± 2 standard deviations and 0.3% beyond ± 3 standard deviations.

Unexpectedly high or low results occur in even the best organised tests and there is a temptation to reject such figures as atypical. Although there are statistical tests which could be applied, it makes good practical sense to reject no result unless there is a proven physical reason for so doing. All results should be treated with suspicion, particularly extreme values, but nothing is rejected unless a faulty instrument, a damaged test piece or similar occurrence, can be shown to be the cause. Even faulty test pieces should not necessarily be a reason for rejection if, for example, such flaws were typical of the product or material, or their presence was significant in making an assessment of the product.

3.5.2 Tests of Significance

If a number of samples are taken from a population and tests made on the test pieces in each sample, the results will show a number of sample means distributed about the true population mean and a number of sample standard deviations which are distributed about a population standard deviation. Hence, for any one sample we wish to know how near its mean and standard deviation lie to the true population mean and standard deviation or, more usefully, what reliance we can place on these estimates of the true values.

The usual method of expressing the precision of an estimate is by calculating confidence limits, that is the limits which there is a given probability that the true value will be within.

The confidence limits of the mean are given by:

$$\bar{x} \pm \frac{ts}{\sqrt{n}}$$

where s is the standard deviation, n the number of test pieces and t is a factor which varies with the degree of confidence required and the value of n. Values of t are given in Table 3.1 for 5% probability (i.e. 1 in 20 chance) of the true mean being outside this range and 1% probability (1 in 100 chance) of the true mean being outside this range. The value is taken at number of 'degrees of freedom', $v = n - 1$, using the double-sided table.

If there are two sets of results, for example if the same test has been carried out on two materials, it is of particular interest to know whether the means of the two sets of results are significantly different. The parameter calculated is the least significant difference. If the difference between the means exceeds this there is only a small chance (depending on the probability level chosen) that the two sets of data are from the same population, i.e. there is a good chance that the populations are different. The least significant difference is calculated from:

$$ts\sqrt{\frac{1}{n_1}+\frac{1}{n_2}} \left(ts\sqrt{\frac{2}{n}} \quad \text{if each set contains the same number of results} \right)$$

s is the pooled standard deviation calculated from:

$$s = \sqrt{\frac{(n_1 - 1)s_1^2 + (n_2 - 1)s_2^2}{n_1 + n_2 - 2}}$$

t can be found from Table 3.1 but the number of degrees of freedom in that table is $(n_1 + n_2 - 2)$. Again the double-sided columns are used.

The single-sided table is only used when it is necessary to state whether a population mean is greater (or smaller) than the lower (higher) confidence limit, e.g. it is 95% certain that the population mean is greater than $x - (ts|\sqrt{n})$. t is then taken from the 5% column of the single-sided table.

Sometimes differences in variability as measured by the standard deviations are of as much or more interest as the difference between

TABLE 3.1

PROBABILITY POINTS OF STUDENT'S t DISTRIBUTION

Single-sided test			*Double-sided test*		
v	1%	5%	v	1%	5%
1	31.8	6.31	1	63.7	12.7
2	6.96	2.92	2	9.92	4.30
3	4.54	2.35	3	5.84	3.18
4	3.75	2.13	4	4.60	2.78
5	3.36	2.01	5	4.03	2.57
6	3.14	1.94	6	3.71	2.45
7	3.00	1.89	7	3.50	2.36
8	2.90	1.86	8	3.36	2.31
9	2.82	1.83	9	3.25	2.26
10	2.76	1.81	10	3.17	2.23
11	2.72	1.80	11	3.11	2.20
12	2.68	1.78	12	3.05	2.18
13	2.65	1.77	13	3.01	2.16
14	2.62	1.76	14	2.98	2.14
15	2.60	1.75	15	2.95	2.13
16	2.58	1.75	16	2.92	2.12
17	2.57	1.74	17	2.90	2.11
18	2.55	1.73	18	2.88	2.10
19	2.54	1.73	19	2.86	2.09
20	2.53	1.72	20	2.85	2.09
21	2.52	1.72	21	2.83	2.08
22	2.51	1.72	22	2.82	2.07
23	2.50	1.71	23	2.81	2.07
24	2.49	1.71	24	2.80	2.06
25	2.48	1.71	25	2.79	2.06
26	2.48	1.71	26	2.78	2.06
27	2.47	1.70	27	2.77	2.05
28	2.47	1.70	28	2.76	2.05
29	2.46	1.70	29	2.76	2.05
30	2.46	1.70	30	2.75	2.04
40	2.42	1.68	40	2.70	2.02
60	2.39	1.67	60	2.66	2.00
120	2.36	1.66	120	2.62	1.98
∞	2.33	1.64	∞	2.58	1.96

means. This might be the case if the performances of two instruments giving similar means were being compared. The ratio

$$\frac{s_1^2(\text{larger } s)}{s_2^2(\text{smaller } s)}$$

is calculated.

The two standard deviations are significantly different (at the given probability level) if this ratio is greater than the appropriate F value given in Table 3.2. The numbers of degrees of freedom corresponding to s_1 and s_2, respectively, are $n_1 - 1$ and $n_2 - 1$.

These tests for significance are only valid if the distribution of data is, or is nearly, normal. The test for significant differences of means using pooled standard deviations strictly should only be used if the F ratio test above has not shown the standard deviations to be significantly different.

3.5.3 Analysis of Variance

It has been emphasised earlier in the chapter that variability arises from many sources and it is frequently desirable to attribute proportions of the total measured variability to the individual sources. Analysis of variance is a technique used to isolate and assess the significance of the components of the total variability which are due to each source and is based on the fact that the sum of the variances (not standard deviations) due to independent factors contributing to the total variance, is equal to the total variance. The general principle is that possible sources of variability are identified, an analysis of variance table is constructed giving the sums of squares of deviations from the mean due to each component together with the relevant number of degrees of freedom and the mean squares, and then F ratio tests performed on the mean squares.

The only rapid way to make this easily intelligible is by means of worked examples and there can be as many different examples as there can be designs of experiment. Hence, there is no possibility of giving comprehensive coverage here but examples can be found in the textbooks referenced earlier. A very simple example applicable to a rubber testing situation is given in BS 5324.[21] The analysis of an experiment is completely interwoven with the design of that experiment and many examples of experiment design together with analysis are given by Davies.[3]

3.5.4 Correlation

This section is concerned with assessing the degree of correlation between two independent variables and the description of one variable in terms

TABLE 3.2

5% POINTS OF THE F DISTRIBUTION

v_2 \ v_1	1	2	3	4	5	6	7	8	9	10	12	15	20	24	30	40	60	120	∞
1	161.4	199.5	215.7	224.6	230.2	234.0	236.8	238.9	240.5	241.9	243.9	245.9	248.0	249.1	250.1	251.1	252.2	253.3	254.3
2	18.51	19.00	19.16	19.25	19.30	19.33	19.35	19.37	19.38	19.40	19.41	19.43	19.45	19.45	19.46	19.47	19.48	19.49	19.50
3	10.13	9.55	9.28	9.12	9.01	8.94	8.89	8.85	8.81	8.79	8.74	8.70	8.66	8.64	8.62	8.59	8.57	8.55	8.53
4	7.71	6.94	6.59	6.39	6.26	6.16	6.09	6.04	6.00	5.96	5.91	5.86	5.80	5.77	5.75	5.72	5.69	5.66	5.63
5	6.61	5.79	5.41	5.19	5.05	4.95	4.88	4.82	4.77	4.74	4.68	4.62	4.56	4.53	4.50	4.46	4.43	4.40	4.36
6	5.99	5.14	4.76	4.53	4.39	4.28	4.21	4.15	4.10	4.06	4.00	3.94	3.87	3.84	3.81	3.77	3.74	3.70	3.67
7	5.59	4.74	4.35	4.12	3.97	3.87	3.79	3.73	3.68	3.64	3.57	3.51	3.44	3.41	3.38	3.34	3.30	3.27	3.23
8	5.32	4.46	4.07	3.84	3.69	3.58	3.50	3.44	3.39	3.35	3.28	3.22	3.15	3.12	3.08	3.04	3.01	2.97	2.93
9	5.12	4.26	3.86	3.63	3.48	3.37	3.29	3.23	3.18	3.14	3.07	3.01	2.94	2.90	2.86	2.83	2.79	2.75	2.71
10	4.96	4.10	3.71	3.48	3.33	3.22	3.14	3.07	3.02	2.98	2.91	2.85	2.77	2.74	2.70	2.66	2.62	2.58	2.54
11	4.84	3.98	3.59	3.36	3.20	3.09	3.01	2.95	2.90	2.85	2.79	2.72	2.65	2.61	2.57	2.53	2.49	2.45	2.40
12	4.75	3.89	3.49	3.26	3.11	3.00	2.91	2.85	2.80	2.75	2.69	2.62	2.54	2.51	2.47	2.43	2.38	2.34	2.30
13	4.67	3.81	3.41	3.18	3.03	2.92	2.83	2.77	2.71	2.67	2.60	2.53	2.46	2.42	2.38	2.34	2.30	2.25	2.21
14	4.60	3.74	3.34	3.11	2.96	2.85	2.76	2.70	2.65	2.60	2.53	2.46	2.39	2.35	2.31	2.27	2.22	2.18	2.13
15	4.54	3.68	3.29	3.06	2.90	2.79	2.71	2.64	2.59	2.54	2.48	2.40	2.33	2.29	2.25	2.20	2.16	2.11	2.07
16	4.49	3.63	3.24	3.01	2.85	2.74	2.66	2.59	2.54	2.49	2.42	2.35	2.28	2.24	2.19	2.15	2.11	2.06	2.01
17	4.45	3.59	3.20	2.96	2.81	2.70	2.61	2.55	2.49	2.45	2.38	2.31	2.23	2.19	2.15	2.10	2.06	2.01	1.96
18	4.41	3.55	3.16	2.93	2.77	2.66	2.58	2.51	2.46	2.41	2.34	2.27	2.19	2.15	2.11	2.06	2.02	1.97	1.92
19	4.38	3.52	3.13	2.90	2.74	2.63	2.54	2.48	2.42	2.38	2.31	2.23	2.16	2.11	2.07	2.03	1.98	1.93	1.88
20	4.35	3.49	3.10	2.87	2.71	2.60	2.51	2.45	2.39	2.35	2.28	2.20	2.12	2.08	2.04	1.99	1.95	1.90	1.84
21	4.32	3.47	3.07	2.84	2.68	2.57	2.49	2.42	2.37	2.32	2.25	2.18	2.10	2.05	2.01	1.96	1.92	1.87	1.81
22	4.30	3.44	3.05	2.82	2.66	2.55	2.46	2.40	2.34	2.30	2.23	2.15	2.07	2.03	1.98	1.94	1.89	1.84	1.78
23	4.28	3.42	3.03	2.80	2.64	2.53	2.44	2.37	2.32	2.27	2.20	2.13	2.05	2.01	1.96	1.91	1.86	1.81	1.76
24	4.26	3.40	3.01	2.78	2.62	2.51	2.42	2.36	2.30	2.25	2.18	2.11	2.03	1.98	1.94	1.89	1.84	1.79	1.73
25	4.24	3.39	2.99	2.76	2.60	2.49	2.40	2.34	2.28	2.24	2.16	2.09	2.01	1.96	1.92	1.87	1.82	1.77	1.71
26	4.23	3.37	2.98	2.74	2.59	2.47	2.39	2.32	2.27	2.22	2.15	2.07	1.99	1.95	1.90	1.85	1.80	1.75	1.69
27	4.21	3.35	2.96	2.73	2.57	2.46	2.37	2.31	2.25	2.20	2.13	2.06	1.97	1.93	1.88	1.84	1.79	1.73	1.67
28	4.20	3.34	2.95	2.71	2.56	2.45	2.36	2.29	2.24	2.19	2.12	2.04	1.96	1.91	1.87	1.82	1.77	1.71	1.65
29	4.18	3.33	2.93	2.70	2.55	2.43	2.35	2.27	2.22	2.18	2.10	2.03	1.95	1.90	1.85	1.81	1.75	1.70	1.64
30	4.17	3.32	2.92	2.69	2.53	2.42	2.33	2.25	2.21	2.16	2.09	2.01	1.93	1.89	1.84	1.79	1.74	1.68	1.62
40	4.08	3.23	2.84	2.61	2.45	2.34	2.25	2.18	2.12	2.08	2.00	1.92	1.84	1.79	1.74	1.69	1.64	1.58	1.51
60	4.00	3.15	2.76	2.53	2.37	2.25	2.17	2.10	2.04	1.99	1.92	1.84	1.75	1.70	1.65	1.59	1.53	1.47	1.39
120	3.92	3.07	2.68	2.45	2.29	2.17	2.09	2.02	1.96	1.91	1.83	1.75	1.66	1.61	1.55	1.50	1.43	1.35	1.25
∞	3.84	3.00	2.60	2.37	2.21	2.10	2.01	1.94	1.88	1.83	1.75	1.67	1.57	1.52	1.46	1.39	1.32	1.22	1.00

of another. A simple way of assessing the degree of linear correlation is to calculate the correlation coefficient from:

$$r = \frac{\sum_{i=1}^{n} (x_i - \bar{x})(y_i - \bar{y})}{\sqrt{\sum_{i=1}^{n} (x_i - \bar{x})^2 \sum_{i=1}^{n} (y_i - \bar{y})^2}}$$

where r is the correlation coefficient and x_i and y_i are the values of the two variables.

r can have values between -1 and $+1$; values near to $+1$ signify very good positive correlation, values near to -1 very good negative correlation whereas values near to 0 indicate poor correlation. The value of r which could be said to indicate significant correlation depends on the numbers of pairs of values of x and y used and care needs to be taken in interpreting the result, particularly as, even if the value of r does not indicate a good linear relationship, there may be some more complicated function which gives a better fit.

If the correlation is significant its numerical form can be obtained by the technique known as regression analysis which consists of estimating the best 'fitting' line through the plotted pairs of values of x and y. A linear regression analysis, i.e. fitting a straight line, minimises the sum of squares of deviations of y about the straight line $y = a + bx$ by a suitable choice of a and b.

The calculations are:

$$b = \frac{\sum_{i=1}^{n} (x_i - \bar{x})(y_i - \bar{y})}{\sum_{i=1}^{n} (x_i - \bar{x})^2}$$

$$a = \bar{y} - b\bar{x}$$

The confidence limits of the predicted value of y corresponding to any given values of x, say X, are given by:

$$\pm ts \sqrt{\frac{1}{n} + \frac{(X - \bar{x})^2}{\sum_{i=1}^{n} (x_i - \bar{x})^2}}$$

where s is the standard deviation of the distances of points in the y direction from the line and is calculated from:

$$s = \sqrt{\frac{\sum_{i=1}^{n} (y_i - a - bx_i)^2}{n-2}}$$

n is the number of pairs of readings and t is found from Table 3.1 at point $(n-2)$.

All the dangers of extrapolation apply to using the above equation to predict values outside of the experimental region covered.

In practice the value of regression analysis is much increased if the technique is extended to fitting higher orders of equations (e.g. quadratic) and especially by extending to the analysis of more than two independent variables, which is called multiple regression. In the very brief account of statistical methods being given here, it is not possible to cover this.

3.6 DESIGN OF EXPERIMENTS

All the clever analysis in the world will not compensate for poor experiment design and planning. In particular, it is no use screaming for the statistician to sort out the mess after the testing has been done. The moral, I suggest, is to keep the complexity of an experiment within your design and analysis capability, unless a statistician's help is available, in which case he must be called in at the very beginning.

A great part of experiment planning is 'common sense'; taking care to eliminate as many unwanted sources of variability as possible whilst retaining all the important factors which should be studied. There are numerous well-known tricks which can be applied, such as testing aged and unaged test pieces at the same time to avoid variability due to different operators or machines on different days. It is invaluable to include a 'standard' material as a control, and errors due to specimen preparation in interlaboratory tests can be minimised by central preparation. In the latter case there may be advantage in randomisation of the test pieces or, depending on the reason for the work, a completely opposite form of selection of test pieces may be needed. The same principle applies to blending of batches—does batch to batch variability need to be studied or as far as possible, eliminated? Each stage, each step, each facet of the experiment should be consciously planned at the same time as planning how the analysis of results will be made at the end. The two are inseparable.

It follows that statistical principles play as great a part in design as they do in analysis and indeed the design of experiments is a most important part of statistics in general. Experimental design is dealt with in most of the text books referenced earlier and there are many volumes more particularly devoted to this subject; for example, references 3, 46 and 47. In the field of developing and evaluating test methods and specifications useful comment on the role of statistical design has been made by Werrimont[48] and experimental design from the viewpoint of standards committees developing tests has been presented by Youden.[49] The documents on precision statements referenced earlier are very relevant to the design of interlaboratory tests. Derringer[50] had considered the role of statistical experimental design in testing and problem solving with examples all taken from the rubber industry.

Unfortunately, when one delves into experimental design there appear to be more designs than there are experiments, but it is, however, possible to introduce simply some of the principles which apply generally. First, there is the problem of an uncontrolled variable which varies with time; a reasonable example would be testing materials for abrasion resistance using several test pieces of each. A perfectly valid approach would be to randomise all the test pieces but it most definitely would not be a good idea to test all the test pieces of the first material followed by the second material and so on, because the abradant is probably slowly losing its cutting power.

To reduce the effects of time trends a 'randomised block design' is often used. For example, if five different fillers were to be compared and three batches of compound using each filler were to be tested they could not be all mixed at the same time and there could well be a systematic trend with time. The mixing would be divided into three blocks, each block consisting of one batch using each filler and with the order of using the five fillers randomised:

Block	*Batch No.*	*Filler*
1	1	B
	2	D
	3	C
	4	E
	5	A

Block	*Batch No.*	*Filler*
2	6	B
	7	A
	8	C
	9	E
	10	D
3	11	A
	12	E
	13	B
	14	C
	15	D

In a fatigue test on a machine with seven stations and with seven materials A–G to compare, using seven test pieces of each, a 'Latin square' design could be used to eliminate the variation due to position on the machine and between different runs:

Station	*Run number*						
	1	2	3	4	5	6	7
1	A	B	C	D	E	F	G
2	B	E	A	G	F	D	C
3	C	F	G	B	D	A	E
4	D	G	E	F	C	B	A
5	E	D	B	C	A	G	F
6	F	C	D	A	G	E	B
7	G	A	F	E	B	C	D

Such block designs can also be used to select test pieces from a number of sheets or mixes.

A 'Youden square', which perhaps should be a Youden rectangle, is a form of incomplete Latin square. If we had five laboratories taking part in a testing programme and wished to give each five test pieces and had five sheets of rubber (A–E), each of which yielded four test pieces, we could distribute the test pieces as follows so as to minimise the effect of different sheets and position in the sheet:

Position in sheet	*Laboratory*				
	1	2	3	4	5
1	A	E	D	C	B
2	B	A	E	D	C
3	C	B	A	E	D
4	D	C	B	A	E

The same design could be applied, for example, to arranging five runs of five materials on a machine with four stations.

An extremely common situation is that in which the effects of more than one variable need to be evaluated; for example, the effect on compression set of amounts of accelerator and sulphur and cure time. If it was decided to conduct an experiment with two different levels of accelerator, two levels of sulphur and two cure times, the 'classical approach' would be to vary one factor at a time, i.e. set the accelerator and sulphur level and test after two different cure times, then varying the accelerator and sulphur for the other levels in turn. The disadvantage of this approach is that if, as is likely, the effect of cure time was different at different accelerator or sulphur levels, we would not readily find the best combination. Changing one factor affecting the response to another factor is called interaction and is obviously very important in practice. Interaction can be studied using a factorial design of experiment in which the factors are changed together. For the example given the test sequence would be:

Experiment	*Accelerator level*	*Sulphur level*	*Cure time*
1	A	S	Y
2	B	S	Z
3	A	R	Z
4	B	R	Y
5	A	S	Z
6	B	S	Y
7	A	R	Y
8	B	R	Z

The results can be analysed using the analysis of variance technique, and similar factorial experiments can be constructed for other numbers of factors and levels of each factor.

The subject of experiment design and analysis has only been touched on here and for a proper understanding of the statistical principles and techniques involved it is necessary to consult the textbooks referenced earlier. Simple examples are included in BS 5324[21] and numerous designs together with analysis are given by Davies.[3]

REFERENCES

1. Fisher, R. A. (1973). *Statistical Methods for Research Workers*, 15th Edn, Hafner.
2. Davies, O. L. and Goldsmith, P. L. (1976). *Statistical Methods in Research and Production*, Longman.

3. Davies, O. L. (1978). *The Design and Analysis of Industrial Experiments*, Longman.
4. Cooper, B. E. (1969). *Statistics for Experimentalists*, Pergamon.
5. Jardine (1975). *Statistical Methods for Quality Control*, Heinemann.
6. BS 2846. Statistical Interpretation of Data, Part 1. 1975. Routine Analysis of Quantitative Data.
7. BS 2846: Part 2, 1981. Estimation of the Mean: Confidence Limit.
8. BS 2846: Part 3, 1975. Determination of a Statistical Tolerance Interval.
9. BS 1846: Part 4, 1976. Techniques of Estimation and Tests Relating to Means and Variances.
10. BS 2846: Part 5, 1977. Power of Tests Relating to Means and Variances.
11. BS 2846: Part 6, 1976. Comparison of Two Means in the Case of Paired Observations.
12. BS 2846: Part 7, 1984. Tests for Departure from Normality.
13. ISO 2602, 1980. Estimation of the Mean—Confidence Interval.
14. ISO 3207, 1975. Determination of a Statistical Tolerance Interval.
15. ISO 2854, 1976. Techniques of Estimation and Tests Relating to Means and Variances.
16. ISO 3494, 1976. Power of Tests Relating to Means and Variances.
17. ISO 3301, 1975. Comparison of Two Means in the Case of Paired Observations.
18. ISO DIS 5479. Tests for Departure from Normality.
19. BS 5532, 1978. Statistics—Vocabulary and Symbols.
20. ISO 3534, 1977. Statistics—Vocabulary and Symbols.
21. BS 5324, 1976. Guide to Application of Statistics to Rubber Testing.
22. BS 5214: Part 1, 1975. Testing Machines for Rubbers and Plastics, Constant Rate of Traverse Machines.
23. ISO 5893, 1985. Rubber and Plastics Test Equipment—Tensile, Flexural and Compression Types (Constant Rate of Traverse)—Description.
24. BS 6460: Part 1, 1983. Specification for the General Requirements for the Technical Competence of Testing Laboratories.
25. BS 6460: Part 2. To be published.
26. ISO Guide 25, 1982. General Requirements for the Technical Competence of Testing Laboratories.
27. BS 5781, 1979. Measurement and Calibration Systems.
28. Bryson, J., Drake, L., Hall, W. and Thomas, O. (1982). *Bibliography on Laboratory Accreditation*, National Bureau of Standards.
29. ISO/IEC Guide 43, 1984. Development and Operation of Laboratory Proficiency Testing.
30. McCormick, C., ACS Rubber Division Meeting, Cleveland, Oct. 1979, Paper 15.
31. ISO TC45 Standard Practice, Determination of Precision for Test Method Standards, 1984.
32. ISO 5725, 1981. Determination of Repeatability and Reproducibility by Interlaboratory Tests.
33. BS 5497: Part 1, 1979. Guide for the determination of repeatability and reproducibility for a standard test method.
34. Stiehler, R. D., ASTM, STP553, 1974.

35. Leong, Y. S. and Loke, K. M. International Rubber Conference, Kuala Lumpur, 1975.
36. Juran, J. M. (Ed.) (1962). *Quality Control Handbook*. McGraw-Hill.
37. Caplen, R. H. (1978). *A Practical Approach to Quality Control*. Business Books.
38. Drury, C. G. and Fox, J. G. (1975). *Human Reliability in Quality Control*. Taylor and Francis.
39. *BS Handbook 22, Quality Control*.
40. Wain, B. J., RAPRA Members Report No. 26, 1979.
41. Thorn, A. D. and Wain, B. J., RAPRA Members Report No. 80, 1982.
42. Wain, B. J., RAPRA Members Report No. 58, 1980.
43. Wain, B. J., RAPRA Members Report No. 85, 1983.
44. Heap, R. D. (1965). *Trans. IRI*, **41**(3).
45. Barker, L. R. and Smith, J. F. (1985). *Polym. Test.*, **5**(6).
46. Cochran, W. G. and Cox, G. M. (1956). *Experimental Designs*. John Wiley and Sons.
47. Cox, D. R. (1958). *Planning of Experiments*, John Wiley and Sons.
48. Werrimont, G. (Sept. 1969), Materials Research and Standards.
49. Youden, W. J. (Nov. 1961). Materials and Standards.
50. Derringer, G. C., ACS Rubber Division Meeting, Cleveland, Oct. 1979, Paper 17.

Chapter 4

Preparation of Test Pieces

Except for work on complete products, a test piece must be formed before the test can be carried out. In many cases the test piece can be directly moulded but, particularly when tests on finished products are concerned, the specimens need to be cut and/or buffed to some particular geometric shape. It is convenient to consider separately, first the mixing and moulding leading up to a vulcanised test piece or test sheet and secondly the preparation of test pieces from moulded sheets or products. The preparation of test pieces for tests on raw rubber and unvulcanised compounds will be considered integrally with those tests in Section 6.

4.1 MIXING AND MOULDING

Processing variables can affect to a very great extent the results obtained on the vulcanisate and in fact a great number of physical tests are carried out in order to detect the result of these variables, for example state of cure and dispersion. In a great many cases tests are made on the factory prepared mix or the final product as it is received, but where the experiment involves the laboratory preparation of compounds and its vulcanisation it is sensible to have standard procedures to help reduce as far as possible sources of variability. Such procedures are provided by BS 1674[1] which covers both mills and internal mixers of the 'Banbury' or 'Intermix' type, and also procedures for compression moulding.

A two-roll mill is specified fairly precisely in terms of dimensions, nominally 150 mm diameter × 300 mm long, and temperature control is required to be within $\pm 5\,°C$. A simple procedure is given for determining the clearance between rolls to ± 0.01 mm. However, temperatures and clearances to be used are not specified but must be taken from particular material specifications or, presumably, agreed between parties concerned. This applies also to details of the mixing schedule.

It is doubtless next to impossible to give such details for a reasonably wide range of formulations and hence, if no material standard is being used, the onus is on the operator to devise and record a reproducible procedure. The tolerance allowed on mass of ingredients is fairly tight at 0.25% or 10 mg whichever is the greater and there is a limit on the difference between the sum of the masses of the ingredients and the final mass of the mixed batch of 0.3% for a gum mix and 0.6% for a filled mix.

The ISO equivalent is ISO 2393[2] which differs in several important respects from the British standard. The British standard permits a wider range of mill roll speeds, on the basis that extensive laboratory comparisons, which are not referenced, have shown that within this range there is only limited effect on the properties of the vulcanised mix. Some dimensions and tolerances are appreciably different in the two standards although not in most cases to such a degree as to cause significant differences in the final mix. ISO 2393 covers mill mixing only and not internal mixers, which seems a serious omission considering that internal mixers are now used for most purposes in a great many, perhaps a majority, of laboratories.

The format of BS 1674 is similar for internal mixers as for the mill, in that mixer parameters are fairly closely defined but no precise schedules are given. Tolerances on ingredient and batch masses are given but no tolerances on temperature control. There is less experience of standards for the use of internal mixers than for mills and probably further parameters will be specified in future versions. Nevertheless, this part of BS 1674 provides a very sound basis for obtaining reproducible mixes provided the detailed schedule for any particular mix is properly standardised and adhered to.

The British standard specifies that Mooney viscosity be determined on the completed mix whereas in many instances in practice it would also be desirable that measurements be made using a cure meter. ISO makes no recommendation at all. The conditions and time of storage between mixing and vulcanisation can affect the properties of the vulcanisate and hence in the British standard storage in the dark in a dry atmosphere is specified. The range of time allowed is rather large at a minimum of 2 h, a preferred maximum of 24 h and an absolute maximum of 72 h. It is not made absolutely clear that the mix is cooled to room temperature before the beginning of the storage period. ISO 2393 allows only between 2 and 24 h at one of the standard temperatures of ISO 471,[3] but in addition makes provision for remilling.

It should be noted that the intention of these standards is to provide conditions and procedures which will lead to reproducible mixes, which is all that can be hoped for considering that in general the results from laboratory mills and mixers are not identical with those obtained with full-sized factory equipment.

BS 1674 and ISO 2393 specify cavity moulds for the compression moulding of sheets but the British standard also details the frame type of mould which is widely used in practice. Furthermore, BS 1674 is intended to cover moulds to produce a variety of test pieces and gives an example of a stepped two thickness sheet mould; whereas ISO 2393 is restricted to sheets for cutting tensile dumb-bells. However, the international standard details a mould for ring tensile test pieces which is omitted from BS 1674 on the basis that such test pieces are rarely used in Britain.

As with the mixing section, there are several differences of detail between the two standards including the tolerances on temperature control of the press platens. Generally the British standard is more detailed which, considering the necessity of giving attention to detail to reduce moulding variability, is advantageous.

The most important parameters are the time and temperature of moulding and both standards specify close limits, $\pm 0.5°C$, on the latter. ISO 2393 requires only that the mould is loaded and unloaded as quickly as possible whereas BS 1674 allows 45 s for each of these operations—a detail of difference which could be important.

ASTM standardisation follows the same pattern as ISO and BSI standards in that there is one standard D3182[4] covering the mixing and moulding equipment and general procedures to be adopted, with detailed mixing schedules being given in relevant material standards. D3182 is, like the British equivalent, a variation on the ISO document with considerable rewording, a number of differences in detail and certain additions. The ASTM standard includes the Banbury type of internal mixer but not the Intermix. As regards moulds, D3182 is very similar to ISO 2393 but omits ring test pieces.

Considering that there is perhaps no absolutely correct procedure for mixing and moulding it is not surprising that there is not universal agreement. The essential is that reproducible test pieces are produced and, as the standards make clear, this can only be achieved by applying the tightest possible control on equipment, times, temperatures and procedures.

4.2 CUTTING FROM SHEET

Although it is debatable whether mixing and moulding are strictly part of testing, particularly as these processes are often not under the control of the tester, there is no doubt that the preparation of test pieces from vulcanised sheet or products is part of the testing process. The most common operation is cutting or stamping from sheet, by which means the vast majority of test piece shapes can be produced.

To stamp, for example, a dumb-bell from sheet requires only a die and a press, although a hammer has been known to replace the latter. There has been a tendency in the past to treat stamping as so simple an operation as to merit little attention, despite the fact that the accuracy of the final test result depends very considerably on the accuracy with which the test piece was prepared. The necessary dimensions of the die are given in the relevant test method standard, for example ISO 37[5] for tensile properties but ISO 4661[6] deals specifically with the preparation of test pieces from vulcanised rubber. This standard is being revised editorially and will be numbered ISO 4661:Part 1 to allow the publication of a Part 2 which deals with the preparation of samples for chemical tests. The British equivalent is BS 903:Part A36[7] and is identical to the ISO standard.

The first requirement is that the test piece should be dimensionally accurate but this is not dealt with in ISO 4661, the necessary tolerances and dimensions remaining a subject for the individual test method. The important dimensions can be conveniently checked on a cut test piece using a projection microscope but the dimensions of the cut test piece will not necessarily be identical with the dimensions of the die because of the pressure of the blade deforming the rubber. In the majority of tests it is the test piece dimensions which are those specified.

It is essential that cutters are very sharp and free from nicks or unevenness in the cutting edge which would produce flaws in the test piece. This is especially important for tests involving the measurement of strength, where a flaw would produce premature failure. Even with the sharpest cutter there is a tendency for the cut edges of the test piece to be concave and it is normal to restrict stamping to sheet no thicker than 4 mm as the 'dishing' effect becomes more severe as the thickness increases. Thicker sheet is cut more successfully using a rotating cutter. Dies for stamping can be of two types, fixed blade and changeable blade. A suitable design for the cutting edge profile of a fixed edge blade type is given in ISO 4661 and the standard also points out the necessity for the die to be suitably rigid and the desirability of some form of test piece ejection system. If there is no automatic ejection system some care has to be taken not to damage the cutting edge of the die or the test piece whilst prodding with whatever sharp object has come to hand. Changeable blade type cutters make use of sharpened strips of the steel rather like long single-edged razor blades. These have the obvious advantage of being very sharp when new and are simply replaced when blunt. They are commonly used for simple shapes such as parallel sided strips but, although very successful dumb-bell cutters can be made in this manner,[8] such dies do not appear to be commercially available.

ISO 4661 does not give any details of the press which should be used with the dies for stamping operations and this probably confirms the finding[9] that the particular design of press is not important as long as it operates smoothly and vertically to the test piece surface. A hammer is unlikely to do this! In practice, quite a variety of presses are to be found and, although the choice is largely a matter of personal preference, there are several points which can be considered. Automatic sample ejection has been mentioned, but this is not very easy to combine with rapid interchange of die shapes. Some toggle action presses require rather more force to operate than is convenient for routine use. Recoil types can be operated very rapidly but are found by some people to be difficult to use. For general use there is a lot to be said for the screw action type operated by a large hand-wheel. Motorised presses are only worthwhile if the volume of work is very large.

Rotary cutters can be used to produce discs or rings from thin sheet and are necessary for sheet above about 4 mm thick to prevent distortion.

Generally, such cutters are used on vertical drilling machines and may consist of either annular or part annular blades. A number of designs have been tried including the incorporation of a second blade simultaneously cutting a large diameter disc. No particular design is referenced in ISO 4661 nor is any recommendation given as to suitable speeds of rotation. However, further information can be found in certain test method standards regarding the preparation of the test piece required for that particular test.

The cutting of rubber is made much easier if a lubricant is applied to either the rubber or the cutting blade. A lubricant which has no effect on the rubber mustbe used and a weak solution of detergent in water has been found suitable. It is not normally necessary to lubricate for stamping operations but it is virtually essential when using a rotating cutter.

The effect of blunt cutters on tensile strength was investigated as long ago as 1934 by van Wijk[10] and later by Scott[11] who found that blunt knives lowered tensile strength on ring test pieces by 8%. Chipped cutters could have a greater effect and it is hence essential that only sharp blades are used which, for fixed blade cutters, means frequent sharpening. It would appear that nothing is more simple than to obtain a sharp cutter but it cannot be over emphasised that many low results and cases of poor reproducibility are caused by blunt or chipped cutting dies. People take them for granted but they need hours of attention and sharpening is a very skilled job. This can be done by the manufacturer or by workshop personnel, but only rarely is the necessary facility and expertise available in the laboratory. A technique suitable for the laboratory has been described by Ennor[12] which uses shaped stones in a vertical drilling machine and this procedure is reproduced in ISO 4661. Experience at RAPRA has shown that drilling machines generally revolve too slowly and better results may be obtained using the high speed routerof a plastics test specimen machining apparatus.

It should be noted that the procedure of using cylindrical stones with the die mounted on a tilted base is inaccurate on the curved parts of the die.

4.3 TEST PIECES FROM FINISHED PRODUCTS

It is obviously desirable to make tests wherever possible on the actual finished product rather than on specially prepared test pieces which may have been produced under rather different conditions. Apart from the difficulty of having sufficient bulk in the product to obtain standard test pieces, extra operations may be involved which are time consuming and are likely to lead to lower test results because of destruction of the moulded surface. However, these difficulties can often be overcome satisfactorily by the use of miniaturised test pieces and by careful use of cutting and buffing apparatus.

The additional operations which may be necessary to obtain a test piece from a finished product are cutting from a large block and the reduction of thickness or removal of irregularities.

In practice the cutting from a large product is often carried out in an arbitrary fashion using a variety of knives and hammers. ISO 4661[6] and BS 903:Part A36[7] cover cutting and buffing from products as well as from

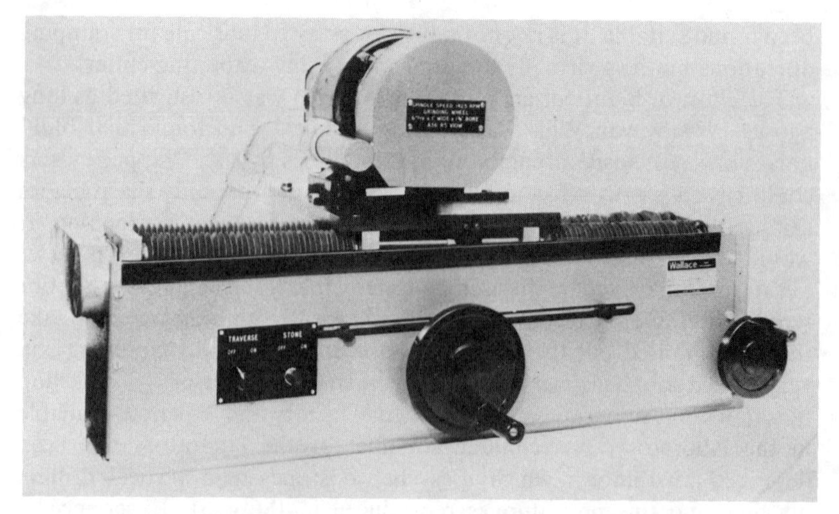

FIG. 4.1. Semi-automatic test piece buffing apparatus.

test sheets and makes reference to rotating knife equipment which can simply mean a powerful bacon slicer and this is indeed very effective if the product can be gripped efficiently. The standard also makes reference to skiving machines based on leather slitting machinery. These are precision machines and not in very widespread use because of high cost. They are however extremely efficient and the best equipment is capable of slitting a visiting card.

Buffing is most effective for the removal of surface irregularities including those left by cutting operations or moderate reduction of thickness. Although it should not be used to remove large quantities of material, when cutting is both quicker and less damaging, rather more than just cloth impressions can conveniently be buffed away. The particular disadvantage of buffing is that heat is generated which may cause significant degradation of the rubber surface and hence the best results are obtained when heat build-up is minimised.

The effect of buffing on tensile properties has been studied by Morley and Scott[13] using a buffing machine with manual test piece feed. They concluded that strength is most affected on soft rubbers, by as much as 15%, whilst for a tyre tread type the drop was about 5% when a smooth surface had been obtained by careful buffing with an 80 grit wheel. Later tests[14] conducted using a machine with automatic test piece feed and specially chosen open structure abrasive wheels (Fig. 4.1) confirmed that tensile strength and elongation at break could be significantly lowered and that modulus may be increased. The magnitude of the effect varied for different rubbers. The more automatic apparatus was considerably safer and more convenient to use as well as being easier for unskilled operators.

Morley and Scott's work suggested that although it was necessary to obtain a smooth finish the depth of cut was not important. However, ISO 4661 places a maximum of 0.2 mm on depth of cut on the assumption (probably erroneous) that heat build-up is minimised by making several light cuts. The standard specifies both abrasive wheel and abrasive band types of buffing machines, claiming that the latter produces less heat build-up. No mandatory parameters for the machines are given but grit sizes and surface speeds are suggested. ASTM D3183[15] deals only with cutting test pieces from other than standard test sheets but the content is very similar to the relevant parts of ISO 4661.

More recently, James and Gilder[16] have made an extremely careful and detailed study of the effects of both buffing and slitting in comparison to moulded surfaces, taking into account the grain from milling operations, grain from the splitting or buffing operation, degree of carbon black distribution and degree of cure.

They concluded that their results confirm previous work in that buffing, when carried out very carefully, gives tensile results which compare reasonably with those from moulded sheets but that the differences vary with the compound. Grain and degree of cure were more dominant than the texture produced by buffing and the former may well account for some of the results obtained by Brown and Jones. For some soft rubber compounds buffing would be better than slitting. They found slitting to be generally as good, and perhaps a little better than buffing, but may be difficult to use outside of the hardness range 55–85 IRHD.

Whichever method of preparation is used they suggest that the surface lines induced by the preparative technique run parallel to the dumb-bell axes and that treating one side of a sheet is as effective as treating both. They also suggest the use of the ratio of tensile strength to elongation at break to indicate whether grain is likely to have a dominant influence on the test results.

It is extremely unlikely that the conditions for buffing and slitting given in the standards are the only conditions which yield good results and there is no one best procedure for all circumstances. Certainly the cutting or buffing operations may significantly affect the measured properties but this can be minimised by careful choice of conditions and procedures.

Furthermore, apparatus is available which allows these operations to be carried out with reasonable convenience.

As would be expected, James and Gilder report that the degree of cure can be crucial. This may be an important factor if cure varies through a thick product or sheet but is especially likely to be an influence when results from sheets and products are compared. A study covering a number of different products[17] revealed very significant and sometimes alarming differences indicating that in many instances the difference in results from sheets and products will be much more due to processing differences than to test piece preparation. This provides considerable support for the school of thought which strongly advocates taking test pieces from the product despite the extra difficulties.

The effect of test piece dimensions on results will be mentioned on several occasions in later chapters in relation to particular tests but it can also be made as a general consideration. If the test piece obtainable from a product is not of standard dimensions this is very likely to significantly influence the result obtained.

REFERENCES

1. BS 1674, 1976. Equipment and General Procedure for Mixing and Vulcanising Test Mixes.
2. ISO 2393, 1973. Rubber Test Mixes—Preparation, Mixing and Vulcanisation—Equipment and Procedures.
3. ISO 471, 1983. Standard Atmospheres for the Conditioning and Testing of Rubber Test Pieces.
4. ASTM D3182–82. Standard Practice for Rubber Materials, Equipment and Procedures for Mixing Standard Compounds and Preparing Standard Vulcanised Sheets.
5. ISO 37, 1977. Determination of Tensile Stress–Strain Properties of Vulcanised Rubber.
6. ISO 4661, 1977. Preparation of Test Pieces.
7. BS 903:Part A36, 1978. Preparation of Test Pieces.
8. Delabertauche, C. and Dean, S.K (April 1968). *J. IRI.*
9. RAPRA. Unpublished information.
10. van Wijk (1934). *Kautschuk*, **10**.
11. Scott, J.R. (April 1944). *J. Rubb. Res.*, **13**(4).
12. Ennor, J.L. (April 1968). *J. IRI.*
13. Morley, J.F. and Scott , J.R. (Oct. 1946). *J. Rubb. Res.*
14. Brown, R.P. and Jones, W.L. (Feb. 1972). *RAPRA Bulletin.*
15. ASTM D3183–84. Preparation of Pieces for Test Purposes from Products.
16. James, D.I. and Gilder, J.S. RAPRA Members Report No. 87, 1984.
17. Brown, R.P. (1980). *Polym. Test*, **2**(1).

Chapter 5

Conditioning and Test Atmospheres

The properties of rubber depend, often to a considerable extent, on its history before test and the atmospheric conditions under which the test was carried out. That is, the results are affected by the age of the rubber, the conditions such as temperature and humidity under which it was stored, any mechanical deformation before test and temperature and humidity at the time of the test. Hence, to produce consistent results it is essential that these factors are controlled within suitable limits.

It is usual to divide the period before test into storage and conditioning; where conditioning refers specifically to the process of bringing the test pieces to the required conditions of temperature and perhaps humidity immediately before test, and storage refers to the period before this back to the time of vulcanisation.

5.1 STORAGE

The properties of vulcanised rubbers change most rapidly immediately after vulcanisation and later, assuming that no accelerating influences are present, the changes become so slow as to be negligible over a period of, say, a few weeks. Hence, it is desirable that a minimum period is allowed between vulcanisation and testing. This minimum period is inevitably arbitrary, but has been standardised in ISO 1826.[1]

The essence of ISO 1826 is now reproduced in most ISO and British standards in words such as:

For all test purposes the minimum time between vulcanisation and testing shall be 16 h. For non-product tests the maximum time between vulcanisation and testing shall be 4 weeks and for evaluations intended to be comparable the tests should, as far as possible, be after the same time interval. For product tests, whenever possible, the time between vulcanisation should not exceed 3 months. In other cases tests shall be made within 2 months of the date of receipt of the product by the customer.

The 16 h period must be treated as an absolute minimum because there is evidence that in some cases several days are necessary before properties have stabilised. There must be a maximum storage period if results are to be representative of the unaged material and again the period quoted is somewhat arbitrary. There is obviously some difficulty in legislating for the storage period on a product when the date of manufacture is unknown but the standard wording makes a very reasonable attempt to cover this case.

Regardless of the time of storage, it is necessary that the rubber is not subjected during this period to high temperatures or other conditions likely to cause deterioration. These include ozone and other chemicals. Most current test method standards take this rather for granted and protection from light is the only condition normally found, for example in various parts of BS 903. The actual temperature and humidity during storage are not critical as conditioning takes place afterwards, but sensible limits would be between 10 and 30°C and below 80% relative humidity. In addition, different rubbers must be separated such that there is no migration of constituents. Special attention needs to be given when the surface condition of the test piece is important, for example ozone or paint staining tests. Although intended to cover the long-term storage of rubber products, it is relevant to call attention to BS 3574, Storage of Vulcanised Rubber.[2]

5.2 CONDITIONING

Virtually all test methods specify a conditioning period, prior to test, in a 'standard atmosphere'. The terms atmosphere, conditioning atmosphere, test atmosphere and reference atmosphere, which are really self-explanatory are defined in ISO 558.[3] ISO 554[4] is a general standard specifying standard atmospheres but there is a separate standard ISO 471[5] (BS 903:Part A35 is identical) which specifically covers temperatures and humidities for conditioning and testing rubber test pieces. The standard conditions are: (a) 23°C and 50% relative humidity, and (b) 27°C and

65% humidity. The latter condition is intended for use in tropical countries.

Where control of temperature only is required, this is either 23 °C or 27 °C, and a further atmosphere where neither temperature nor humidity need be controlled is defined as 'prevailing ambient temperature and humidity'. A note is given to draw attention to the atmosphere 20 °C and 65% relative humidity used for textiles, which may be used if necessary where the product is a composite of textile and rubber. A list of preferred sub-normal and elevated temperatures is also given, which would be used when testing was carried out at other than the normal ambient temperatures.

The normal tolerances are ±2 °C on temperature and ±5% on relative humidity; however, provision is made for closer tolerances if required. For certain tests where the result is very sensitive to temperature change ±1 °C would be specified.

When both temperature and humidity are controlled the standard conditioning time is a minimum of 16 h and where temperature only controlled at 23 °C or 27 °C, a minimum of 3h. At the sub-normal and elevated temperatures it is simply specified that the time should be sufficient for the test piece to reach equilibrium with the environment and no advice is given as to how this time may be estimated. Tables of approximate times required to reach equilibrium in both air and liquid media have been given[6] for a wide range of temperatures and various test piece geometries, and specific instructions are given in some test method standards. Reference 6 is reproduced in the appendix to this chapter.

Generally, 3 h in air is more than sufficient to reach equilibrium at the normal temperatures of 23 °C and 27 °C whatever the test piece geometry. Usually rather shorter times would be used at the sub-normal or elevated temperatures and it is important to note that, whilst a minimum time is required, an excessive time at an elevated temperature may cause significant ageing before test.

It is generally assumed that humidity is not important in most rubber tests and hence conditioning in an atmosphere with control of temperature only is usually specified. Control of humidity is considered necessary in certain cases, however; for example, testing latex rubber and electrical tests. In many instances the 16 h minimum conditioning period will not be sufficient for equilibrium to have been reached, especially with relatively thick test pieces. Hence, all that this conditioning can hope to achieve is to bring test pieces having similar dimensions into more nearly comparable conditions than they would otherwise be. To reach complete moisture equilibrium would in many cases take several days and for thicker test pieces probably weeks.

Inevitably there are certain special cases, for example after accelerated ageing tests ISO and British standards specify conditioning for between 16 h and 6 days which is a stipulation akin to the minimum and maximum storage periods after vulcanisation. The 6 days maximum is on the basis that deteriorated samples may deteriorate further relatively rapidly. Evidence of this does not seem to have been published but there can be little doubt that after exposure to liquids the subsequent delay before testing will be critical because of drying or further chemical attack. This is catered for in ISO and British standards by specifying either testing immediately or after drying at 40 °C and conditioning for 3 h at 23 °C. Another special case is where there has been preparation other than moulding, for example buffing or cutting. In the latest draft ISO tensile testing standard it is specified that testing shall be between 16 h and 72 h after buffing which is based on the evidence of Morley and Scott[7] who found that buffed test pieces show a gradual drop in tensile strength and elongation at break with time. Yet another exception is after mechanical conditioning when at least one standard specifies 16 h–48 h between mechanical conditioning and testing. It becomes apparent that despite standards for storage and conditioning times it is essential to study each test method very carefully to ensure that the exact procedure specified is followed. In this context it should be noted that standards are not consistent as to whether the test piece is cut before or after the conditioning period. Although in most cases it would not matter which was done there will be cases when the result could be affected.

5.3 TESTING CONDITIONS

The object of conditioning is to bring the test piece as nearly as possible into equilibrium with a standard atmosphere and it is reasonable that the test atmosphere should be identical with the conditioning atmosphere. ISO 471 states that this should be the case unless otherwise specified but includes a note to the effect that test pieces conditioned in one atmosphere may be tested in a less rigorous atmosphere in cases where the changes do not affect the results. The most common application of relaxing the testing conditions is after conditioning at 23 °C and 50% relative humidity to test at 23 °C without humidity control and this is perfectly sound practice if the test is performed quickly. It is generally not sound practice to condition at sub-normal or elevated temperature and then test at 23°C unless the test piece is very bulky and the test is made extremely rapidly.

Most testing is carried out in one of the normal standard atmospheres but, as has been mentioned, ISO 471 gives a list of preferred sub-normal and elevated temperatures. This list covering the range of interest for rubber testing is taken from the much wider range given in the general ISO document on preferred test temperatures ISO 3205.[8] It should be noted that the tolerances given in ISO 471 do not completely agree with those in ISO 3205, the tighter tolerances in the former reflecting the increased temperature dependence of rubber compared to many other materials. The preferred temperatures (°C) from ISO 471 are:

−80; −70; −55; −40; −25; −10; 0;
40; 55; 70; 85; 100; 125;
150; 175; 200; 225; 250; 275; 300

Plus, of course, 23°C and 27°C.

The equivalent list given in ASTM D1349[9] is not quite identical, −75 being given instead of −70 and −80°C, and 275 and 300°C being omitted. Additionally, the equivalents in °F are given. This ASTM standard also provides a very abbreviated version of ISO 471 in that it quotes the standard atmosphere 23°C and 50% relative humidity.

ASTM also has a standard, D832,[10] covering conditioning for low temperature testing for which there is no international equivalent. This document explains the underlying theory of testing rubbers for the effects of low temperatures and gives advice on conditioning times.

5.4 APPARATUS FOR CONDITIONING

It would not be appropriate to attempt to deal in any detail here with the specialist subject of environmental enclosures and air conditioning. However, considering the lack of any reference to apparatus in standards such as ISO 471, it is of value to discuss briefly certain aspects of the subject.

5.4.1 Air-Conditioned Rooms

It is rather difficult to operate a rubber physical testing laboratory without air conditioning of the room in terms of temperature. Despite this, a vast number of laboratories do not have this facility, presumably on the grounds that the cost involved is too high. In fact, if the room is reasonably well isolated by doors, not made totally of glass and not of excessive size the cost of installing air conditioning for temperature control using self-contained cooling units is surprisingly low. Complete air conditioning of both temperature and humidity is inevitably more expensive and in most cases not necessary for rubber testing. Relatively few tests call for humidity control and when it is required it is usually possible to use humidity cabinets. Humidity is much more important for testing plastics. Although it is necessary to consult an expert when considering the installation of air conditioning, it may be noted that there is a British standard covering the design of controlled atmosphere laboratories.[11]

5.4.2 Enclosures

In principle the usual type of circulating air laboratory oven can be used for conditioning test pieces when temperature only is controlled. However, for temperatures near to ambient, enclosures equipped with cooling coils are absolutely necessary and frequently it is convenient to use cabinets which also have the facility for the control of relative humidity.

Two types of humidity cabinet are in common use, salt-tray cabinets and moisture-injection type which are covered by BS 3718[12] and 3898,[13] respectively. The salt-tray type of cabinet is very much the simpler of the two types being essentially a temperature controlled enclosure in which the humidity is controlled by the use of saturated salt solutions. Despite the relative simplicity such enclosures must be designed and operated with care if accurate conditions are to be realised and much useful information will be found in the detail of BS 3718. An ISO standard pertaining particularly to plastics,[14] R483, covers similar ground, but in addition describes the use of glycerol solutions to replace the saturated salt solutions. Both standards give tables listing suitable salts to cover a range of humidities. At the condition of most interest, 23°C and 50% relative humidity, sodium dichromate is just within the tolerance of $\pm 5\%$ but no salt is listed to give $50 \pm 2\%$. R483 claims that a glycerol solution will achieve this tolerance if its refractive index is maintained between 1.444 ± 0.002.

The rather more sophisticated injection type of humidity cabinet uses a humidity sensitive device, frequently a wet and dry bulb hygrometer, with mercury contact thermometers to control the injection of moisture into the cabinet from a reservoir. Humidity levels are rather more easily changed with this type of apparatus and some types have the means to cycle both humidity and temperature in a prescribed manner, so extending the range of tests which can be carried out. BS 4864[15] describes closures with a greater operating range than those of BS 3898, including cycling and control of pressure, but this type of apparatus is really more pertinent to accelerated ageing than conditioning as defined here.

5.4.3 Hygrometers

The most useful standard hygrometer is still the wet and dry bulb type which should be used in conditions where air is circulating around the hygrometer at a velocity of not less than 3 m/s. Hygrometric tables for use with wet and dry bulb instruments are given in BS 4833[16] which also contains a bibliography. There are nowadays a large number of different types of hygrometer available commercially and the principles behind many of these are described in *The Instrument Manual*.[17] Simple hair or paper hygrometers can be useful because of their size, and are relatively inexpensive; they are, however, very often inaccurate and must be calibrated frequently.

5.4.4 Thermometers

The ordinary mercury-in-glass thermometer as covered by BS 593[18] is in such common use that it is rather badly taken for granted. In practice, much of the variability associated with testing at a set temperature can be traced to the misuse of thermometers. They should be calibrated frequently, carefully inspected for separation of the mercury, and immersed to the correct depth. The worst errors are usually found with low temperature thermometers (not covered by BS 593) and hence particular care should be taken when conditioning or testing at sub-zero temperatures. Other standards covering thermometers are ISO R653[19] and ISO R654[20] and the British equivalent BS 5074[21] for precision thermometers and BS 1704[22] for general purpose thermometers.

There are of course many types of temperature-measuring instrument in use apart from the liquid-in-glass thermometer and the same principles apply as regards calibration and careful use. Again, extensive descriptions of the various types can be found in *The Instrument Manual*.[17] There is also a guide to selection and use of thermometers, BS 1041.[23–26]

5.4.5 Apparatus for Elevated and Sub-Normal Temperatures

Generally, conditioning at elevated or sub-normal temperatures indicates that the test will be carried out at that same temperature and normally will be carried out in the same enclosure as used for conditioning. That is, the conditioning enclosure forms part of the testing apparatus and hence is likely to take many forms depending on the nature of the test in question. Comment on the types of enclosure available is given in the *RAPRA Guide to Rubber and Plastics Test Equipment*[27] and the requirement for particular tests will be discussed in the relevant section of this book. The ASTM publication on low temperature conditioning[10] has been mentioned but there is also an ISO standard,[28] ISO 3383, which gives general directions for achieving elevated or sub-normal temperatures for rubber testing.

ISO 3383 is useful in that it lists the various types of chamber construction and heat transfer media which may be used and specifies a number of general performance requirements. It is not in any way a detailed document and much of its content is of an elementary nature. It does not specify tolerances on temperature control and it would seem unlikely that its publication will enable the apparatus sections of test method standards to be simplified. An annexe gives advice on the time required for test pieces to reach equilibrium but the information is not in a very convenient form. How the times were derived is not given and they do not wholly agree with the tables referenced previously.[6] A revision of ISO 3383 has been agreed in which the annexe is taken from reference 6.

5.5 MECHANICAL CONDITIONING

It is known, for example from the work of Mullins,[29,30] that vulcanised rubbers containing fillers have their stress/strain curve semi-permanently changed when they are deformed. In particular, there is a reduction in the stiffness measured at any elongation below that to which the rubber has been previously stretched, as illustrated in Fig. 5.1. Repeated stretchings produce successively smaller effects, indicating an approach to an equilibrium stress/strain curve. The effect is not permanent but recovery to the original stress/strain curve may be very slow, even at elevated temperatures.

This effect of pre-stressing is due to physical breakdown of some structure of the filler/rubber composite, its exact nature being unimportant as regards testing procedure. It is self-evident, however, that if a rubber during service is subjected to repeated deformations testing should be carried out after pre-stressing rather than in the initial state where the result may be changed by the effect of the unstable structure.

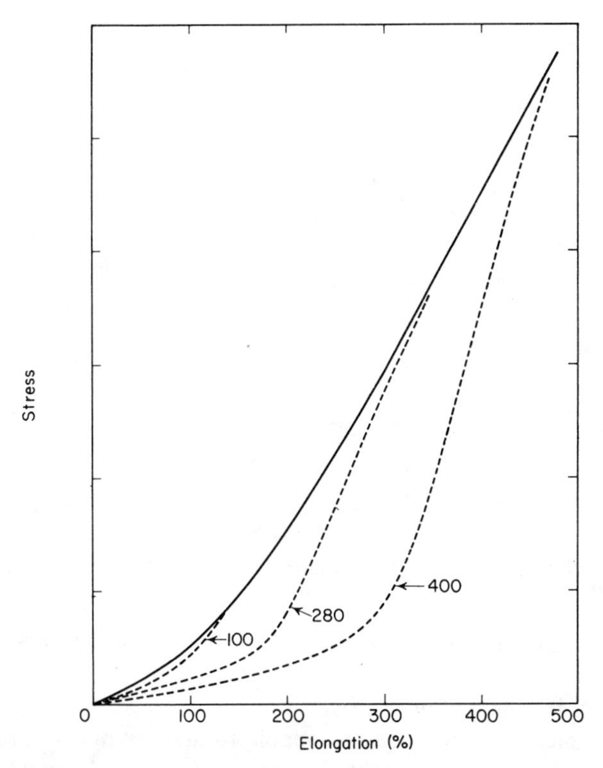

FIG. 5.1. Stress/strain curves for carbon black vulcanisate showing effect of previous stretching. —— before stretching; ---- after stretching to elongation (%) shown.

Pre-stressing or mechanical conditioning takes place automatically in many forms of dynamic test but is very rarely incorporated into other test procedures. This is doubtless not due to ignorance but because mechanical conditioning is an inconvenience and the effect is probably not so important in quality control work. Attempts to incorporate mandatory mechanical conditioning into ISO standards have met with resistance for the reasons above and also because trials with certain particular tests have failed to produce evidence that the effect is large enough to be significant. One cannot help but feel that a more systematic study of the phenomenon covering a wide range of current test methods would be valuable.

There would be little to gain from tabulating an exhaustive survey of standards where mechanical conditioning has been included, but certain

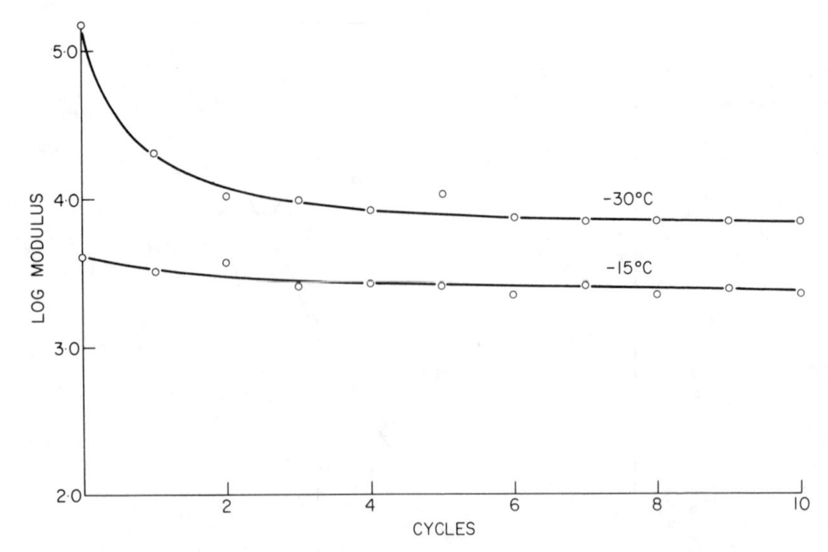

FIG. 5.2. Effect of mechanical conditioning on modulus at low temperatures.

cases which have been given attention are worth noting. ISO 2856, General Requirements for Dynamic Testing[31] suggests that dynamic measurements should only be made after at least six cycles of stress have been applied to the test piece. It would be reasonable to assume that in many cases rather more than six cycles would be needed to approach equilibrium but in any continuous dynamic test any changes would be self-evident. In the simplest form of dynamic test—rebound resilience as covered by the Lupke pendulum in BS 903:Part A8[32]—it is specified that impacts on the same spot are repeated until a constant reading is maintained for three consecutive impacts. The effect of mechanical conditioning on low temperature testing is well illustrated by results obtained using the RAPRA dynamic low temperature tester[33] where the test piece is slowly cycled in tension and the stress monitored. Figure 5.2 shows results for an SBR compound which illustrates the increasing effect of mechanical conditioning as the temperature is lowered.

REFERENCES

1. ISO 1826, 1981. Time Interval between Vulcanisation and Testing.
2. BS 3574, 1963. Storage of Vulcanised Rubber.

3. ISO 558, 1980 Conditioning and Testing—Standard Atmospheres—Definitions.
4. ISO 554, 1976. Standard Atmospheres for Conditioning and/or Testing of Test Pieces.
5. ISO 471, 1983. Standard Temperatures, Humidities and Times for the Conditioning and Testing of Test Pieces.
6. Brown, R.P. and Hands, D. (April, 1973). *RAPRA Members J.*
7. Morley, J.F. and Scott, J.R. (Oct. 1946). *J. Rubb. Res.*, **15**(10).
8. ISO 3205, 1976. Preferred Test Temperatures.
9. ASTM D1349-78. Standard Temperatures and Atmospheres for Testing and Conditioning.
10. ASTM D832-79. Rubber, Conditioning for Low Temperature Testing.
11. BS 4194, 1967 (1984). Design Requirements and Testing of Controlled Atmosphere Laboratories.
12. BS 3718, 1964 (1984). Laboratory Humidity Ovens (Non-Injection Type).
13. BS 3898, 1965 (1984). Laboratory Humidity Ovens (Injection Type).
14. ISO R483, 1966. Methods for Maintaining Constant Relative Humidity in Small Enclosures by Means of Aqueous Solutions.
15. BS 4864, 1973. Design and Testing of Enclosures for Environmental Testing.
16. BS 4833, 1972. Hygrometric Tables for Use in the Testing or Operation of Environmental Enclosures.
17. Miller, J.T. (Ed.) (1975). *The Instrument Manual*, United Trade Press.
18. BS 593, 1974 (1981). Laboratory Thermometers.
19. ISO R653, 1980. Long Solid Stem Thermometers for Precision Use.
20. ISO R654, 1980. Short Solid Stem Thermometers for Precision Use.
21. BS 5074, 1974. Short and Long Stem Thermometers for Precision Use.
22. BS 1704, 1951. General Purpose Thermometers .
23. BS 1041:Part 2.1, 1985. Guide to Selection and Use of Liquid in Glass Thermometers.
24. BS 1041:Part 3, 1969. Industrial Resistance Thermometry.
25. BS 1041:Part 4, 1966. Thermocouples.
26. BS 1041:Part 5, 1972. Radiation Pyrometers.
27. Brown, R.P. (Ed.) (1979). *RAPRA Guide to Rubber and Plastics Test Equipment*.
28. ISO 3383, 1976. General Directions for Achieving Elevated or Sub-Normal Temperatures.
29. Mullins, L. (1948). *Trans. IRI* , **23**, 280.
30. Mullins, L., RABRM Research Memo R. 342, 1948.
31. ISO 2856, 1981. Elastomers—General Requirements for Dynamic Testing.
32. BS 903:Part A8, 1963. Determination of Rebound Resilience.
33. Eagles, A.E. and Fletcher, W. ASTM Sp. Publication 553 Symposium Philadelphia, June 1973.

APPENDIX

THERMAL EQUILIBRIUM TIMES FOR NON-AMBIENT TESTING

If a product is used at non-ambient temperatures, it is logical that it should be tested under these same service conditions. However, elevated or sub-normal temperature testing inevitably requires more complex, and hence more expensive apparatus and normally takes a considerably longer time to carry out. In this respect the conditioning period is a crucial consideration and is often the limiting factor to efficient testing. This is of course of particular importance in quality control testing where both time and money may be of paramount importance.

If the conditioning time used is too short the effect of temperature change on the material will be under-rated, and furthermore the variability of the test results is liable to increase. Alternatively if excessive periods of conditioning are allowed, not only is the test made more expensive but it may become more difficult to distinguish time-dependent effects on the material. Generally technologists have used a rule of thumb method to arrive at conditioning times, frequently biased by financial considerations. Their difficulty has been the almost complete lack of available information and the complication of working with a multitude of different test piece geometries. Relatively few standards are available in which conditioning times are specified and the derivations of the figures quoted are never given. For example in BS 903:Part A2, Tensile Stress/Strain Properties of Rubber it states 'If the test is to be carried out at a temperature other than 20°C the test pieces shall be conditioned at the test temperature immediately prior to testing for a period sufficient to reach substantial temperature equilibrium'.

What are required are the times for the centre of the test piece to come within the tolerance allowed at the temperature in question, calculated for a variety of materials, each in liquid and gaseous media over a range of temperature. Furthermore the figures will be required for all the test piece geometries in normal use. The absence of these figures, or in some cases, inspired guesses at the true values being substituted, has been both because of the lack of basic thermal data and the difficulty of deriving the required information. Thermal property data is certainly still rather sparse but considerable advance has been made in the technique of calculating temperature distributions during heating or cooling.

In this report the methods described by Hands[1] have been extended and applied to standard test pieces for rubbers and plastics subjected to step temperature changes in conditioning chambers and baths. Results are tabulated for three materials in each of three basic geometries which cover almost any test piece configuration in use. The times to equilibrium are given over a range of temperatures for both air and oil heat transfer media.

Test Pieces and Conditions

To make individual calculations for every test piece in current use would clearly be impractical. Fortunately nearly all test pieces used for rubber and plastics fall into three basic geometries. These are cylinders, characterised by diameter and length, flat sheets characterised by thickness and flat strips. A test piece may be considered to approximate to an infinite sheet when both its length and width are greater than four times its thickness, or to an infinite strip when its length only is greater than four times its thickness. The principal form of test piece which does not obviously fall into one of these three geometries is the dumb-bell used in tensile testing. However for our purpose dumb-bells may be considered as flat strips. For most purposes it is the central parallel part of a dumb-bell which is of importance and this width may be taken as the strip width. If the tab ends of the dumb-bell are to be considered as well then the overall width would be used.

Generally, when a measurement at a non-ambient temperature is to be made, the test piece, at room temperature, is introduced into a cabinet preset at the test temperature. It must be expected that the introduction of a test piece will alter the temperature of the cabinet to some extent but in theory at least the test piece is subjected to a step change of temperature. If time is required for the cabinet to regain its set temperature then this must be considered additional to the times given in the tables. The equilibrium times given have been calculated for step changes from 20°C, and normal variations of room temperature around 20°C may be neglected. The range from −50°C to +250°C was chosen to cover the test temperatures most commonly specified.

Tolerances allowed on temperatures of test are commonly ±1°C or ±2°C. Consequently the equilibrium times calculated are those to reach within 1°C of the set temperature which should be adequate for all normal circumstances.

The majority of temperature controlled cabinets used for testing use either air or a liquid as the heat transfer media, although occasionally test pieces may be in contact with metal platens. Separate figures are quoted in the tables for both air and liquid media. Although a variety of liquids are used their heat transfer properties have been considered to be very similar.

Although it is essential that the test piece is given sufficient time to reach equilibrium, in practice the conditioning time is not critical to the nearest minute and consequently all times in the table have been rounded up to the next highest multiple of five minutes.

Calculation of Equilibrium Times

To calculate the time needed for a sample to reach a given temperature we require values for the thermophysical properties of the sample material. To give the tables as wide an application as possible we have considered three classes of materials, crystalline plastics, amorphous plastics and rubbers, and assumed typical values for each group. The chosen are given below.

	Crystalline plastic	*Amorphous plastic*	*Rubber*
Thermal conductivity cal/cm s°C	8×10^{-4}	4×10^{-4}	5×10^{-4}
Thermal diffusivity cm²/s	1.2×10^{-3}	0.9×10^{-3}	1.0×10^{-3}

We also assumed a value for the surface heat transfer coefficient in air of 5×10^{-4} cal/cm²s°C, and in oil of 1.8×10^{-2} cal/cm²s°C.

Temperature Equilibrium Tables

Three tables are presented, one for each of the three basic geometries considered, and for each basic geometry results are tabulated for a range of dimensions. In each case the times to reach equilibrium are given for the three polymer types in both air and oil media over the temperature

CYLINDER

Diameter (mm)	Height (mm)	Temp. (°C)	Time in 1°C off—Equilibrium (min): Rubber, In air	Rubber, In oil	Crystalline, In air	Crystalline, In oil	Amorphous, In air	Amorphous, In oil
64	38	−50	130	75	135	60	130	80
		0	95	60	100	45	95	65
		50	105	65	115	50	105	70
		100	130	80	140	60	130	85
		150	145	85	155	65	145	90
		200	155	90	165	70	155	95
		250	160	90	170	75	160	100
40	30	−50	75	35	85	30	75	40
		0	55	30	60	25	55	35
		50	60	30	70	25	60	35
		100	75	35	85	30	75	45
		150	85	40	95	35	85	45
		200	90	45	100	35	90	50
		250	95	45	105	40	90	50
37	10.2	−50	35	10	40	10	35	10
		0	25	10	30	10	25	10
		50	30	10	35	10	25	10
		100	35	10	40	10	35	10
		150	40	10	45	10	35	10
		200	40	10	50	10	40	15
		250	45	15	50	10	40	15
32	16.5	−50	45	15	50	15	45	20
		0	35	15	40	10	30	15
		50	35	15	45	15	35	15
		100	45	20	55	15	45	20
		150	50	20	60	15	50	20
		200	55	20	65	20	50	20
		250	55	20	65	20	55	25
29	25	−50	50	20	60	20	50	25
		0	40	15	45	15	40	20
		50	45	20	50	15	40	20
		100	55	25	60	20	50	25
		150	60	25	70	20	55	25
		200	65	25	70	25	60	30
		250	65	25	75	25	65	30
28.7	12.7	−50	35	10	40	10	35	15
		0	25	10	30	10	25	10
		50	30	10	35	10	30	10
		100	35	15	45	10	35	15
		150	40	15	50	10	40	15
		200	45	15	50	15	40	15
		250	45	15	55	15	40	15

Diameter (mm)	Height (mm)	Temp. (°C)	Time in 1°C off—Equilibrium (min): Rubber, In air	Rubber, In oil	Crystalline, In air	Crystalline, In oil	Amorphous, In air	Amorphous, In oil
25	20	−50	40	15	50	15	40	20
		0	30	15	35	10	30	15
		50	35	15	40	10	35	15
		100	45	15	50	15	40	20
		150	45	20	55	15	45	20
		200	50	20	60	15	50	20
		250	50	20	60	15	50	20
25	8	−50	25	5	30	5	25	10
		0	20	5	20	5	20	5
		50	20	5	25	5	20	5
		100	25	5	30	5	25	10
		150	30	10	35	5	25	10
		200	30	10	35	10	30	10
		250	30	10	35	10	30	10
25	6.3	−50	20	5	25	5	20	5
		0	15	5	20	5	15	5
		50	20	5	20	5	15	5
		100	20	5	25	5	20	5
		150	25	5	30	5	20	5
		200	25	5	30	5	25	5
		250	25	5	30	5	25	5
13	12.6	−50	20	5	25	5	20	5
		0	15	5	20	5	15	5
		50	20	5	20	5	15	5
		100	20	5	30	5	20	10
		150	25	10	30	5	25	10
		200	25	10	30	5	25	10
		250	25	10	35	5	25	10
13	6.3	−50	15	5	20	5	15	5
		0	10	5	15	5	10	5
		50	15	5	15	5	15	5
		100	15	5	20	5	15	5
		150	20	5	20	5	15	5
		200	20	5	25	5	20	5
		250	20	5	25	5	20	5
9.5	9.5	−50	15	5	5	5	15	5
		0	10	5	5	5	10	5
		50	15	5	5	5	10	5
		100	15	5	5	5	15	5
		150	20	5	5	5	15	5
		200	20	5	5	5	15	5
		250	20	5	5	5	20	5

TABLE 2
FLAT SHEETS

Thickness (mm)	Temperature (°C)	Time in 1°C off—Equilibrium (min): Crystalline plastic		Amorphous plastic		Rubber	
		Air	Oil	Air	Oil	Air	Oil
25	−50	115	80	145	100	135	90
25	0	80	65	105	85	95	75
25	50	90	70	120	90	110	80
25	100	115	80	150	100	140	90
25	150	130	85	165	105	155	95
25	200	135	85	180	110	160	100
25	250	140	90	185	115	170	105
15	−50	60	30	80	40	70	35
15	0	40	25	55	30	50	30
15	50	45	30	65	35	60	30
15	100	60	30	80	40	75	35
15	150	65	35	90	40	80	40
15	200	70	35	95	40	85	40
15	250	75	35	100	45	90	40
10	−50	35	15	50	20	45	15
10	0	25	15	35	15	30	15
10	50	30	15	40	15	35	15
10	100	40	15	50	20	45	20
10	150	40	15	55	20	50	20
10	200	45	15	60	20	55	20
10	250	45	20	60	20	55	20
8	−50	30	10	40	15	35	10
8	0	20	10	30	10	25	10
8	50	25	10	30	10	30	10
8	100	30	10	40	15	35	10
8	150	35	10	45	15	40	10
8	200	35	10	45	15	40	15
8	250	35	15	50	15	45	15
5	−50	20	5	25	5	20	5
5	0	15	5	20	5	15	5
5	50	15	5	20	5	20	5
5	100	20	5	25	5	20	5
5	150	20	5	25	5	25	5
5	200	20	5	30	5	25	5
5	250	20	5	30	10	25	5
3	−50	10	5	15	5	15	5
3	0	10	5	10	5	10	5
3	50	10	5	15	5	10	5
3	100	10	5	15	5	15	5
3	150	15	5	15	5	15	5
3	200	15	5	20	5	15	5
3	250	15	5	20	5	15	5
2	−50	10	5	10	5	10	5
2	0	5	5	10	5	10	5
2	50	5	5	10	5	10	5
2	100	10	5	10	5	10	5
2	150	10	5	10	5	10	5
2	200	10	5	15	5	10	5
2	250	10	5	15	5	10	5
1	−50	5	5	5	5	5	5
1	0	5	5	5	5	5	5
1	50	5	5	5	5	5	5
1	100	5	5	5	5	5	5
1	150	5	5	5	5	5	5
1	200	5	5	5	5	5	5
1	250	5	5	10	5	5	5
0.2	−50	5	5	5	5	5	5
0.2	0	5	5	5	5	5	5
0.2	50	5	5	5	5	5	5
0.2	100	5	5	5	5	5	5
0.2	150	5	5	5	5	5	5
0.2	200	5	5	5	5	5	5
0.2	250	5	5	5	5	5	5

TABLE 3
FLAT STRIPS

Width (mm)	Thickness (mm)	Temp. (°C)	Time in 1°C off—Equilibrium (min) Crystalline plastic		Amorphous plastic		Rubber	
			Air	Oil	Air	Oil	Air	Oil
25.4	12.7	−50	50	10	40	15	45	15
		0	35	10	30	10	30	10
		50	40	10	35	15	35	10
		100	55	10	40	15	45	15
		150	60	15	45	15	50	15
		200	60	15	50	15	50	15
		250	65	15	50	20	55	15
	10	−50	45	10	35	10	35	10
		0	30	5	25	10	25	10
		50	35	10	30	10	30	10
		100	45	10	35	10	35	10
		150	50	10	40	10	40	10
		200	50	10	40	15	40	10
		250	55	10	40	15	45	10
	9.5	−50	40	10	30	10	35	10
		0	30	5	25	10	25	10
		50	35	10	25	10	30	10
		100	40	10	35	10	35	10
		150	45	10	35	10	40	10
		200	50	10	40	10	40	10
		250	50	10	40	10	40	10
	6.5	−50	30	5	25	5	25	5
		0	20	5	15	5	20	5
		50	25	5	20	5	20	5
		100	30	5	25	5	25	5
		150	35	5	25	5	30	5
		200	35	5	25	5	30	5
		250	40	5	30	10	30	5
	5.0	−50	25	5	20	5	20	5
		0	20	5	15	5	15	5
		50	20	5	15	5	15	5
		100	25	5	20	5	20	5
		150	30	5	20	5	20	5
		200	30	5	20	5	25	5
		250	30	5	25	5	25	5

Width (mm)	Thickness (mm)	Temp. (°C)	Time in 1°C off—Equilibrium (min) Crystalline plastic		Amorphous plastic		Rubber	
			Air	Oil	Air	Oil	Air	Oil
	3.0	−50	15	5	10	5	15	5
		0	10	5	10	5	10	5
		50	15	5	10	5	10	5
		100	15	5	10	5	15	5
		150	20	5	15	5	15	5
		200	20	5	15	5	15	5
		250	20	5	15	5	15	5
	2.0	−50	10	5	10	5	10	5
		0	10	5	5	5	10	5
		50	10	5	10	5	10	5
		100	10	5	10	5	10	5
		150	15	5	10	5	10	5
		200	15	5	10	5	10	5
		250	15	5	10	5	10	5
	1.0	−50	5	5	5	5	5	5
		0	5	5	5	5	5	5
		50	5	5	5	5	5	5
		100	5	5	5	5	5	5
		150	10	5	5	5	5	5
		200	10	5	5	5	5	5
		250	10	5	5	5	5	5
15.0	15.0	−50	45	10	35	15	35	10
		0	35	10	25	10	30	10
		50	35	10	30	10	30	10
		100	45	10	35	15	40	10
		150	50	10	40	15	40	15
		200	55	15	40	15	45	15
		250	55	15	45	15	45	15
12.7	12.7	−50	35	10	30	10	30	10
		0	25	5	20	10	25	10
		50	30	10	25	10	25	10
		100	40	10	30	10	30	10
		150	40	10	35	10	35	10
		200	45	10	35	10	35	10
		250	45	10	35	10	40	10

(continued)

TABLE 3—*contd.*

FLAT STRIPS

Width (mm)	Thickness (mm)	Temp. (°C)	Time in 1°C off—Equilibrium (min) Crystalline plastic		Amorphous plastic		Rubber	
			Air	Oil	Air	Oil	Air	Oil
12.7	10.0	−50	35	5	25	10	25	10
		0	25	5	20	5	20	5
		50	25	5	20	5	20	5
		100	35	5	25	10	30	10
		150	35	10	30	10	30	10
		200	40	10	30	10	30	10
		250	40	10	30	10	35	10
	9.5	−50	30	5	25	10	25	10
		0	25	5	20	5	20	5
		50	25	5	20	5	20	5
		100	35	5	25	10	25	10
		150	35	10	30	10	30	10
		200	40	10	30	10	30	10
		250	40	10	30	10	35	10
	6.5	−50	25	5	20	5	20	5
		0	20	5	15	5	15	5
		50	20	5	15	5	15	5
		100	25	5	20	5	20	5
		150	30	5	20	5	25	5
		200	30	5	25	5	25	5
		250	30	5	25	5	25	5
	5.0	−50	20	5	15	5	15	5
		0	15	5	10	5	15	5
		50	15	5	15	5	15	5
		100	20	5	15	5	20	5
		150	25	5	20	5	20	5
		200	25	5	20	5	20	5
		250	25	5	20	5	20	5
	3.2	−50	15	5	10	5	15	5
		0	10	5	10	5	10	5
		50	15	5	10	5	10	5
		100	15	5	10	5	15	5
		150	15	5	15	5	15	5
		200	20	5	15	5	15	5
		250	20	5	15	5	15	5

Width (mm)	Thickness (mm)	Temp. (°C)	Time in 1°C off—Equilibrium (min) Crystalline plastic		Amorphous plastic		Rubber	
			Air	Oil	Air	Oil	Air	Oil
	3.0	−50	15	5	10	5	10	5
		0	10	5	10	5	10	5
		50	10	5	10	5	10	5
		100	15	5	10	5	10	5
		150	15	5	10	5	10	5
		200	15	5	10	5	10	5
		250	15	5	10	5	10	5
	2.0	−50	10	5	10	5	10	5
		0	10	5	5	5	5	5
		50	10	5	5	5	5	5
		100	10	5	10	5	10	5
		150	10	5	10	5	10	5
		200	10	5	10	5	10	5
		250	10	5	10	5	10	5
	1.52	−50	10	5	5	5	5	5
		0	5	5	5	5	5	5
		50	5	5	5	5	5	5
		100	10	5	5	5	5	5
		150	10	5	5	5	10	5
		200	10	5	10	5	10	5
		250	10	5	10	5	10	5
	1.0	−50	5	5	5	5	5	5
		0	5	5	5	5	5	5
		50	5	5	5	5	5	5
		100	5	5	5	5	5	5
		150	5	5	5	5	5	5
		200	10	5	5	5	5	5
		250	10	5	5	5	5	5
4.0	12.7	−50	20	5	15	5	15	5
		0	15	5	10	5	10	5
		50	15	5	10	5	10	5
		100	20	5	15	5	15	5
		150	20	5	15	5	15	5
		200	20	5	15	5	15	5
		250	20	5	15	5	20	5

	3.0	−50	15	5	10	5	10	5
		0	10	5	10	5	10	5
		50	10	5	10	5	10	5
		100	15	5	10	5	10	5
		150	15	5	10	5	15	5
		200	15	5	15	5	15	5
		250	20	5	15	5	15	5
	2.0	−50	10	5	10	5	10	5
		0	10	5	5	5	5	5
		50	10	5	5	5	10	5
		100	10	5	10	5	10	5
		150	10	5	10	5	10	5
		200	15	5	10	5	10	5
		250	15	5	10	5	10	5
	1.0	−50	5	5	5	5	5	5
		0	5	5	5	5	5	5
		50	5	5	5	5	5	5
		100	5	5	5	5	5	5
		150	10	5	5	5	5	5
		200	10	5	5	5	5	5
		250	10	5	5	5	5	5
6.35	12.7	−50	25	5	20	5	20	5
		0	20	5	15	5	15	5
		50	20	5	15	5	15	5
		100	25	5	20	5	20	5
		150	30	5	20	5	25	5
		200	30	5	20	5	25	5
		250	30	5	25	5	25	5
	10.0	−50	25	5	15	5	20	5
		0	15	5	15	5	15	5
		50	20	5	15	5	15	5
		100	25	5	20	5	20	5
		150	25	5	20	5	20	5
		200	25	5	20	5	20	5
		250	30	5	20	5	25	5
	6.5	−50	20	5	15	5	15	5
		0	15	5	10	5	10	5
		50	15	5	10	5	15	5
		100	20	5	15	5	15	5
		150	20	5	15	5	15	5
		200	25	5	15	5	20	5
		250	25	5	20	5	20	5
	5.0	−50	15	5	15	5	15	5
		0	15	5	10	5	10	5
		50	15	5	10	5	10	5
		100	15	5	15	5	15	5
		150	20	5	15	5	15	5
		200	20	5	15	5	15	5
		250	20	5	15	5	15	5
	10.0	−50	15	5	15	5	15	5
		0	15	5	10	5	10	5
		50	15	5	10	5	10	5
		100	15	5	15	5	15	5
		150	20	5	15	5	15	5
		200	20	5	15	5	15	5
		250	20	5	15	5	15	5
	6.5	−50	15	5	10	5	10	5
		0	10	5	10	5	10	5
		50	10	5	10	5	10	5
		100	15	5	10	5	10	5
		150	15	5	15	5	15	5
		200	20	5	15	5	15	5
		250	20	5	15	5	15	5
	5.0	−50	15	5	10	5	10	5
		0	10	5	10	5	10	5
		50	10	5	10	5	10	5
		100	15	5	10	5	10	5
		150	15	5	10	5	10	5
		200	15	5	10	5	15	5
		250	15	5	10	5	15	5
	3.0	−50	10	5	10	5	10	5
		0	10	5	5	5	5	5
		50	10	5	5	5	10	5
		100	10	5	10	5	10	5
		150	10	5	10	5	10	5
		200	15	5	10	5	10	5
		250	15	5	10	5	10	5
	2.0	−50	10	5	5	5	5	5
		0	5	5	5	5	5	5
		50	10	5	5	5	5	5
		100	10	5	5	5	10	5
		150	10	5	10	5	10	5
		200	10	5	10	5	10	5
		250	10	5	10	5	10	5
	1.0	−50	5	5	5	5	5	5
		0	5	5	5	5	5	5
		50	5	5	5	5	5	5
		100	5	5	5	5	5	5
		150	5	5	5	5	5	5
		200	5	5	5	5	5	5
		250	5	5	5	5	5	5

range −50°C to +250°C. It must be remembered that the times given have been rounded up to the nearest five minutes and represent the time for the centre of the test piece to come within one degree of the oven temperature when subjected to a step change from 20°C. For intermediate temperature changes the time for the next highest temperature in the table should be used.

Cylinders
Cylindrical test pieces are frequently used in compression stress/strain and compression set tests and also for some resilience and abrasion tests. In the table cylinders have been characterised by diameter and length, specific sizes having been chosen to correspond with currently used standard test pieces. For example the 28.7mm × 12.7mm cylinder is that specified in BS903:Part A4, Compression Stress/Strain.

Flat Sheets
Flat sheets refer to test pieces where both length and width are large compared with thickness. This covers, for example, specimens for falling weight impact strength measurements and sheets used for hardness and indentation tests. Because the thickness used is often not closely specified results are given for a range of thicknesses up to 25mm and for any intermediate thickness the time given for the next thickest sheet in the table should be used.

Flat Strips
This is the most important table because the majority of specimens for mechanical tests are of this form. There are, for example, the flexural tests and many impact tests and, the largest group of all, the tensile tests. Dumb-bells, although shaped, are essentially of strip form and may be treated for most purposes as strips of the same width as the central parallel portion. For example, the Type 1 dumb-bell of BS903:Part A2 (Fig. 301.9 of BS2782) would be taken as a 6mm wide strip. Also, ring test pieces may be considered as strips with dimensions equal to the cross sections of the ring. Results are given for a range of widths up to 25mm. Above this the strip could be treated as a flat sheet. At each width a range of thicknesses are tabulated which includes some specific to a particular test. Again, for any intermediate dimensions the time for the next largest strip should be used. It should also be noted that the terms width and thickness are completely interchangeable.

References
1. Hands, D. (July 1971). *RAPRA Technical Review*, No. 60.

Chapter 6

Tests on Unvulcanised Rubbers

Most of this book is concerned with tests on vulcanised rubber which is the state in which the consumer receives the product. The consumer is generally only interested in the properties of the vulcanised product whereas the supplier is also concerned about the properties of the raw and compounded but unvulcanised rubber. His interest is in the control of his raw materials as regards their processing qualities and the likelihood of them producing satisfactory vulcanisates. Consequently, the tests are of immediate concern in the rubber factory and great emphasis has been placed on the development of methods and apparatus which provide very rapid quality control data. Processibility is a particularly ill-defined term and rubber frequently manages to exhibit a processing quirk which does not show up in laboratory tests. Thus there is continuing effort to develop tests which more realistically predict processing behaviour. As evidence of the effort some 70 papers on the subject were abstracted over the last five years.

The principal properties requiring measurement are viscoelastic flow behaviour and curing characteristics. Cure rate or, more generally, curing characteristics is a reasonably precise concept but viscoelastic flow behaviour is used here as a convenient term to cover plasticity, viscosity and other parameters which collectively are in fact 'processibility'. Plasticity measurements were the first processibility tests but as it became apparent that the complicated flow behaviour of rubber demanded more searching tests of its viscoelastic behaviour so the scope of tests was extended. A comprehensive review of processibility testing has been given by Norman and Johnson.[1]

6.1 STANDARD METHODS FOR PARTICULAR POLYMERS

A number of international standards have been published which give specific test mix recipes and evaluation procedures for particular polymers. These are recorded in references 2–10 and similar British and ASTM methods also exist. These standards refer to the general test methods and standard methods of preparation which are discussed below but include additional detail relevant to the polymer in question. They also include very limited testing on the vulcanisates. The main purpose of these standards is to provide a basis for comparison and evaluation of particular polymers as regards their processing and vulcanisation characteristics; their scope does not extend to the general physical properties of vulcanisates.

6.2 SAMPLE PREPARATION

Mixing and vulcanisation have been considered under the general heading of preparation of test pieces in Chapter 4. Where compounded but unvulcanised rubbers are to be tested, the same standard mixing procedures will be relevant together with further details relevant to particular polymers as referred to in Section 6.1 above.

Additional, general procedures have been standardised for the tests discussed in this chapter. ISO 1796[11] gives very simple procedures for homogenising a sample by milling under specified conditions. Although the title and the procedure indicate that the method is intended for raw rubber, the list of tests which may be carried out on the homogenised sample includes rate of cure. The object of such sample preparation is to ensure that the sample is homogeneous and to get it into a suitable shape for test. Inevitably this must involve working on the mill which will alter the characteristics of the rubber. The standard conditions given in ISO 1796 are intended to minimise as far as possible the effects of milling whilst giving a satisfactory sample and, although other conditions may be equally suitable, it is essential that the same method is rigorously followed in comparative tests.

After homogenising, ISO 1796 indicates how test portions should be taken for various tests and gives a sequence of operations which should be followed when making a number of different tests.

BS 6315[12] gives essentially the same procedures together with the instructions on sampling raw rubber in bales which is given in ISO 1795.[13]

6.3 VISCOELASTIC FLOW BEHAVIOUR

Plasticity can be defined as 'ease of deformation' so that a highly plastic rubber is one that deforms or flows easily. Viscosity is the resistance to plastic deformation or flow and hence the inverse of plasticity. It is defined as shear stress/shear rate. Unfortunately the terms are often used indiscriminately, for example the result of a test may be in units of stiffness but is called plasticity. Unvulcanised rubbers are not totally plastic or viscous but exhibit some elastic behaviour and 'plasticity tests' have been devised which measure the elastic as well as the plastic component of deformation. Consequently, when such terms as plasticity and viscosity are used care should be taken to ascertain exactly what is meant by them.

Before describing instruments and test methods currently used it is desirable to briefly consider some of the aspects of the flow or rheological properties of unvulcanised rubbers to draw attention to the difficulties and limitations associated with these methods.

Figure 6.1 shows the possible flow curves for two rubbers (curves B and C) together with that for a material exhibiting Newtonian flow (i.e. shear rate proportional to stress) (curve A). The approximate shear rates realised in various rubber processing operations are also noted. Firstly rubber does not exhibit simple Newtonian flow characteristics nor can its behaviour be accurately represented by a power law as this would also

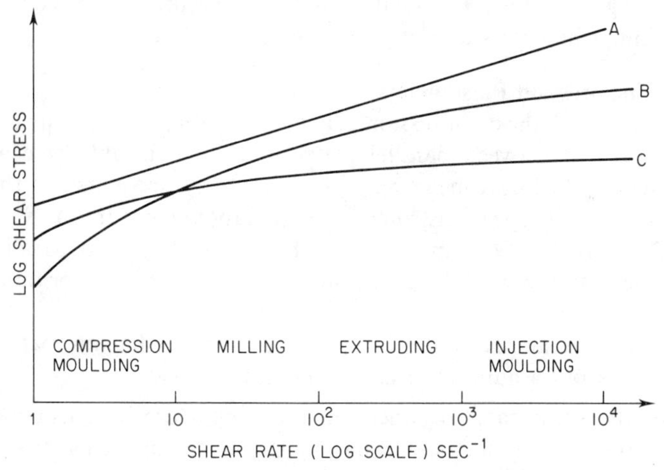

FIG. 6.1. Shear stress/shear rate curves.

give a straight line on a log–log plot. Also the viscosity will change drastically with temperature. The important consequence is that the flow properties of rubber cannot be represented by a single measurement.

Furthermore, any measurement of flow properties should be made at the shear rate of interest. In the example, rubber B has a lower shear stress at low strain rates but has the higher stress of the two at very high rates. Unfortunately the shear rates encountered in rubber processing operations can be very high ($10^4\,s^{-1}$) whereas many conventional plasticity tests operate at much lower rates (roughly in the range 0.0025 to $1\,s^{-1}$). A plasticity test operating at a shear rate of less than $1\,s^{-1}$ is quite likely to yield results which do not correlate with injection moulding behaviour.

Filled rubbers can undergo profound changes in plasticity as a result of storage or of deformation. Storage leads to the formation of filler/filler and filler/rubber 'structure' which is more or less broken down by subsequent deformation. During processing the 'structure' will be broken down by the rapid shearing and considerable heat generated. Even if a laboratory test applies sufficient rapid shearing to break down the structure there may be difficulty in dissipating the heat quickly.

It is clear that to get an accurate idea of how a rubber behaves in the nip of a mill or the die of an extruder is a difficult problem which has not been solved by the existing plasticity tests. Hence, plasticity tests have mostly been used as a check on the uniformity of repeat batches. As Fig. 6.1 shows this may not be satisfactory as no difference in behaviour under the test conditions does not mean that there is not a difference under processing conditions. Fortunately, in practice repeat batches of basically very similar materials do not yield intersecting flow curves.

6.3.1 Compression Plastimeters

The principle of the compression plastimeter is very simple—the test piece is compressed between parallel plates under a constant force and the compressed thickness measured. This simplicity accounts for the early adoption of this type of instrument and its subsequent continued popularity. The work of Williams[14] led to the first widely used parallel plate instrument and eventually to various modified forms all working on the same principle.

Apart from simplicity the compression principle has no real inherent advantages but a number of disadvantages:

(a) the shear rates produced in the rubber are low, usually below $0.1\,s^{-1}$ although somewhat higher (up to almost $1\,s^{-1}$) in the so-called rapid plastimeters.

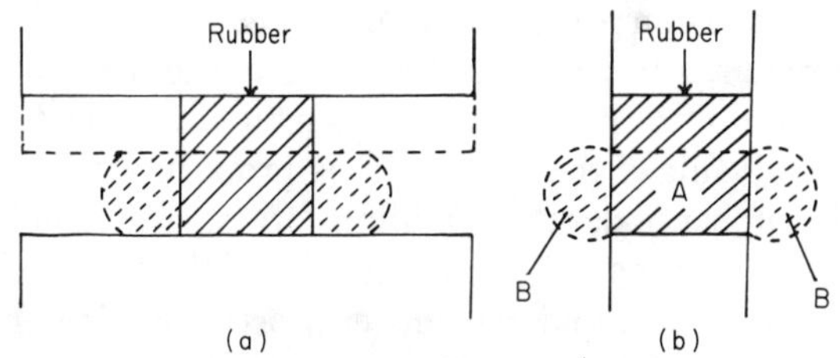

FIG. 6.2. Forms of parallel plate compression test for plasticity. (a) 'Plate' test; (b) 'disc' test. Broken lines and shading show position of upper plate and shape of test piece after compression.

(b) The rubber is not deformed sufficiently to break down any structure effects in black reinforced materials.
(c) The flow produced by compression is extremely complex, the shear rate is not uniform throughout the test piece and changes during the course of the test. Consequently it is virtually impossible to deduce fundamental rheological parameters of the rubber.

Nevertheless, compression plastimeters have been found very useful for routine testing, particularly of uncompounded rubber, where only basically similar materials are compared.

There are basically two forms of parallel plate compression plastimeter: (a) with both compression plates much larger than the test piece (Fig. 6.2(a), 'plate' test; and (b) with one or both plates of approximately the same diameter as the test piece (Fig. 6.2(b), 'disc' test).

In the plate test the test piece area increases and hence the pressure decreases as the rubber spreads out, whereas in the disc test the test piece area remains effectively constant because the excess material (B in Fig. 6.2(b)) is outside the compression zone A. However, although the compression pressure remains constant the shear stresses in the rubber vary as its thickness is decreased.[15,16] A more important advantage of the disc test is that the result is less affected by variations in test piece volume;[17,18] of the order of $\pm 5\%$ can be allowed in the disc method as against $\pm 1\%$ in the plate method. On the other hand, the initial test piece shape factor (ratio of height to diameter) influences the result more in the disc test than in the plate test.[19] Consequently there is an advantage in precompressing the test piece to constant thickness before commencing the test proper. Precompression has another advantage in that the thinner test piece can be brought to the test temperature more quickly and this is the basis of the various 'rapid' plastimeters.

The basic compression plastimeter principle can be modified by measuring the force required to compress the test piece to a given thickness in a given time. This was the principle adopted in, for example, the Defometer and it has the advantage that this force is proportional to the effective viscosity of the rubber under the conditions of test, although this viscosity is an average for the range of shear rates throughout the rubber.

ISO 2007[20] specifies a rapid plastimeter procedure using an instrument with one platen either 7.3, 10 or 14 mm diameter and the other platen 'of larger diameter than the first' (i.e. disc type method). The size of the first platen is chosen such that the measured plasticity is between 20 and 85. The depth of the smaller platen should be 4.5 mm and not 3.2 mm as stated in the standard. This should have been changed in the light of work demonstrating that the larger size gave better reproducibility. At least one commercial instrument was physically changed. The test piece is cut with a punch which will give a constant volume of 0.40 ± 0.04 cm,^{3} the thickness being approximately 3 mm and the diameter approximately 13 mm. The test piece is precompressed to a thickness of 1 ± 0.01 mm within 2 s and heated for 15 s. The test load of 100 N is then applied for 15 s when the test piece thickness is measured. The usual temperature of test is 100 °C and the result is expressed as the thickness of the test piece at the end of the test in units of 0.01 mm and called the 'rapid plasticity number'. The Wallace rapid plastimeter and presumably other commercial instruments conform to this specification but it would be sensible to check with the manufacturers. A technically identical method is given in BS 1673:Part 3.[21]

ASTM D926[22] specifies a plate type test based on the Williams plastimeter with plates 4 cm in diameter. The test piece is 2.00 ± 0.02 cm^3 in volume and can conveniently be a cylinder 16 mm diameter and 10 mm thick. As discussed above, a close tolerance on volume is necessary for this type of plastimeter. The test piece is preheated for 15 min (the temperature of test is usually 70 °C or 100 °C) and compressed under a force of 49 N. The thickness of the compressed test piece is measured in mm and this value multiplied by 100 quoted as the plasticity number. No one standard time of compression is given in the standard but times between 3 and 10 min are suggested, which must of course be reported.

The ASTM method also gives procedures for measuring the recovery of the test piece after removal of the load. In procedure A the test piece is removed from the plastimeter, allowed to cool for 1 min and its height measured. The 'recovery value' is reported as the difference between plasticity number and recovered height. In procedure B the test piece is compressed, not under a fixed load, but to a fixed height of 5 mm for 30 s. It is then allowed to recover for 5 min at the test temperature and its height measured. The recovery value is in this case the increase in height above the 5 mm in multiples of 0.01 mm.

The measurement of recovery is intended to be a measure of the elastic component of softness but it is rather debatable whether the strain rates and recovery times in the ASTM procedure yield results relevant to processing conditions. Recovery at room temperature as in procedure A must be liable to lead to variable results. ISO is now producing a similar standard method.

Koopmann[23] has adopted the Defo compression test to produce figures for viscosity, shear rate dependence, elasticity (recovery) and a coefficient of elasticity representing the shear rate dependence of the elastic behaviour. His procedure is based on compressing several test pieces under different loads followed by recovery for the same length of time, yielding curves for both the loading and recovery phases of test piece height against time. The shear rate dependence coefficient is calculated on the basis that at the shear rates of the test a power law relationship was sufficiently closely followed for the raw rubber tested. By repeating the compression and recovery sequence on the same test piece he also deduced a rheological fatigue factor.

6.3.2 Plasticity Retention Index

Most ageing tests are carried out on vulcanised compounds, but there has been a need to assess the oxidative effects of storage on natural rubber. Various accelerated procedures using ovens or infra-red lamps have been used with visual assessment of deterioration. Now a more satisfactory procedure based on the measurement of plasticity after oven ageing has been standardised as ISO 2930[24] and the result is known as the plasticity retention index.

The plasticity measurements are made before and after ageing with a parallel plate compression instrument in accordance with ISO 2007.[20] Very close control of temperature is required (± 0.2°C) together with a defined rate of air throughput and specially designed ovens are used. Test pieces are aged at 140°C for 30 ± 0.25 min and cooled to room temperature before making the plasticity measurement. The plasticity retention index (PRI) is calculated from:

$$\frac{\text{aged rapid plasticity number}}{\text{unaged rapid plasticity number}} \times 100$$

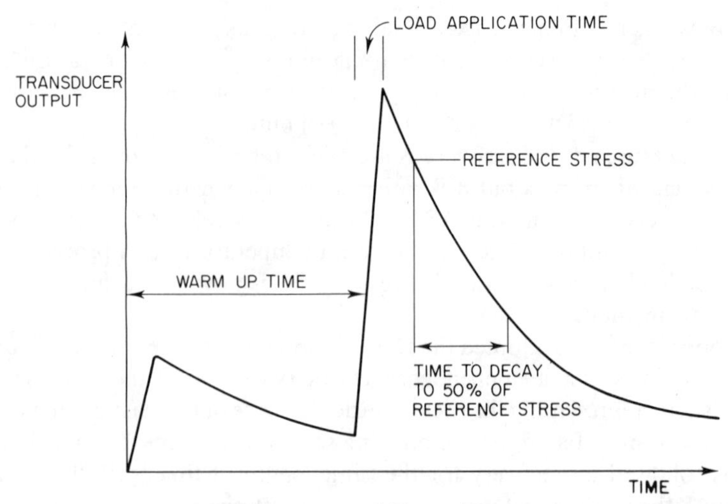

FIG. 6.3. Processibility tester trace.

The same method is given in BS 1673: Part 3[21] and in ASTM D3194.[25] In the ASTM standard the procedure for rapid plasticity measurement is given as there is no ASTM equivalent of ISO 2007 to reference, and a larger tolerance on oven temperature (±0.5°C) and hence greater variability is allowed.

6.3.3 Stress Relaxation

The use of a parallel plate plastimeter to determine both softness and recovery is a simple way of obtaining a measure of both the viscous and elastic components on deformation behaviour, albeit under conditions somewhat removed from those met during processing. An alternative approach is to measure the stress relaxation in a test piece and this is the basis of an instrument described by Moghe[26] which operates in shear, and also the Monsanto Stress Relaxation Processibility Tester (SRPT) recently developed at RAPRA , which operates in compression.

The SRPT and its operation has been described by Norman[27] and examples of its application also given by Berry and Sambrook,[28] and Leblanc.[29] Amsden[30] has given a description, examples of use and a comparison with other plastimeters. The same test piece as used in the Wallace rapid plasticity test is precompressed and heated between parallel platens. The test piece is then rapidly further compressed by a standard amount and the resultant stress monitored by a transducer located in one platen. The change of stress with time is shown schematically in Fig. 6.3.

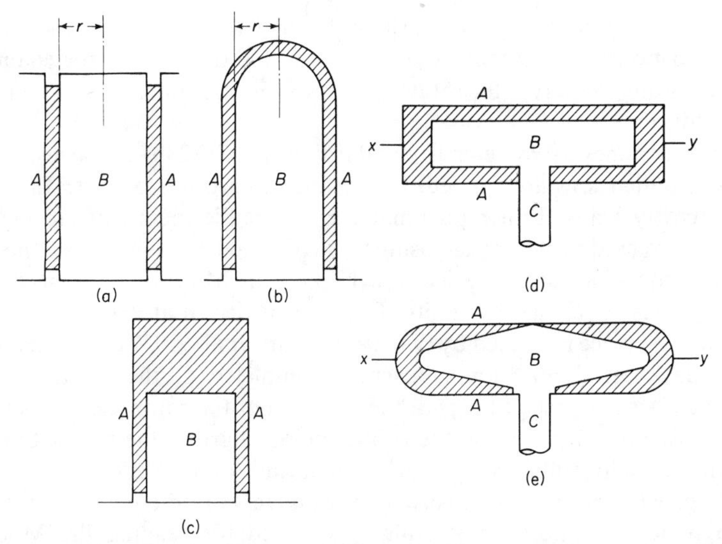

FIG. 6.4. Rotation plastimeter geometry. (a), (b) and (c) Coaxial cylinder types; (d) and (e) concentric disc types. ((d) is the Mooney geometry.) A is usually the stator and B the rotor, C is a rotating shaft and r is the cylinder radius (much larger than the clearance between A and B). xy indicates the mid-plane along which the chamber can be opened for filling.

The instrument digitally displays the reference force which is a measure of viscosity and the time for the force to decay to a fraction (usually $\frac{1}{2}$) of this reference force which is a measure of the viscoelastic behaviour. Hence the instrument rapidly provides in digital form fundamental processibility data and the complete relaxation curve can be recorded if required.

The instrument is perhaps best thought of, as presented by Amsden, as a very superior type of compression plastimeter. It is rapid, automatic, uses a small test piece and because it gives a measure of the viscoelastic response as well as viscosity it can be expected to detect differences in processing characteristics not apparent from the standard plasticity and Mooney tests.

6.3.4 Rotation Plastimeters

A number of plastimeters of this type have been used for rubbers, often for research purposes, but one instrument, the Mooney viscometer, is extensively used for routine quality control purposes. The principle of the Mooney is shown in Fig. 6.4. The rotor turns at a constant rate inside a closed cavity containing the test piece so that a shearing action takes place between the flat surfaces of the rotor and the walls of the chamber. The torque required to rotate the rotor is monitored by a suitable transducer.

The Mooney viscometer is standardised in ISO 289.[31] The rotor and cavity dimensions are defined, as are anti-slip grooves on both the rotor and cavity walls. Either pneumatic or hydraulic means of closing the cavity are used providing a closing force of 11500 ± 500 N during the test. The torque indicating device is calibrated in Mooney units such that 8.3 ± 0.02 N m equals 100 units. The rotor is driven at 2 rev/min.

The test piece is formed by two discs of rubber about 59 mm in diameter and thickness about 6 mm sufficient to completely fill the die cavity. One of the discs is pierced to permit the insertion of the rotor stem. It is usual to allow 1 min for heating the rubber before starting the motor but this is not actually sufficient to reach equilibrium and longer heating times may give better agreement between viscometers of different construction. No preferred time after starting is specified for reading the Mooney viscosity as this is chosen to suit the viscosity/time curve encountered; but commonly 4 min is suitable for many materials and 8 min for butyl.

Typically a Mooney viscosity would be expressed as:

$$50\,\mathrm{ML}\quad(1+4)\quad 100\,^{\circ}\mathrm{C}$$

Where 50 M is the Mooney viscosity, L indicates the use of the large (i.e. standard) rotor, 1 is the preheating time in minutes, 4 is the reading time in minutes and 100 °C is the test temperature.

The same method is given in BS 1673: Part 3[21] and ASTM D1646[32] but in the latter case a procedure for assessing cure characteristics with the Mooney is contained in the same standard (see Section 6.4).

The Delta Mooney (Δ Mooney) test is an extension of the Mooney used on empirical grounds as a general indication of processibility for non-pigmented oil extended emulsion styrene/butadiene rubber. It quantifies the changes that occur in Mooney viscosity with time either as the difference between viscosities recorded at two specified times or as the difference between the minimum viscosity recorded immediately after the commencement of the test and the subsequent maximum viscosity. Both definitions of Delta Mooney are used in the procedure standardised in ASTM D3346[33] and in a draft British standard.

For an instrument used for decades in most rubber laboratories it is a little surprising and even disturbing to find recent attention being devoted to the sources and elimination of considerable differences in results obtained by different laboratories. Investigations have been reported by Kramer[34] and Niemiec.[35]

Kramer found differences of up to 11 MU (Mooney units) between 12 laboratories which he attributed partly to failure to comply with the standards' instructions and partly to vagueness on the part of the standards. In a second trial with selected laboratories and in which the test instruments were subjected to physical analysis, he found a difference of 4.6 MU which he considered mainly due to differences in test piece temperature.

Niemiec discusses in detail the factors which can contribute to variability, which could be considerable, but concludes that at least for a butyl rubber with 8 min running time it is possible to get agreement to within ± 1.5 MU with strict control of all factors.

Nakajima and Harrell[36] have derived expressions for calculating shear stresses from the Mooney torque values to give viscosities in agreement with those obtained from other instruments and also an expression to correct for the edge effects.[37]

The main advantage of rotation plastimeters over compression instruments is that shearing at constant rate can be continued for as long as required so that thixotropic or structure effects can be studied. Rather higher shear rates are possible although the Mooney operates at only about $1\,s^{-1}$ (the shear rate varies across the diameter of the rotor). A practical difficulty is to avoid slippage of the rubber over the metal parts and this is why the Mooney operates with a positive hydrostatic pressure and has grooves cut in the metal surfaces.

Another problem, which limits the shear rates which can be used, is the heat generated during shearing. Calculated and measured temperature rises[38] indicate that at $10\,s^{-1}$ the thickness of rubber must be no more than 1 mm to keep the equilibrium rise to 1 °C. Piper and Scott[39] used a bi-conical shaped rotor which operated at $10\,s^{-1}$ and the apparatus of Bulgin and Wratten[40] with a clearance variable down to 0.25 mm could operate at up to $100\,s^{-1}$. The shearing cone geometry has the advantage over the Mooney disc type that with suitable design the shear rate is fairly uniform rather than varying from zero at the centre to a maximum at the periphery of the rotor .

An alternative geometry to the shearing disc (or shearing cone) type of rotation viscometer is the coaxial cylinder type (Fig. 6.4). The inner cylinder can rotate inside the outer cylinder or the inner cylinder could be the stator. Such a geometry gives a substantially uniform shear rate in the annulus of the rubber provided the clearance between rotor and stator is very small in comparison with the inner cylinder radius. Coaxial cylinder types have proved valuable for research purposes but there is a practical difficulty of maintaining the hydrostatic pressure without introducing friction by the device used to close the gap. Sealing around the rotating shaft in the Mooney is relatively easy because frictional force on the shaft contributes little to the total torque.

Yet another geometry is the cone and plate viscometer. This generally operates without a positive hydrostatic pressure and although often used for plastics melts is not suitable for rubbers because of excess slipping. An example of a commercial cone and plate instrument is the Weissenberg Rheogoniometer and this and many other forms of viscometer are described in a very practical handbook of rheology by Van Waser *et al.*[41]

In principle it is possible to extend the use of rotation plastimeters to measure elastic recovery but this has not been common in practice. The procedure would be to allow the rotor to continue moving under the elastic stresses in the rubber after the drive had been stopped but this is difficult because of friction in the system which would oppose the motion and so invalidate the result. Attempts have been made with the Mooney,[42,43] the shearing cone viscometer[44] and, for plastics, a cone and plate viscometer.[45]

6.3.5 Extrusion Plastimeters and Die Swell

In the extrusion plastimeter, rubber is forced through a small cylindrical die under a known pressure and the volume, extruded in a given time, measured (or at a constant rate and the pressure measured). It is therefore rather similar in action to extruders used in the factory and is indeed often thought of as a test for extrudibility rather than plasticity. It also resembles the classical capillary viscometer, but with rubbers there is uncertainty as to what extent the material slips over the die. Apart from the rate of extrusion it is also possible to visually estimate the quality of the extrusion, for which purpose the cylindrical die could be replaced with one having a more complicated cross section.

It is apparent that the first advantage of the extrusion plastimeter is that it simulates to some extent, the processing operations of extrusion and injection moulding. Apart from this it has the important advantage that much higher shear rates are possible than are normally obtainable in compression or rotation instruments. The shear rate can be comparable with those encountered in processing but heat build-up is not a great problem because passage of the rubber through the die is very rapid. However, the short period of shearing is insufficient to break down thixotropic structures completely.

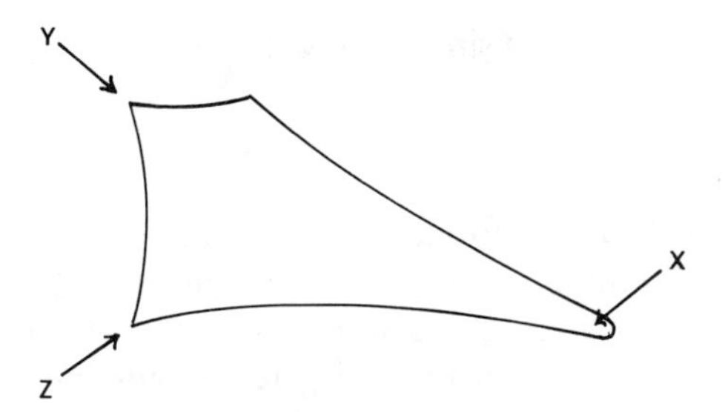

FIG. 6.5. Cross section of the Garvey die.

Some of the earliest plastimeters were of the extrusion type but despite the advantages mentioned none has become anything like as popular as the Mooney and they have not been favoured by international and national standardisation.

However, in recent times it has been increasingly realised that to obtain rheological data relevant to processes such as injection moulding measurements must be made at the higher shear rates obtainable with extrusion plastimeters and in consequence their use is increasing. Basically, extrusion plastimeters are quite simple instruments but many of those on the market have been made sophisticated, and hence expensive, both in respect of the arrangements for applying pressure and controlling temperature and by the addition of microprocessors and even laser technology. Probably the most versatile instrument of the type is the so-called Monsanto Processibility Tester (not to be confused with the SRPT, see Section 6.3.3) which with a pressure transducer at the entrance to the orifice, a microprocessor system and a laser device allows the measurement of viscosity, die swell and stress relaxation. A summary of the uses of extrusion rheometry is given by Colbert[46] and the characterisation of natural rubber mixes is discussed by Halim and Brichta.[47] Applications of the Monsanto Processibility Tester have been given by, for example, Leblanc,[48] Pica *et al.*[49] and Sezna.[50]

Small instrumented extruders, which are available commercially, are also used in the laboratory to measure extrudability. The best known special die is the Garvey die[51] shown in Fig. 6.5 which was designed to show up typical faults that can occur in a mix with poor extrusion characteristics. This die has a cross section including an acute-angled wedge portion (X in Fig. 6.5). Mixes are rated by assigning a numerical value or score to each of the following characteristics: (a) surface, (b) sharpness of the 'edge' X, (c) the two 'corners' Y and Z, and (d) cross-sectional dimensions (die swell). A method of test for extrudability is given in ASTM D2230[52] using a screw type laboratory extruder and a Garvey die.

6.3.6 Mixing Machines

A very different approach to measuring 'processibility' is to use what is effectively a small-scale internal mixer and to monitor the torque required to turn the rotors which gives a measure of the effective viscosity. Such instruments (torque rheometers) are perhaps most appropriate in estimating mixing behaviour and have been used more for plastics than rubbers. Commercial instruments are available, for example the Brabender Plastograph and Plasticorder, and the latter can also be used with an extruder head. The RAPRA Variable Torque Rheometer[53] is another example of the mixer type of instrument and its use with rubber has been reported by Paul.[54] The potential advantages of these instruments are the similarity of their action to full-scale mixing or extrusion equipment together with being able to operate at shear rates appropriate to factory operations. However, because of the difficulty of matching exactly the range of shear rates, etc., which exist in full-size plant, successful scaling-up of the results may be difficult.

6.3.7 Other Processibility Tests

In processing operations the deformation of the rubber is largely in shear but there are circumstances where elongation deformation is important. 'Elongational flow' measurements have been reported by several workers, in which a sample is stretched in uniaxial tension at a constant strain rate. Useful discussions of this type of measurement have been given by Denby,[55] White,[56] Cotten and Thiele,[57] and Ng.[58]

Curemeters will be discussed in Section 6.4 but these instruments, oscillating disc, rotorless and reciprocating paddle types, are forms of plastimeter which measure plasticity before the onset of, as well as during, cure.

The viscoelastic behaviour of uncured rubbers can be characterised by measurement of in-phase and out-of-phase moduli and loss tangent, i.e. by 'dynamic' measurements as will be discussed in Chapter 9. Such measurements have been made with a Rheovibron apparatus[59] and with an 'eccentric rotating discs instrument'.[60]

Electrical properties of compounds have been shown to depend on the quality of mixing and hence methods of monitoring the mixing process based on measurement of resistivity or conductivity have been suggested.[61,62]

Various other mechanical tests are possible. Indentation or penetration by a plunger under load as in the Vicat test for plastics could be applied to unvulcanised rubber; the so-called Humboldt Penetrometer works on this principle and it has been claimed that reproducible plasticity measurements can be obtained with a spring-loaded durometer.[63] An Indentation method using a needle has been described by Kusano *et al.*[64] and details of a Russian dynamic penetration method were circulated to ISO TC45. Such methods would be limited to low shear rates and the test would be too short to break down filler structure. Higher shear rates could be obtained by using a rebound resilience test, although again the time scale is too short to break down filler structure, and Yoshida used the Schob pendulum for this purpose.[65]

6.3.8 Correlation between Plastimeters

With a number of plastimeters in common use, it is inevitable that there is a demand to know the relationship between the readings obtained with them. From the foregoing discussion emphasising the dependence of plasticity results on the shear rate and other conditions of test, it must be clear that the question is not really a sensible one. Any relationship found between two different instruments can only be valid for the compounds used and the particular conditions of test, simply because the flow properties of rubber cannot be defined by a single parameter.

Many workers have studied and published correlations between various types of plastimeter, often to show that they do not agree and to illustrate the superiority of the instrument which supposedly agrees best with processing behaviour. Several comparisons are included in the literature already noted and other examples are shown in Fig. 6.6. Figure 6.6(a) shows the relatively close correlation obtained between two compression instruments, the Wallace Rapid and Williams plastimeters, for materials of similar flow characteristics (plasticised natural rubber). When such rubbers are compared on two basically different instruments (compression and extrusion) the correlation is less good (Fig. 6.6(b)). A similar degree of scatter is shown in Fig. 6.6(c) where a variety of tyre tread and carcase mixes are tested on another pair of basically dissimilar instruments. Similar correlations between Mooney and Rapid Plasticity have been given by Bristow[66] who demonstrates that correlation is improved if initial Mooney values are used.

Baader[67] has published curves showing how the relationship between readings on the Mooney viscometer and a compression (Defo) test varies according to the type of material tested and has also studied[68,69] the

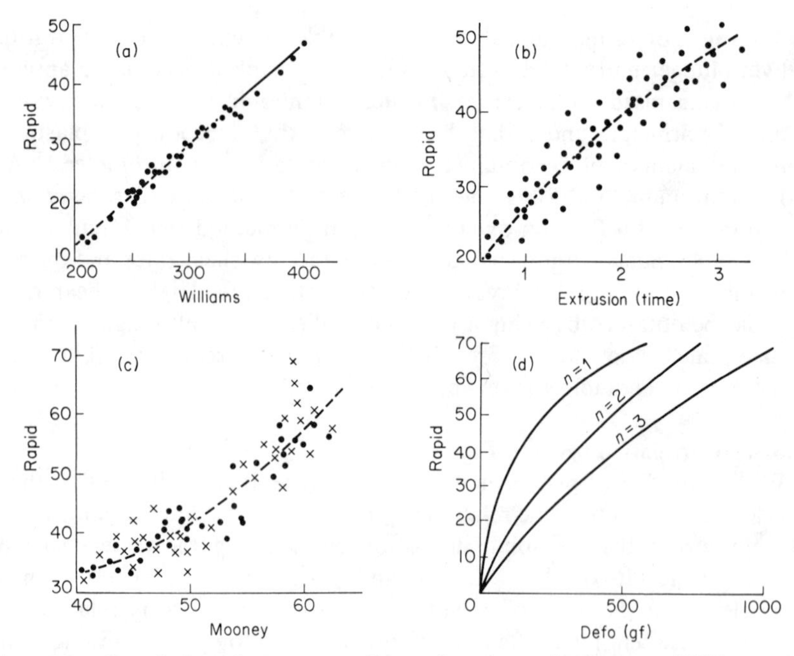

FIG. 6.6. Correlations between readings of various plastimeters. Williams—compression, plate type; 'Rapid'—Wallace rapid, compression, disc type; 'Extrusion'—Griffiths (1926); Defo—compression, disc type.

correlation of Defo with Williams parallel plate compression results. Figure 6.6(d) gives the calculated relationship between two compression tests (Defo and Wallace Rapid) obeying the relation: shear rate proportional to $(\text{stress})^n$ (an oversimplified representation of the flow of rubber), and serves to show how widely the relationship can vary with the characteristics of the material tested, even when both instruments are of basically the same type. Kandyrin *et al.*[70] compared results from a Mooney viscometer with viscosity determined on a gas capillary viscometer and with melt flow index measurements and derived relationships for the rubbers tested. Mooney, Defo, Williams and Wallace Rapid instruments were compared by Brezik.[71]

6.4 SCORCH AND CURE RATE

If a fully compounded rubber is subjected to a plasticity measurement at a high enough temperature and for long enough it will cure and consequently there is not always a clear distinction between a plasticity test and a test for scorch or rate of cure. For example, the Mooney viscometer is used to measure scorch, i.e. the onset of vulcanisation, and an oscillating disc rheometer will measure the plasticity of the compound before the onset of cure as well as the increase in stiffness as curing takes place.

Tests for scorch and rate of cure should be distinguished from tests for degree of cure or optimum cure measured on the vulcanised material. The latter type of test measures the physical properties of test pieces vulcanised for various times, tensile properties, swelling and set measurements being the parameters most commonly used.

The most obvious changes in a rubber mix when vulcanisation sets in are an increase in stiffness and an increase in the elastic component of its viscoelastic deformation. In addition, the ease of solution in common rubber solvents decreases and this has been used with some success as a very simple, if not very convenient, test for natural rubbers. The method is not successful for reinforcing black-filled mixes for example, because these will not dissolve readily even when not scorched, and for synthetic rubbers a range of solvents would be necessary.

Parallel plate compression plastimeters have been quite widely used for measuring rate of cure and in the past methods have been standardised. The test pieces are heated for various times and then tested in the plastimeter. The change in plasticity or recovery, or some combination of these, is then plotted against time of heating to give a 'scorch curve'. The Mooney viscometer offers a more convenient way of measuring scorch and even rate of cure, and a standard method is given in ISO 667.[72] This is essentially the method of ISO 289 continued until the viscosity reaches 40 MU above its minimum value. The minimum viscosity, the times to reach 5 MU and 35 MU above minimum viscosity, and the difference in time between reaching 5 and 35 MU above minimum viscosity are reported. The time to reach 5 MU above minimum can be taken as a measure of the time until the onset of vulcanisation and the difference between times to 5 MU and 35 MU above minimum viscosity as an indication of the rate of cure.

The same method is included in BS 1673: Part 3[21] but in a draft revision only the measurement of scorch time (i.e. time to reach 5 MU above minimum viscosity) is retained on the basis that the Mooney is now relatively little used for cure rate measurements because of the increased use of 'curemeters'. The ASTM equivalent is contained in the standard for Mooney viscosity, ASTM D1646,[32] and is the same as the ISO procedure except that alternative viscosity rises of 3 and 18 MU above minimum viscosity are included for when the small rotor is used.

Laboratory measurement of curing characteristics has been somewhat revolutionised by the introduction of so-called 'curemeters' which are now almost universally used for the routine control of fully compounded rubbers. These instruments have been so successful[73] that the use of the Mooney to measure scorch, and physical tests on moulded test pieces, have been much reduced as a control routine. Although curemeters also measure viscosity before the onset of cure they have by no means replaced other types of plastimeter and viscometer for this purpose.

There are basically two types of curemeter in common use: the reciprocating paddle type, as for example the Wallace–Shawbury Curometer and the first Vulcameter, and the oscillating disc type such as the Monsanto Rheometer, with a third type, the rotorless curemeter increasing in popularity. In the reciprocating paddle type a small paddle embedded in the rubber, which is itself enclosed in a heated cavity, is reciprocated. Either the change in amplitude of oscillation at constant force or the change in force to produce constant amplitude is monitored as a measure of change in stiffness. In the oscillating disc type a bi-conical disc is embedded in the rubber in a closed cavity rather in the manner of the Mooney. The disc is oscillated through constant angular displacement and the torque required monitored. The rotorless type is a curemeter in which one half of the die enclosing the test piece, rather than a paddle or disc within the test piece, oscillates or reciprocates. It may be noted that a reciprocating or oscillating motion rather than continuous rotation is used in curemeters to prevent break-up of the test piece once cure is well advanced.

Despite the widespread use of curemeters, progress to international standardisation was relatively slow, partly because of patent difficulties as a result of the virtual monopoly of certain commercial instruments. There are now international standards for the oscillating disc curemeter, ISO 3417,[74] and for rotorless curemeters, ISO 6502.[75] The British standard, BS 1673: Part 10[76] covers both the oscillating disc and the reciprocating paddle type, whilst ASTM D 2084[77] covers only the oscillating disc.

The oscillating disc and reciprocating paddle types are specified in some detail, based on commercially available equipment. Examples of typical curemeter curves are given with details of the various parameters which can be derived from them. By way of illustration the curve for a plateau type cure on an oscillating disc curemeter is shown in Fig. 6.7. Minimum torque, maximum torque or the slope of the curve (cure rate) can be

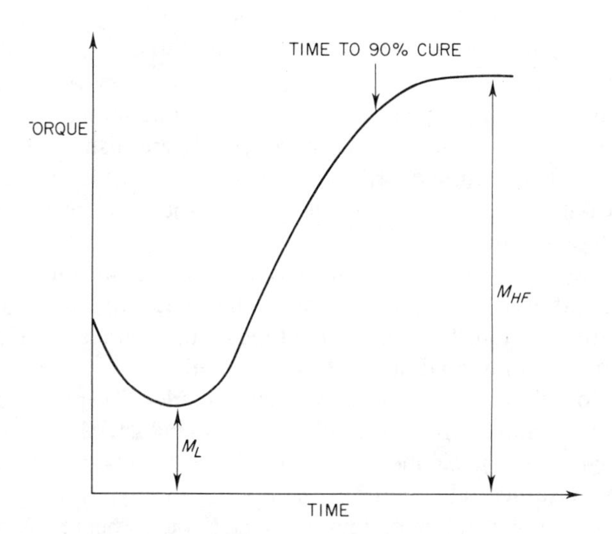

FIG. 6.7. Oscillating disc rheometer trace.

taken but perhaps the most useful single figure is the time to achieve a given degree of cure which is the time for the torque to increase to:

$$\frac{Y}{100}(M_{\mathrm{HF}} - M_{\mathrm{L}}) + M_{\mathrm{L}}$$

where Y is the percentage cure required (usually 90% for a 'practical' cure), M_{HF} is the plateau torque, and M_{L} is the minimum torque.

A similar estimate of cure time is taken from an oscillating paddle curemeter curve.

The standard for rotorless curemeters is far less specific in specifying the apparatus and probably does not encompass the most recent commercial instruments. The general method of use and form of results is the same as the other types but the presentation in ISO 6502 differs considerably from that of ISO 3417. The impression is given that a standard for rotorless meters was premature and was pushed through for parochial reasons.

The development of curemeters including reference to many of the commercial instruments has been discussed by, for example, Hofmann[78] and Bennett.[79] Further details of apparatus can be found in manufacturers' literature which sometimes also includes examples of factory use.

There is little doubt that the oscillating disc curemeter has achieved the greater popularity compared to the reciprocating paddle type because of high precision and convenience in use but at the expense of higher purchasing price. Norman[80] gives a valuable discussion of the problems with curemeters, pointing out that there is no one level of cure which gives optimum values for all physical properties and listing problems as non-uniform temperature distribution, possible slip of the test piece over rotor or cavity, porosity and no satisfactory procedure for dealing with a 'marching modulus'.

One advantage claimed for the paddle type instrument is that the rubber test piece reaches test temperature more quickly and there is less temperature variation. There must be a time lag and Hands and Horsfall[81] have used an isothermal apparatus to obtain basic cure rate data and have developed a mathematical cure model for predicting cure distributions in non-isothermal conditions, as in industrial processes. The rotorless curemeter is claimed, for example by Hands *et al.*[82] to more approximate isothermal conditions than both other types because of the absence of an unheated rotor and a thin test piece, hence giving more accurate predictions for fast curing materials.

Clearly curemeters will not always agree due to their differing thermal characteristics. Because the Mooney is often used to measure scorch it is worth noting a comparison between Mooney and curemeter search made by Bristow[83] in which he found poor correlation for higher curing temperatures.

As mentioned earlier, the degree of cure of vulcanised material can be estimated by various physical tests. A rapid non-destructive procedure would have considerable value and a procedure using an 'ultra high frequency apparatus' has been described.[84]

6.5 TACK

Tack is best thought of as the ability of two pieces of rubber to stick when pressed together and is sometimes called auto-adhesion. It is hence not quite the same as stickiness or adhesion, which generally involves sticking or adhering to a second material. High tack can be a nuisance when handling sheets of rubber but is very important in the manufacture of articles built up from separate pieces of uncured mix.

Despite its importance, the measurement of tack has not attracted the attention of standards committees and satisfactory tests have been few and far between. Recently ISO TC45 started work on a method for tack but abandoned it through lack of interest. In essence tack measurements consist simply on pressing together two pieces of rubber and then measuring the force required to separate them. In practice, the results are likely to be very variable and it is essential that the pressure applied when joining the test pieces and the time for which it acts are carefully controlled; also that measurements are distributed over a fairly large area of the rubber sheet in question.

A large number of instruments for measuring tack have been devised, some portable for use on the factory floor and some for laboratory operation. Many of these have not been produced commercially and are not used outside of the factory of origin. Bussemaker[85] has reviewed many tack measuring instruments, discussing their mode of operation, advantages and disadvantages. One fairly precise apparatus available commercially is the Monsanto Tel Tack which uses two strip test pieces pressed together at right angles to give a known contact area. After a set contact time under known load, both of which could be varied, the two test pieces are separated in direct tension at a fixed rate of grip travel and the maximum force noted. If required, one test piece can be replaced by a metal strip so that the difference between adhesion and auto-adhesion can be measured.

It should be remembered that tack is very much influenced by the surface condition of the rubber as well as dwell time and pressure. This includes contamination and the surface roughness of the test piece.

6.6 OTHER TESTS

Shrinkage on moulding has been discussed in Section 7.2.5. When uncured rubber is deformed, as during milling, the elastic recovery of the rubber will cause a change in dimensions when the deforming force is removed. Such shrinkage can be measured by cutting a piece of known dimensions from the sheet whilst still on the mill and remeasuring these dimensions after a period of rest.

Contact methods of measuring surface roughness (see Section 7.2.3) are not likely to be successful with uncured rubber because of its softness. It is unlikely that roughness needs to be known very precisely and a simple method has been given by Orlovskii *et al.*[86] The volume of a disc is calculated using the overall thickness measured on top of any irregularities and compared to the true volume measured by a liquid displacement method.

Green strength, the strength of the unvulcanised rubber compound, has been more particularly defined as a property of the bulk of the rubber which is optimised when the breaking strength minus the yield strength is a maximum. More often the term is used vaguely without proper definition. Interest in a standard method has recently arisen in ISO TC45 and a draft proposal is in circulation where green strength is defined as resistance to tensile deformation or fracture.

Green strength can be estimated by testing strip or dumb-bell test pieces on a tensile machine to obtain a stress/strain curve and this is the procedure used in the ISO draft. The results will be dependent on temperature and strain rate. A review of green strength has been given by McDuff.[87]

REFERENCES

1. Norman, R.H. and Johnson, P.S. (1981). *Rubb. Chem. Tech.,* **54**(3).
2. ISO 1658, 1973. Natural Rubber. Test Recipes and Evaluation of Vulcanisation Characteristics.
3. ISO 2302, 1978. Rubber, Isobutylene–Isoprene (IIR)—Evaluation Procedures.
4. ISO 2303, 1983. Rubber, Isoprene (IR)—Non Oil-Extended, Solution Polymerised Types—Test Recipe and Evaluation of Vulcanisation Characteristics.
5. ISO 2322, 1981. Rubber, Styrene–Butadiene (SBR)—Emulsion Polymerised—Test Recipe and Method of Evaluation.
6. ISO 2475, 1975. Rubber, Chloroprene (CR)—General Purpose Types—Evaluation Procedures.
7. ISO 2476, 1980. Rubber, Butadiene (BR) Solution Polymerised Types—Test Recipe and Evaluation of Vulcanisation Characteristics.
8. ISO 4097, 1980. Rubber, Ethylene–Propylene–Diene (EPDM) Non Oil Extended Raw General Purpose Types—Evaluation Procedures.
9. ISO 4658, 1980. Rubber, Acrylonitrile–Butadiene (NBR)—Test Recipe and Evaluation of Vulcanisation Characteristics.
10. ISO 4659, 1981. Rubber, Raw Styrene Butadiene (Carbon Black or Carbon Black and Oil Masterbatches)—Test Recipe and Method of Evaluation.
11. ISO 1796, 1974. Raw Rubber in Bales—Sampling.
12. BS 6315, 1980. Sampling and Sample Preparation of Raw Rubber.
13. ISO 1795, 1974. Raw Rubber in Bales—Sampling.
14. Williams, I. (1924). *Ind. Eng. Chem.*, **16**, 362.
15. Scott, J.R. (1931). *Trans. IRI*, **7**, 169.
16. Scott, J.R. (1935). *Trans. IRI*, **10**, 481.
17. Scott, J.R. RABRM Lab. Circular 172, 1940.
18. Herne, H., Hine, D.J. and Wright, M. RABRM Research Report 70, 1952.
19. Scott, J.R., RABRM Lab. Circular 134, 1935.
20. ISO 2007, 1981. Determination of Plasticity—Rapid Plastimeter Method.
21. BS 1673: Part 3, 1969. Methods of Physical Testing.

22. ASTM D926-83. Plasticity and Recovery (Parallel Plate Method).
23. Koopmann, R. (1985). *Polym. Test.*, **5**(5).
24. ISO 2930, 1981. Rubber, Raw Natural—Determination of Plasticity Retention Index (PRI)
25. ASTM D3194-84. Rubber from Natural Sources—Plasticity Retention Index (PRI).
26. Moghe, S.R. (1976). *Rubb. Chem. Tech.*, **49**(2), 247.
27. Norman, R.H. (1980). *Plast. Rubb. Int.*, **5**(6), 243.
28. Berry, J.P. and Sambrook, R.W., Paper 28, Rubbercon 77, May 1977.
29. Leblanc, J.L. (1980). *Eur. Rubb. J.*, **162**(2), 20.
30. Amsden, C.S. (1985). *Polym. Test.*, **5**(1), 45.
31. ISO 289, 1985. Determination of Mooney Viscosity.
32. ASTM D1646-81. Viscosity and Vulcanisation Characteristics (Mooney Viscometer).
33. ASTM D3346-81. Processibility of SBR (Styrene—Butadiene Rubber) with the Mooney Viscometer.
34. Kramer, H. (1981). *Polym. Test.*, **2**(2), 107.
35. Niemiec, S. (1980). *Polym. Test*, **1**(3), 201.
36. Nakajima, N. and Harrell, E.R. (1979). *Rubb. Chem. Tech.*, **52**(1), 9.
37. Nakajima, N. and Harrell, E.R. (1979). *Rubb. Chem. Tech.*, **52**(5), 962.
38. Whorlow, R.W. RABRM Res. Memo. R347, 1949; *J. Rubb. Res.*, **20**, 39 (1951).
39. Piper, G.H. and Scott, J.R. (1945). *J. Sci. Instrum.*, **22**, 206.
40. Bulgin, D. and Wratten, R. (1953). *Proc. 2nd Inst. Congress on Rheology*, Butterworth.
41 Van Waser, J.R., Lyons, J.W., Kim, K.Y., and Colwell, R.E. (1963). *Viscosity and Flow Measurement*, Interscience.
42. Taylor, R.H., Fielding, J.H. and Mooney, M. (1947). *Rubb. Age, NY*, **61**, 567.
43. Koopmann, R. and Kramer, H. Paper 012 Conference Internationale du Caoutchouc, Paris, June 1982.
44. Wharlow, R.W. RABRM Res. Memo. R375, 1951.
45. Batchelor, J. RAPRA Res. Report 180, 1970.
46. Colbert, G.P. (April 2, 1979). *Rubb. and Plast. News.*
47. Halim, O.A. and Brichta, A.M. (1983). *Rubb. World*, **188**(5).
48. Leblanc, J.L. (1981). *Plast. Rubber Process. Appln*, **1**(2).
49. Pica, D., Barket, R., Rice, P. and Ma, C.C. (1979). *Rubb. World*, **180**(4).
50. Sezna, J.A., ACS Rubber Division Meeting, Indianapolis, May 1984, Paper 22.
51. Garvey, B.S., Whitlock, M.H. and Freeze, J.A. (1942). *Ind. Eng. Chem.*, **34**, 1309.
52. ASTM D2230-83. Extrudability of Unvulcanised Compounds.
53. Paul, K.T. (Feb. 1972). *RAPRA Bulletin.*
54. Paul, K.T. (March 1973). *RAPRA Members J.*
55. Denby, J.M., (1971). *Polym. Eng. Sci.*, **11**, 433.
56. White, J.L., (1977). *Rubb. Chem. Tech.*, **50**, 163.
57. Cotten, G.R. and Thiele, J.L., ACS Rubber Division Meeting, Montreal, May 1978, Paper 4.

58. Ng, T.S. (1983). *Kaut. u. Gummi. Kunst.*, **36**(1).
59. Nakajima, N., and Collins, E.A., (1975). *Rubb. Chem. Tech.*, **48**, 69.
60. Macosko, C.W. ACS Rubber Division Meeting, Minneapolis, April 1976, Paper 49.
61. Khaskhachikh, A.O. Dozortser, M.S. and Plotkin, S.L. (1977). *Int. Polym. Sci. Tech.*, **4**, 4.
62. Usacher, S.V., Zakhorkin, O.A. and Zakharov, N.D. (1977). *Int. Polym. Sci. Tech.*, **4**, 5.
63. Muramatsu, K., Furukawa, J. and Yamashitu, S. (1961). *Nippon Gomu Kyokaishi*, **34**, 997.
64. Kusano, T., Murakami, K. and Yamakawa, T. (1981). *Int. Polym. Sci. Tech.*, **8**, 4.
65. Yoshida, J. (1952). *Chem. Abs.*, **46**, 3786.
66. Bristow, G.M. (1982). *NR Technol.*, **13**, Part 3.
67. Baader, T. (1954). *Kaut. u. Gummi*, **7**, 263WT.
68. Baader, T. (1953). *Kaut. u. Gummi*, **6**, 210WT, 235WT, 258WT.
69. Baader, T. (1954). *Kaut. u. Gummi*, **7**, 15WT.
70. Kandyrin, L.B., Al'tzitser, U.S., Anfimov, B.N. and Kuleznev, V.N. (1977). *Kauch. i. Rezina*, No. 4, 18.
71. Brezik, R. (1984). *Int. Polym. Sci. Tech.*, **11**(12).
72. ISO 667, 1981. Determination of Cure Rate—Shearing Disk Method.
73. Dunn, J.R., Wood, A.G. and Burn, P.S. Interscandinavian Rubber Meeting, Copenhagen, May 1976, SGF Publication 48.
74. ISO 3417, 1977. Measurement of Curing Characteristics with the Oscillating Disc Curemeter .
75. ISO 6502, 1983. Measurement of Vulcanisation Characteristics with Rotorless Curemeters.
76. BS1673: Part 10, 1977. Measurement of Pre-vulcanising and Curing Characteristics by Means of Cure Meters.
77. ASTM D2084-81. Vulcanisation Characteristics Using Oscillating Disc Cure Meter.
78. Hofmann, W. (Jan. 1974). *Europ. Rubb. J.*, 41.
79. Bennett, F.N.B. (Sept. 1973). *Europ. Rubb. J.*, 34.
80. Norman, R.H. (1980). *Polym. Test.*, **1**(4).
81. Hands, D. and Horsfall, F., ACS Rubber Division Meeting, Houston, Oct. 1983, Paper 11.
82. Hands, D., Norman, R.H. and Stevens, P. A new curemeter. Paper presented at Rubbercon Conference, Birmingham 1984.
83. Bristow, G.M. (1982). *NR Technol.*, **13**(1).
84. Laboratoire de Recherche et de Controlle du Caout. Publication, Étude d'un appareil de measure UHF en continu du degré de reticulation de polymeres, 1982.
85. Bussemaker, O.K.F. (Dec. 1964). *Rubb. Chem. Tech.,* **37**, 1178.
86. Orlovskii, P.N., Lukomskoya, A.I., Tsydzik, M.A. and Bogotova, S.K. (1960). *Sov. Rubb. Technol.,* No. 7, 20.
87. McDuff, K. (1978). *Green Strength—A Review*, RAPRA.

Chapter 7

Density and Dimensions

When putting physical properties into groups it is a little difficult to know where to place density and dimensional measurements. So, for want of any better arrangement, the two are lumped together here to form their own section. There is the obvious link between the two in that density can be derived from a knowledge of dimensions plus mass; but also they are both measurements which are used as an essential part of other physical tests. For example, density being used to calculate volume loss in an abrasion test or as an integral part of volume change measurements. There are very few tests which do not at some point involve the measurement of dimensions. Density and dimensions also have a certain link in the factory, both being important as regards the costing of products. Density is very commonly used as a simple but effective quality control check on batches of compounded rubber as a guard against gross errors; and dimensional checks on products is one of the common operations of routine inspection. Hence, both in the laboratory and in the factory, density and dimensional measurement have a particular position due to their frequent usage. Measurements that are made every day have a habit of being taken for granted and this can certainly happen to the measurement of dimensions, resulting in unnecessary errors. When one considers that in, for example, the determination of tensile strength, any error in the measurement of the cross section results directly in an equivalent percentage error in the strength measurement, it is reasonable to devote considerable attention to the seemingly simple matter of measuring the width and thickness.

7.1 DENSITY

Density is defined as mass per unit volume, whereas relative density is the mass of the substance compared to the mass of an equal volume of a reference substance (usually water) and is hence dimensionless. Relative density used to be commonly known as specific gravity but this term is now deprecated and should not be used.

In practice, the method of measurement often involves the determination of the relative density to water but the density of water is assumed to be 1 Mg/m^3. Furthermore, the determination is often made by observation of gravitational forces but for convenience the forces are expressed in mass units.

For most purposes the density of a rubber is quoted to 0.01 Mg/m^3 and the commonest method of determination is by weighing in air and water. The standard procedure is given in ISO 2781,[1] method A and specifies a test piece weighing a minimum of 2.5 g which can be of any shape as long as the surfaces are smooth and there are no crevices to trap air. It is stated that duplicate tests shall be made which presumably means two test pieces rather than two tests on the same piece. The test piece is weighed in air and then in water using a balance accurate to 1 mg, which is usually wrongly interpreted as a balance reading to 1 mg. A top pan balance is not suitable. It is permissible to wet the test piece with a liquid such as methylated spirit before weighing in water and this is indeed common practice. The water then needs to be changed relatively frequently because of contamination by the alcohol. The best way of suspending the test piece is by means of a very fine filament, the weight of which can be included in the zero adjustment of the balance and its volume in water can be ignored. However, if smaller than standard test pieces are used the effect of the filament could be significant.

If the rubber is less dense than water a less dense liquid of known density could be substituted but it is more usual to attach a sinker to the test piece. The sinker can conveniently be a small piece of lead, but using an item like a paper clip to suspend the test piece leads to complications as it will only be partly submerged. The weight of the sinker in water must be measured and it is a common error among new assistants to make this weighing in air.

ISO 2781 also details a procedure (method B) for use when it is necessary to cut the sample into small pieces to avoid trapped air, as might happen with narrow bore tubing. The test piece comprises a number of small smooth pieces within the size 4 mm × 4 mm × 6 mm. These are weighed in a density bottle both with and without the remaining space filled with water. The bottle is also weighed without rubber both empty and filled with water. This is a more tedious procedure than method A and most people would prefer to go to great lengths to obtain a large test piece free from air bubbles rather than resort to the density bottle, especially as even with the test piece cut up trapped air can still be a problem.

The British equivalent to ISO 2781 is BS 903:Part A1[2] which is identical to the international method. Rather surprisingly ASTM does not appear to have a specific method for density at the present time. There is however a section on density in the method on chemical analysis (D297)[3] and also a separate method for determining the density of rubber chemicals (D1817).[4]

It is not often that density is required to be known to a greater accuracy than provided by the standard methods discussed above, but greater accuracy can be provided by use of a density column as standardised for plastics testing in, for example, ISO R1183.[5] The principle of the method is that two miscible liquids of different densities can be run into a container such that a uniform density gradient from the bottom to the top of the container results. This column can then be calibrated by floats of known density which will come to rest at the depth in the column where their densities equal that of the immediate surrounding liquid. Small test specimens of rubber are then introduced into the column in the same manner and allowed to come to rest, their height in the column measured and their density deduced from a calibration graph. With care a column will last several months and the range of density in a single column would not normally be greater than 0.2 Mg/m^3 but could be as little as 0.02 Mg/m^3. Ten minutes is suggested as the minimum time to allow test pieces to come to equilibrium but a large number of samples can be tested at one time and only a very small sample is required. A typical single column apparatus is shown in Fig. 7.1.

A useful procedure for checking if test pieces lie within certain limits of density is given in R1183 method C whereby two liquids of different but known densities are prepared; to be within the known limits a test piece must sink in one liquid and float in the other. This can be employed, for example, to rapidly sort parts made in two materials which have been mixed up. A further variation[6] is titration of a heavier liquid into a lighter liquid until the test piece just floats.

Because density is often used as a quality control check on batches of rubber compound, there has been a necessity to make measurements

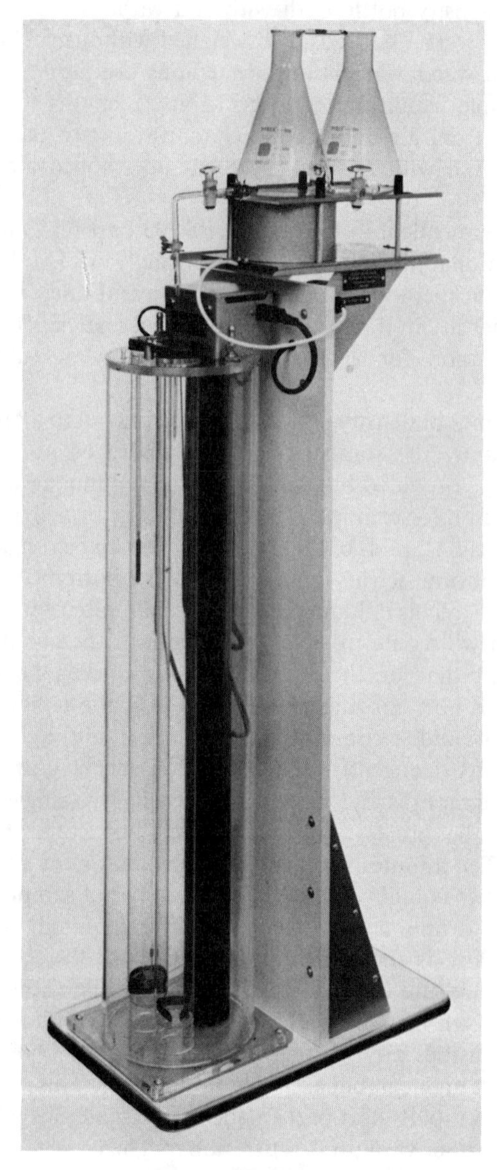

FIG. 7.1. Density column.

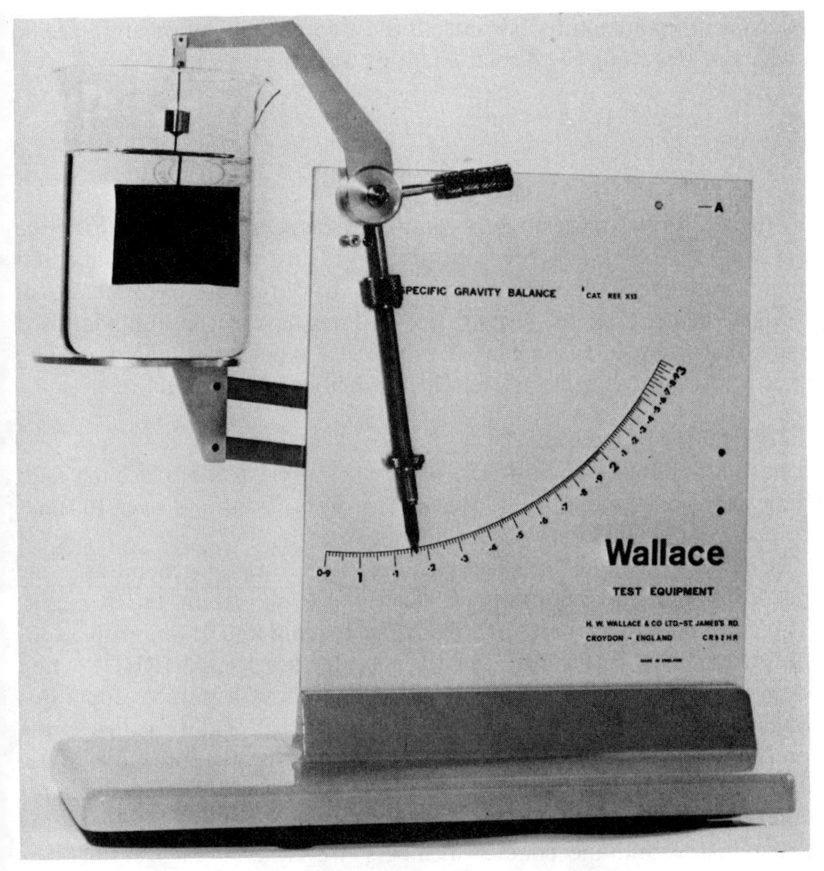

FIG. 7.2. Direct reading density balance.

essentially in accordance with standards such as ISO 2781 but making the determinations as rapidly as possible. Hence various designs of 'specific gravity balance' are in existence which to varying degrees automate the process. In one form of apparatus, shown in Fig. 7.2, the practical steps of weighing in air and water are taken but the result may then be read directly from a scale calibrated in density.

Complete automation using the principle of weighing a moulded test piece of known volume was achieved in an apparatus developed at RAPRA which also automatically measured hardness. More recently an instrument manufacturer has introduced a completely automatic 'specific gravity tester' using the displacement of water principle.

7.2 DIMENSIONS

Virtually all physical test methods involve the measurement of test piece dimensions and it has been the usual practice for each test method standard to specify the apparatus and means of making the dimensional measurements. Over the years the procedures have become standardised such that it was reasonable to produce a separate standard dealing specifically with dimensions, to which other standards could refer.

7.2.1 Standard Methods

The international standard, ISO 4648,[7] has three sections dealing with dimensions less than 30 mm, dimensions over 30 mm and over 100 mm, respectively. BS 903: Part A38[8] is identical.

In the first section three methods are given, one for general use, one non-contact method and one specifically for compression set test pieces. The general method specifies a dial gauge reading to 0.01 mm with a foot pressure of 20 ± 3 kPa (10 ± 2 kPa for hardness below 35 IRHD) acting on a plane flat foot. An annexe gives a diagram of a suitable apparatus involving a weight and a dial gauge with a lock such that the gauge spring pressure does not bear on the rubber. The more usual approach is to use an ordinary dial gauge minus the return spring. The method intended for compression set test pieces is similar except that the force on the foot is 850 ± 30 mN and the contact members are either domed surfaces or 12.5 mm radius or a spherical contact of 6.35 mm diameter and a raised platform of 9.5 mm diameter.

This use of curved surfaces for compression set is based on the fact that after compression, particularly with non-lubricated test pieces, the rubber may well have concave surfaces. This does not happen if the test piece is lubricated and it is expected that the compression set standard will change to specifying an ordinary flat contact foot.

It is fairly obvious that the use of a different foot pressure would, with a soft deformable material such as rubber, produce a different result. Hence, it is not permissible to use a dial gauge with a return spring, calipers or a travelling microscope when this standard method is specified. The errors resultant on using different pressures have been reported by Clamroth and Dobroschke.[9]

These two methods are essentially intended for the measurement of thickness and there has been considerable debate as to how many readings should be taken and what form of average should be used. ISO 4648 specifies at least three readings, taking the median for general use; and one central reading for compression set. However, for a test such a tensile strength there is merit in the notion that it is not the median thickness but the minimum thickness which is required. Practical difficulties probably make the adoption of the mean or median most sensible but it can be expected that not all test methods will be totally in agreement for some time.

Dimensions such as the width of a dumb-bell or the depth of a nick in a tear specimen will be less than 30 mm but could not be measured with the dial gauge; because of the virtual impossibility of applying a known pressure, such measurements must be made in an essentially 'contactless' manner. For low precision, calipers or a rule may suffice but for readings to 0.01 mm a travelling microscope or projection microscope is most suitable and this is specified in ISO 4648 although no details of the apparatus are given. Projection microscopes also find use in examining profiles and for rapid swelling tests (see Section 16.2.2).

For the measurement of dimensions greater than 30 mm ISO 4648 simply specifies vernier calipers reading to 0.25 mm, with the requirement that the test piece shall not be strained. Again there seems no reason why a travelling microscope could not be used and would perhaps be necessary if precision better than 0.25 mm was required. Although projection microscopes do not generally cover dimensions much greater than 30 mm, they can be used with a suitable jig to measure change in dimensions of a large test piece. The last method of ISO 4648 specifies a rule or tape for measuring dimensions greater than 100 mm with an accuracy of 1 mm.

No attempt will be made in this section to consider all the separate measurement clauses to be found in current test method standards. Until the ISO standard for the measurement of dimensions has become well established each test method will have its own procedure and there will not be universal agreement on detail. The essentials are to distinguish between a non-contact measurement and one applying a specified pressure, in the latter case to use the correct standard pressure, and to measure within the accuracy limits specified.

7.2.2 Non-Standard Methods

Although not standardised for use with rubbers, the method of determination of gravimetric thickness is used for thin plastic film and could have use with rubber film or coatings. In the method of BS 2782[10] a square test piece of 100 cm^2 area is weighed in air and water to give mass and density and its area found by measurement of dimensions. From this the average or gravimetric thickness can be calculated.

On-line inspection is a form of testing and in this context dimensional measurements are those most often made. Apart from gauges, micrometers and so on, there are various optical, electrical, nuclear and other methods which may have advantages in production circumstances. Descriptions of the use of such techniques are extensively covered in the literature, for example non-contact thickness measurements on plastics have been reviewed,[11] and a number of references can be found in a review of developments in rubber testing.[12] Unfortunately there is not space here to describe or review the individual techniques. Methods having use in the laboratory have been reviewed in connection with swelling measurements.[13,14]

There are inevitably a great number of special circumstances connected with rubbers where an unusual type of dimensional measurement is required and it may be of value to make reference to cases which have been reported in the literature. Very often a great deal of dimensional information can be found by means of microscopy—such an important subject in its own right as to be in no way considered as a branch of physical testing. One would expect to employ a microscope to determine the thickness of a wax film on the surface of a rubber but a particular technique employing a cigarette paper has been described.[15] Methods for determining footprint area of tyres have been given[16,17] and also an interlaboratory study of tread depth measurements.[18] Back to the measurement of area, a device has been described to measure the contact area between a gasket and a shaft[19] as has an apparatus for determining the surface area of polymer shavings.[20]

These references are of course by no means exhaustive; many more particular cases of dimensional measurement will have been described and a number of methods of interest will be mentioned in later chapters in conjunction with particular physical tests.

7.2.3 Surface Roughness

It is not often necessary to measure the surface roughness of rubber test pieces or products and no standard methods exist. If measurements are attempted, either mechanical profiling as is standard with metals or possibly optical reflectance methods would be used. It has been suggested[21] that the surface geometry of rubber can be assessed by observing the length of shadow thrown by an opaque body. Generally, it is necessary to turn to methods established for metals but only those which can be adapted to take account of the much greater deformability of rubber will be of potential value.

In rubber testing the surface finish of metals is of importance, for example on mould surfaces and compression set plates. The ISO standard for surface finish is 468[22] and the British equivalent BS 1134.[23] This British standard is divided into two parts, the first concerning the method and instrumentation; the second forms a general explanation and is hence a good introduction to the subject. The parameter most often used to grade the roughness of a surface is *Ra* (previously known as *CLA*), the mean deviation of the surface profile above and below the centre line. For example, for compression set the arithmetic mean deviation (*Ra*) of the compression plates must be better than 0.2 μm.

7.2.4 Extensometry

The measurement of extension (or other mode of deformation) is an essential part of several tests, notably tensile or compression stress/strain properties and also thermal expansion. The precision required must be specified in the individual test method and is unlikely to be the same as that required for test piece dimensions. The method of measurement will also be to a considerable extent dependent on the test in question and particular techniques may in some cases be given. Hence, the requirements for particular tests will be discussed in the relevant sections in later chapters.

7.2.5 Dimensional Stability

Generally, vulcanised rubber is dimensionally very stable, which probably explains the lack of standard test methods for this property. (Thermal expansion and swelling in liquids are properties considered in their own right and not normally thought of as being measures of dimensional stability.) This is a different situation to that which exists with plastics where a number of dimensional stability tests are in existence. If a measure of dimensional change is required, the appropriate dimensions of a suitable sized test piece can be measured before and after an ageing treatment by any of the methods mentioned in this chapter.

Although mould shrinkage, i.e. the reduction in size of cooled moulded articles compared to the mould dimensions, is principally a matter of thermal expansion it is usual to make a direct measure of shrinkage by measuring a standard moulded test bar. There must be any number of 'standard' moulds used in various factories for this purpose and the main essential is that the required accuracy can be obtained. For example, to detect 0.1% shrinkage on a 10 cm bar requires a measurement to 0.1 mm. Shrinkage data has been given by Juve and Beatty[24] as well as a procedure for calculating shrinkage for different formulations.

7.2.6 Dispersion

The dispersion of compounding ingredients, particularly carbon black, in the rubber can have a large effect on physical properties and a measure of dispersion can be used to judge the efficiency of mixing. The estimation of degree of dispersion is effectively a dimensional measurement using microscopy techniques and is just one example of the value of microscopy for fault diagnosis in rubber products.

Dispersion measurements are normally made on cured rubber although it is possible to prepare test pieces from some uncured materials. Probably the most widely used techniques are those described carefully by Medalia and Walker[25] and which form the basis of ASTM D2663.[26]

Examination of a torn surface with reflected light gives an overall picture of dispersion and is a useful rapid test in the control laboratory. To examine the dispersion of fine agglomerates of carbon black it is necessary to microtome sections using a freezing stage on a sledge microtome. The sections are examined by transmitted light and either the percentage of agglomerates is estimated by counting or reference made to a standard chart.

Alternative approaches to the measurement of dispersion include the use of a stylus to measure roughness,[27] electrical resistivity[28] and the analysis of the dark field image produced by a reflected light microscope.[29] Persson[30] has reported an improvement to the ASTM microscope method by using split field microscopy. Methods given by Janzen and Kraus,[31] based on optical density are for determining dispersability of blacks and not dispersion in rubber compounds.

Cembrola[32] has compared microscope, stylus and resistivity methods and concludes that no one method is universally the best.

REFERENCES

1. ISO 2781, 1981. Determination of Density.
2. BS 903: Part A1, 1980. Methods of Testing Vulcanised Rubber. Determination of Density.

3. ASTM D297-81. Rubber Products—Chemical Analysis.
4. ASTM D1817-66. Chemicals—Density.
5. ISO R1183, 1970. Methods for Determining the Density and Relative Density of Plastics Excluding Cellular Plastics.
6. Gal, O.S. and Premovic, P. (1980). *Angew. Makromol. Chem.*, **84**, 1.
7. ISO 4648 1978. Determination of Dimensions of Test Pieces.
8. BS 903:Part A38, 1978. Determination of Dimensions of Test Pieces.
9. Clamroth, R. and Dobroschke, P. (1974). *Tech. Notes Rubb. Ind.*, No. 51.
10. BS 2782, Method 631A, 1982. Gravimetric Thickness and Yield of Flexible Sheet.
11. Anon. (July 1969). *Plastics Design and Processing*, 22.
12. Brown, R.P. and Scott, J.R. (1972). *Progress of Rubber Technology*, 36.
13. Brown, R.P. and Jones, W.L. (Feb. 1972). *RAPRA Bulletin.*
14. Brown, R.P. (Aug. 1973). *RAPRA Members J.*
15. Angert, L.G., Mavrina, R.M. and Kuz'minskii, A.S. (Oct. 1970). *Kauch i Rezina*, **28**(10).
16. NBS. (Oct. 1970). *Rubber Age*, **102**(10).
17. Pizer, R.S. and Spinner, S. (Feb. 1970). *Mater. Res. Stand*, **10**(2).
18. Brenner, F.C. and Kondo, A. (May 1974). *Tyre Sci. Tech.*, **2**(2).
19. Yurtsev, N.N. and Kosenkova, A.S. (Dec. 1969). *Kauch i Rezina*, **28**(12).
20. Salloum, R.J. and Eckert, R.E. (May 1972). *J. Polym. Sci., A2,* **10**(5).
21. Tryukova, Yu.P., Travin, G.M. and Gorbunov, O.L. (1972). *Kauch i Rezina*, **31**(7).
22. ISO 468, 1982. Surface Roughness.
23. BS 1134:Parts 1 and 2, 1972. Assessment of Surface Texture.
24. Juve, A.E. and Beatty, J.R. (Oct. 1954). *Rubber World.*
25. Medalia, A.I. and Walker, D.F. Technical Report RG 124 Revised, Cabot Corporation.
26. ASTM D2663-82. Dispersion of Carbon Black.
27. Hess, W.M., Chirico, V.E. and Vegvari, P.C. (1980). *Elastomerics*, **112**(1).
28. Boonstra, B.B. (1977). *Rubb. Chem. Tech.*, **50**(1).
29. Ebell, P.C. and Hemsley, D.A., (1981). *Rubb. Chem. Tech.*, **54**(4).
30. Persson, S. (1978). *Eur. Rubb. J.*, **160**(9).
31. Janzen, J. and Kraus, G. ACS Rubber Division Meeting, Cleveland, Oct. 1979, Paper 3.
32. Cembrola, R.J. (1983). *Rubb. Chem. Tech.* **56**(1).

Chapter 8

Short-Term Stress and Strain Properties

In the first of the chapters devoted to mechanical tests the so-called 'short-term' measurements of stress and strain are grouped together. The term 'static stress/strain' tests is also applied to this group to distinguish them from the dynamic or cyclic tests. Either term is intended to refer to the ordinary stress/strain or force/deformation measurements which are made at somewhat arbitrary chosen strain rates and where the effects of long times and cycling are ignored. Such tests have been studied and standardised for many years and have been of enormous value, particularly in quality control. It must be realised however that they are limited as regards complete mechanical characterisation simply because mechanical properties of rubbers are very dependent on time, temperature and test conditions. There has been on the one hand something of a fanatical belief in, for example, the value of hardness measurements, and, on the other, dismissal of all simple mechanical tests as virtually useless. The truth is, as one might suspect, somewhere in between and the best value is obtained from the simpler tests if care is taken in choosing the most relevant tests, carrying them out under the most relevant conditions and not expecting the result to be necessarily valid when applied to some different conditions.

Stress/strain relationships are commonly studied in tension, compression, shear or indentation. Because in theory all stress/strain relationships except those at breaking point are a function of elastic modulus, it can be questioned as to why so many modes of test are required. The answer is partly because some tests have persisted by tradition, partly because certain tests are very convenient for particular geometry of specimens and because at high strains the physics of rubber elasticity is even now not fully understood so that exact relationships between the various moduli are not known.

It is not necessary to be expert in the theory of rubber elasticity to test rubbers but it is a distinct advantage to be conversant with the main principles. A classic account of the development of modern theory is given by Treloar[1] which, if not digested from cover to cover should be compulsory reading for those concerned with physical testing. Statistical network theory leads to the following relatively simple relationships between stress and deformation:

(a) Simple shear

$$\text{Shear stress} = G\gamma$$

(b) Tension or compression with lubricated ends

Tensile or compressive stress (calculated on initial cross section)

$$= G(\lambda - \lambda^{-2})$$

(c) Equi—biaxial extension

$$\text{Stress} = G(\lambda^2 - \lambda^{-4})$$

where G = shear modulus,
γ = shear strain,
λ = ratio of strained to unstrained length (height).

Rather surprisingly, all these kinds of deformation can be described in terms of a single modulus. This is a result of the assumption that rubber is virtually incompressible (i.e. bulk modulus $\gg$ shear modulus. Young's modulus $E = 3G$ (for filled rubbers the numerical factor may be in fact as high as 4). Indeed these relationships by no means fully describe the complete stress/strain behaviour of real rubbers but may be taken as first approximations. The shear stress relationship is usually good up to strains of 0.4 and the tension relationship approximately true up to 50% extension.

A general relationship for compression with bonded ends[2] is:

$$\text{Compressive stress} = G(\lambda - \lambda^{-2})Z$$

where Z is a shape factor which is a function of the dimensions of the test piece and modulus.

Scott[3] has given the relationship for indentation by a rigid ball (for indentations up to 0.8 of the ball diameter) as:

$$\text{Force} = KGR^{0.65}\ P^{1.35}$$

where K = numerical constant,
R = radius of ball,
P = depth of identation.

For torsion where deformation is essentially in shear the relation for a strip much wider than its thickness is:

$$\text{Torque} = \frac{K_1 w t^3 G \theta}{L}$$

where K_1 = functions of width and thickness of strip,
w = width of strip,
t = thickness of strip,
θ = angle of twist,
L = length of strip.

The fundamental shear and Young's moduli are the slope of the shear or tension/compression stress/strain curve at the origin and the relationship given above is an attempt with theoretical justification to describe the shapes of the stress/strain curves at higher strains. Appreciation of this may avoid confusion between 'the absence of a single modulus figure' for rubber whilst such values are quoted.

It should be noted that the 'rubber technologists modulus', i.e. the tensile stress corresponding to some arbitrary elongation, is not a modulus at all, but many technologists still insist on using the term.

8.1 HARDNESS

A hardness measurement is a simple way of obtaining a measure of the elastic modulus of a rubber by determining its resistance to a rigid indentor to which is applied a force. Indentation involves deformations in tension, shear and compression but, as in the case of a perfectly elastic rubber the moduli controlling these are closely related, it is convenient to regard hardness as depending simply on Young's modulus .

For the various geometry of indentor which might be used the following approximate relationships have been derived for a perfectly elastic rubber:

Ball:

$$P = k_1 \left(\frac{F}{E}\right)^{0.74} R^{-0.48}$$

Flat ended cylinder:

$$P = k_2\left(\frac{F}{E}\right)d^{-1}$$

Cone:

$$P = k_3\left(\frac{F}{E}\right)^{0.5}$$

where $P =$ depth of indentation,
$F =$ indenting force,
$E =$ Young's modulus,
$R =$ radius of ball,
$d =$ diameter of cylinder,

k_1, k_2 and k_3 are constants, k_3 involving the angle of cone.

A truncated cone of fairly small angle behaves roughly like the cylinder.

It would be convenient to have a hardness test where a given difference in indentation always represented the same proportionate difference in modulus (i.e. P a linear function of log E). None of the indentor shapes listed above achieves this with a constant loading force. The cone has the disadvantage of being especially prone to damage, a criticism which also applies to the plunger and truncated cone. Because of this, and the fact that accurate hardened steel balls are readily obtainable, standard tests mostly use a ball indentor, although a truncated cone is used in the Shore durometer .

The indenting force could be applied in three ways:

(a) Application of a constant force, the resulting indentation being measured.
(b) Measurement of the force required to produce a constant indentation.
(c) Use of spring loading, resulting in variation of the indenting force with depth of indentation.

Referring back to the equation given for indentation by a ball, it would appear that method (b) is ideal because the measured force should be proportional to modulus. Modern force transducers would make this measurement very convenient but the method has never been seriously adopted, presumably because when standard methods were being established the measurement of force would have been a severe complication. All pocket instruments use a spring loading system, which does in fact enable a much closer to linear relationship between P and log E to be realised. However, springs are not well thought of as precision measuring elements and the net result is that the internationally recognised reference standards use weights to apply a constant force.

The development of hardness tests over the years is a fascinating subject, there having been an amazing variety of instruments most of which are now a matter of history. For an account of these reference should be made to the work of Soden.[4]

8.1.1 Dead Load Tests

The internationally accepted standard dead load method is given in ISO 48[5] which covers rubbers in the range 30–85 IRHD. For the normal test a ball indentor of either 2.38 mm or 2.5 mm diameter (2.38 mm = 3/32 in) is used, acting under a total force of 5.53 or 5.7 N, respectively. The two conditions have been shown to be equivalent but with complete metrication one would expect the 2.38 mm ball to disappear. The indentation of the ball is made relative to the surface of the test piece and to define this an annular foot of specified dimensions surrounds the indentor and presses on the rubber with a given force. Again, because of the uncertainty of defining the surface position of a soft material, the 'zero' reading of the indentor is taken with a small specified contact force and the indenting force applied in addition. Because the indentation would gradually increase with time due to creep, readings are taken after the arbitrarily fixed time of 30 s after applying the indenting force. The tolerances on dimensions and forces given in ISO 48 have been arrived at after extensive calculations and testing and no significant difference in results would be expected between apparatus within those limits. It should be noted that the indentor must act essentially without friction and in the usual design of apparatus the test piece is gently vibrated in the hope of overcoming any slight residual friction. Dusting the test piece with talc is to reduce any friction between indentor and test piece.

The hardness reading obtained is to some extent dependent on test piece thickness and certainly the effect will be very marked if very thin test pieces are used. The standard thickness is given as 8 to 10 mm but tests may be made on thickness down to 4 mm.

The indentation is usually measured with a dial gauge with its scale directly calibrated in International Rubber Hardness Degrees (IRHD). The IRHD scale owes a lot to the desire to have readings equivalent to the Shore A scale (see Section 8.1.2), which was originally very popular, and to represent increasing hardness by increasing numbers. (Indentation

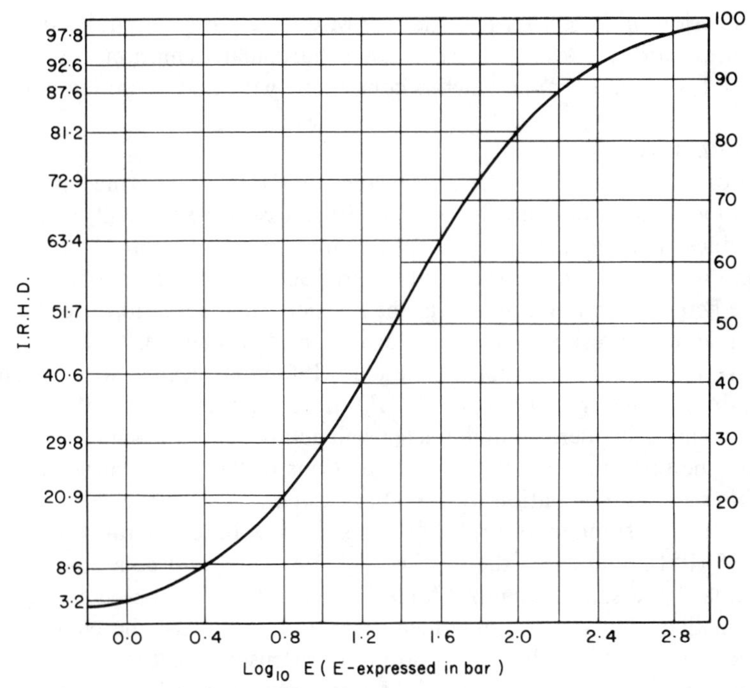

FIG. 8.1. Probit curve relating logarithm of Young's modulus, *E*, and hardness in IRHD.

decreases with increasing hardness.) The basis of the IRHD scale is a probit (integrated normal error) curve relating log(modulus) to hardness (Fig. 8.1) and this curve is reproduced in the standard together with a table of indentation against IRHD. The curve is defined by the value of $\log_{10}$(modulus) at the mid point and the maximum slope of the curve, and results in a scale which is in good agreement with the Shore A scale. The modulus is calculated from the indentation by means of a modification of the formula for indentation by a ball given earlier.

In practice the relationship between hardness and Young's modulus as given by this formula is not realised within a reasonable tolerance and there are other empirical and theoretical formula. Steihler *et al.*[6] demonstrated the dependence of the exponent 1.35 on the radius of the indentor. Yeoh[7] has examined the accuracy of three formulae and suggests that discrepancies can occur simply from the way in which Young's modulus for a rubber is defined and measured. Further discrepancies can be expected with rubbers which are far from perfectly elastic. The conclusion must be that it is most unwise to rely on a modulus derived from a hardness test, at least unless a particular definition is given to modulus and a formula which is compatible with that definition is used.

HIGH AND LOW SCALES

For hardness above 95 and below 30 IRHD the normal standard method is not very satisfactory. In either case a very small change in hardness number results from unit change in indentation. At the high end the indentation needs to be increased relative to the standard test to give better discrimination and at the low end the indentation needs to be decreased to prevent excessive deformation of soft rubbers.

At present there is a separate standard ISO 1400[8] dealing with the high hardness range from 85 to 100 IRHD (note the overlap with ISO 48) by specifying a smaller, 1 mm diameter, ball acting under the same force and with the same foot as in ISO 48. Extensions of the probit curve and the table relating indentation to IRHD are given.

A further standard ISO 1818[9] covers the low hardness range from 10 to 35 IRHD and again the overlap with ISO 48 should be noted. In this standard the same indenting force is used but the ball is increased to 5 mm diameter and in consequence the foot must also be changed to accommodate the ball. The appropriate extensions to the probit curve and table are given. The standard test piece thickness is increased to between 10 to 15 mm with an absolute minimum of 6 mm as even with the increased ball size larger indentations than in ISO 48 are realised.

A draft revision of ISO 48 is in circulation in which the high and low ranges are included. This revision also includes apparent hardness as discussed below.

MICRO TESTS

For testing small products and in particular those having a thickness of less than 4 mm the standard dead load test is altogether too large. Consequently micro tests have been developed and a standard procedure is included in ISO 48. The test is essentially the same as the normal test but uses a 0.395 mm diameter ball with 153 mN total force and an appropriately reduced foot. The standard test piece thickness is 2 mm but ISO 48 points out that for various reasons the micro test will not always give the same result as the normal test even when both are made at their respective standard thicknesses and sometimes these differences can be quite large. Although the micro and normal tests are the same in principle the apparatus used differs considerably. The normal hardness instrument is a fairly straightforward mechanical device (Fig. 8.2) but because of the very small indentations involved and the greater effect of even small amounts of friction, a rather more complex apparatus is necessary for micro tests. The development and evaluation of the micro test has been described by Scott and Soden[10] and the principle of operation of one widely used instrument is shown in Fig. 8.3. The vertical movement of the specimen table is magnified by a 6:1 sliding wedge which compensates directly for the indentation being one sixth of that in the standard test. The table is moved to return the indentor to its position before application of the load, the null point being detected by an AC bridge circuit. The movement of the table effectively measures the indentation but the indentor, being suspended on leaf springs, is not subject to friction and the springs contribute no force in the null position.

APPARENT HARDNESS

It has been stated previously that hardness readings are influenced by test piece thickness. Consequently the term standard hardness refers to measurements made on standard test pieces and measurements on non-standard test pieces are called apparent hardness.

Generally, the apparent hardness will increase as the test piece thickness is reduced because of the effect of compression against the rigid test piece support.

Corrections to hardness readings to compensate for test piece thickness are not entirely satisfactory because the effect varies from rubber to rubber; however, typical readings are shown in Fig. 8.4.

If readings are taken too near the edge of a test piece there will be an 'edge effect' and consequently minimum distances from the edge for various thicknesses are given in ISO 48, 1400 and 1818. An infinite block would of course be an ideal test piece for both the thickness and edge effects but, to say the least, is impractical.

It is often necessary to make hardness measurements on curved surfaces, e.g. rollers or 'O' rings. In the first example the product may be large enough for the hardness instrument to rest upon it whilst in the second it would usually be possible to rest the product on the specimen table. In either case some form of jig is required to locate the test piece and suitable examples and precautions to be taken are described in the British standard for hardness[11] which is discussed later in this section. A comparison of methods for holding small 'O' rings for micro tests has been made[12] which also describes a novel device using metal 'fingers' which proved very satisfactory.

FIG. 8.2. Dead load hardness tester.

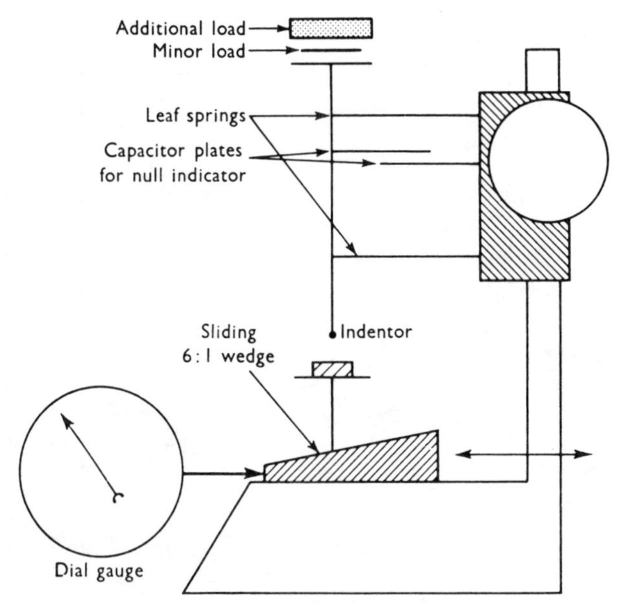

FIG. 8.3. Principle of operation of Wallace micro-hardness tester.

The particular case of the hardness of rollers is covered by draft international standards. DIS 7267[13] is in three parts dealing with the normal dead load method, durometers and the Pusey Jones method respectively. This last method is a very old hardness test which is now virtually never used except on rollers for which it is still very popular, probably mostly for reasons of tradition.

BRITISH AND ASTM STANDARDS

The British standard for hardness[11] is technically very similar to the ISO methods but is presented all in one document with the addition of methods for measuring apparent hardness on curved surfaces, i.e. more in the form of the draft revision of ISO 48. Any of the standard methods could be used for curved surfaces except that it is not possible to use a foot on concave surfaces. For large cylindrical surfaces the hardness tester is either fitted with feet movable in universal joints which rest on the curved surface or the base of the instrument is fitted with two cylindrical rods which rest on the curved surface. The latter method can be used for surfaces with radius of curvature down to 20 mm and is illustrated in the standard. For surfaces having double curvature, only the method using movable feet is suitable. For small products and where the radius of

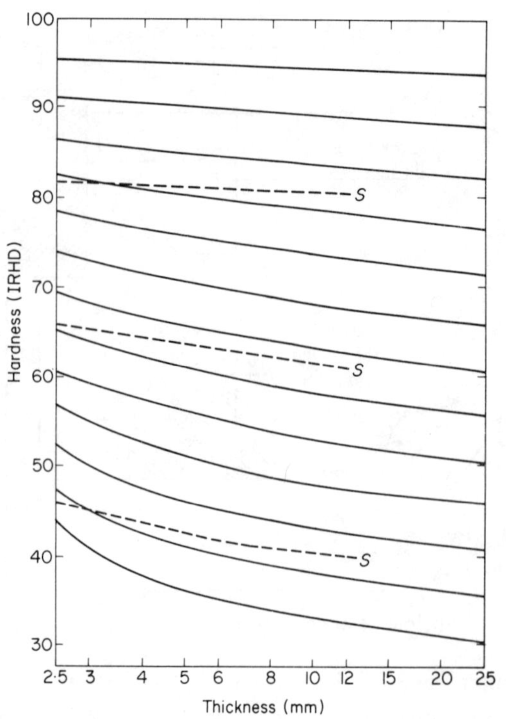

FIG. 8.4. Effect of test piece thickness on hardness reading. —— IRHD; ----S Shore A.

curvature is too small to rest the instrument on the surface the 'test piece' is placed on the base of the instrument as with the standard test piece. The standard specifies that the test piece shall be held so as to prevent bodily movement of the surface and suggests methods to achieve this (see reference 12).

ASTM D1415[14] is essentially the same as ISO 48 but there is no coverage of high and low scales as in ISO 1400 and 1818. Curved surfaces are not covered in D1415 but there is a separate standard D1414[15] for 'O' rings which has a section on measuring hardness. The term apparent hardness is not used and reference is simply made to the micro test and to the use of a durometer (see below) with a jig to aid alignment. The Pusey and Jones test is described in ASTM D531.[16]

8.1.2 Durometer Tests

The term durometer is used for the small 'pocket-type' of hardness meter which, in a myriad of shapes and sizes, has been in common use virtually as long as the rubber industry has existed. The main difference between durometers and the dead load tests is that the former utilise a spring to produce the indenting force. Because of this, and because of variability due to the instruments being hand held, durometers are not as precise or as reproducible as the standard dead load instruments. In one trial[17] the range of variation between five operators on the same rubber averaged 5° for a Shore durometer, 3° for a Wallace durometer but only 1.5° for a dead load instrument.

However, it should be noted that improvement can be made by fixing the durometer to a stand and applying the foot pressure by a weight rather than hand. This defeats the object of having a 'pocket' instrument but in some countries a Shore durometer used in this way is at least as popular as the international dead load method.

For many years there was no move to produce an international standard for durometers but at the time of writing a draft standard[18] is in circulation. It covers the Shore A and D type meters and a meter calibrated in IRHD. The Shore meters are also covered in ASTM D2240[19] and an international standard for plastics ISO 868.[20] The Shore A scale corresponds approximately to the IRHD scale and the D scale can conveniently be used for hard rubbers above about 90 IRHD. The rubber draft specifies a test piece at least 6 mm thick and that the reading is taken 1 s after applying the instrument, although other times are recognised. In practice the 1 s is not strictly adhered to and when testing products all manner of unreasonable test piece geometries are used, resulting in many instances in very large variability. The effect of time has been investigated in some detail.[21]

A British standard for durometers, BS 2719,[22] has existed for some time. The British attitude has been that it is advisable to have a standard giving guidance on the use and calibration of durometers but not to specify the instruments. Hence BS 2719 recognises the existence of many different makes of Shore type instruments and a list is given as an appendix. Details of calibration are given but the actual instruments are not specified, although reference is made to such standards as ASTM D2240 (but not ISO 868).

Most of the currently available durometers referred to in BS 2719 are of the Shore type but one particularly interesting instrument is the Wallace Pocket Meter (shown in Fig. 8.5) which is calibrated directly in IRHD, and which uses a hemispherical indentor. (This type is also covered in the ISO draft.) This eliminates the problem of rapid wear on the truncated cone of the Shore instrument. The Wallace Meter also operates with a

FIG. 8.5. Wallace durometer together with RAPRA standard hardness blocks.

substantially constant spring pressure, varying by 15 gf in a 270 gf mean force as against between 56 and 822 gf in the Shore A.

It is frequently asked how the readings on pocket durometers agree with those of the standard dead load instruments. The answer must be that agreement is at best approximate because, however carefully the durometer is set up, variations due to the operator, the time of application and the thickness of the test piece inevitably result in relatively large scatter of durometer results. Figure 8.6 shows the approximate relationship between Shore A and IRHD and between Shore A and D.

Dead load instruments are intrinsically calibrated by provision of the correct weights and the correct indentor and foot dimensions. Durometers require, in addition to checking dimensions, calibration of the spring over

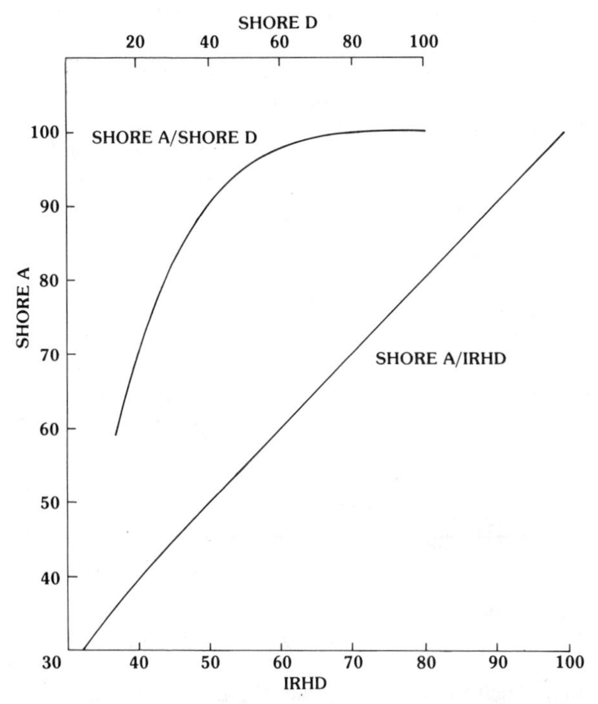

FIG. 8.6. Approximate relationships between Shore A, Shore D and IRHD scales (test piece 10 mm thick).

its complete range. Perhaps a better procedure which checks both dimensions and spring pressure together is to use standard rubbers previously calibrated on a dead load instrument. This method is recommended in BS 2719 as being rapid and applicable to all types of durometer and it also 'calibrates' the operator in terms of the hand pressure exerted.

8.2 TENSILE STRESS/STRAIN

After indentation hardness, the most common type of stress/strain measurement is that made in tension. The ability of rubber to stretch to several times its original length is one of its chief characteristics but it is worth noting that at least as many rubber products are used in compression or shear as are used in tension. Besides being of relevance for products strained in tension, tensile stress/strain properties have been taken, since

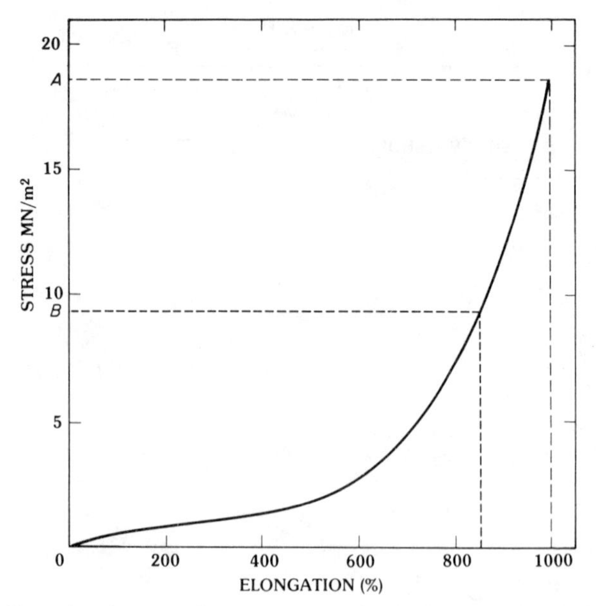

FIG. 8.7. Stress/strain curve for pure gum vulcanisate illustrating the difference in stress between the inside (A) and outside (B) circumstances of a ring test piece.

the beginning of the industry, as a general guide to the quality of a rubber, being sensitive to filler or plasticiser content as well as to mixing and curing efficiency.

Standard methods for determining tensile properties of rubbers have evolved gradually and are now in a well-defined state. Essentially, dumb-bell shaped, or less often ring, test pieces are strained at a constant rate of traverse and force and corresponding extension recorded. The force readings are expressed as stresses by reference to the original cross-sectional area of the test piece.

A typical tensile stress/strain curve for rubber is shown in Fig. 8.7. (This also illustrates the differential strain between the inside and outside of a ring test piece which is discussed later.) It can be seen that there is no linear elastic portion as is usual with, for example, metals, and rubber technologists do not measure a modulus as such but quote the stress at various percentage elongations, commonly at 100%, 200%, etc. If a figure for Young's modulus was required, this could be obtained from a more sensitive measurement of stress and strain in the very early part of the curve.[7]

All the major international and national standards are fundamentally in agreement and it will be convenient to consider the test method section by section.

8.2.1 Form of Test Piece

Two shapes of test pieces are generally used—rings and dumb-bells and both are covered by ISO 37,[23] BS 903:Part A2[24] and ASTM D412.[25] The advantages of rings are that there are no gripping problems as the ring may be mounted on two pulleys and that the elongation is easily measured by monitoring the distance between the pulleys. Their principal disadvantage is that the strain distribution in the ring is not uniform and this will be considered in more detail below.

Dumb-bells, on the other hand, are rather more difficult to grip and the measurement of elongation cannot be taken from grip separation as the strain along the whole test piece is not uniform. However, the stress and strain is uniform throughout the central parallel portion of the dumb-bell and hence the problem of non-uniform strain in ring test pieces is avoided. In addition, by cutting dumb-bells in different directions grain effects can be studied which is not possible with rings.

Largely because of the uneven strain problem, but perhaps also because most laboratories consider rings to be more difficult to cut from sheet, dumb-bells are the more commonly used of the two test pieces.

When a ring is stretched the tensile stress and strain are not uniform over the cross-section but vary from a maximum on the inside circumference to a minimum at the outside. As the ring breaks when the maximum stress equals the breaking stress the force registered at break does not correspond to the true tensile strength. This was pointed out many years ago and the extent of the non-uniformity is shown in the following figures given by Reece in 1935[26] for the standard ring test piece:

E_1	100	200	300	400	500	600	700	800	900	1000
E_2	80	163	247	331	415	499	584	668	752	857

E_1 = % elongation of inside diameter
E_2 = % elongation of outside diameter

The effect of the difference between E_1 and E_2 on the force at break is shown in Fig. 8.7. The approximate 'average' stress is $\frac{1}{2}(A + B)$ which is the apparent tensile strength and may be considerably less than the true tensile strength A. The discrepancy may be as much as 33%[27] and will vary with the steepness of the final part of the stress/strain curve. Although this may not be serious in pure quality control testing, the variation with

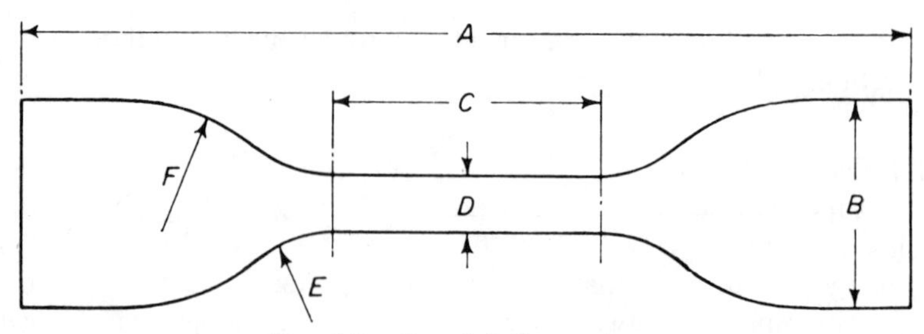

FIG. 8.8. Dumb-bell test piece.

Dimensions	*Type* 1 (*mm*)	*Type* 2 (*mm*)	*Type* 3 (*mm*)
A	115 minimum	75 minimum	35 minimum
B	25 ± 1	12.5 ± 1	6 ± 0.5
C	33 ± 2	25 ± 1	12 ± 0.5
D	$6.0 + 0.4$ -0.0	4 ± 0.1	2 ± 0.1
E	14 ± 1	8 ± 0.5	3 ± 0.1
F	25 ± 2	12.5 ± 1.0	3 ± 0.1

slope of the stress/strain curve could make comparisons between different rubbers invalid.

ISO 37 specifies two rings; the well-known 'Schopper' ring of 44.6 mm internal diameter and 52.6 mm external diameter and a smaller ring of 8 mm internal diameter and 10 mm external diameter. The preferred thicknesses are 4 ± 0.2 mm and 1 ± 0.1 mm, respectively. BS 903:Part A2 has only the large ring whereas ASTM D412 has four rings. These are the two ISO rings plus rings of 50 mm circumference, 1 mm radial width, and 100 mm circumference, 2 mm radial width. Both rings can have a thickness between 1 and 3.3 mm. Whereas it seems unnecessary to have four sizes of rings the two peculiar to ASTM have the advantages of being suitable for standard 2 mm sheet and have been designed so that 50 mm of separation is equal to 100 and 50% elongation respectively.

Considerable effort has been expended on selecting the best dumb-bell shape and size, particularly by ISO Committee TC 45. The type 1 dumb-bell of ISO 37 with a 6 mm wide centre portion and preferably cut from 2 mm thick sheet is now very widely used (Fig. 8.8). The relative width of the centre and ends, and the radii where the two join affect the ease of gripping and the incidence of shoulder breaks and the standard dumb-bell has been selected as the best compromise. The same consideration has been given to the smaller type 2 dumb-bell which has of course the advantage of using less material and for many workers is now the first choice. A need has been found for an even smaller dumb-bell and a type

3 has now been introduced which has an overall length of only 35 mm. The types 1 and 2 are also in BS 903 and the type 3 will be added.

BS 903 is in fact long overdue for revision and when the new version appears it can be expected to be aligned with ISO 37 which itself is, at the time of writing, in the course of revision. The draft includes yet another small dumb-bell of overall length 50 mm. ASTM D412 includes the ISO type 1 but not the types 2 and 3 in the six dumb-bells it lists!

In the interests of standardisation it is desirable to limit as far as possible the variety of test piece sizes allowed. Success in this direction has not always been possible as illustrated by the tensile test pieces detailed in ASTM D412. However, there would be no need to limit dimensions at all if it were not a fact that the size of test pieces can affect the magnitude of the result obtained or at least the variability. In the case of tensile tests, the difference in level between results from rings and dumb-bells has already been mentioned. The variability of the two types of test pieces has been found to be similar. The measured tensile strength has a tendency to decrease with increasing cross-sectional area of the test piece and hence it is desirable to make comparisons only between groups of test pieces of nominally the same type and thickness. The difference between the results from type 1 and 2 dumb-bells is not normally significant but Bartenev and Gul[28] found the tensile strengths of a butadiene–styrene rubber to be 21, 27 and 37 kg/mm^2 for test pieces 2.2, 1.2 and 0.4 mm thick, respectively, and in an ISO comparison[29] there were highly significant differences between miniature dumb-bells and the ISO type 2 using thicknesses from 0.7 mm to 2 mm. A careful study of the phenomenon has been made by Nazeni.[30] The ISO and British standards call for a preferred thickness of 2 ± 0.2 mm which implies that results are critically dependent on thickness. An unpublished report to ASTM found no significant difference over the range 1.3–3.3 mm but this was based on one type of material only.

Ring test pieces can be made by four methods: (a) stamping from sheet, (b) cutting with revolving knives from sheet, (c) cutting from tube on a lathe, and (d) by moulding. Method (a) is not really satisfactory for a 4 mm × 4 mm cross section and method (b) is probably the most widely used. Cutting from tube has the inherent inconvenience of the need to first obtain the tube, and in any moulding method care must be taken to avoid interference from flash. Considerable detail of a rotating cutter for rings is given in ASTM D412 whereas ISO 37 only gives an apparatus for small rings—which are considered especially difficult to cut.

Dumb-bells can usually only be stamped or moulded but the latter method is little used. Stamping is satisfactory using sharp dies if the thickness is restricted to 3 mm for a type 1 dumb-bell and 2.5 mm for a type 2.

Any imperfections in the cut edge are potentially sites for premature failure. Patrikeev *et al.*[31] have investigated the effect of artificially introduced nicks as well as the effect of test piece width. Their paper is difficult to follow but apparently the critical depth of a nick (above which strength is effected) ranges from 0.1 to 0.9 mm for different rubbers! They also found a strong dependence of strength on test piece width, the magnitude again depending on rubber type.

8.2.2 Measurement of Test Pieces

The measurement of dimensions is covered in Section 7.2. It is, however, necessary to stress the importance of accurate measurement of the small cross-sectional area of tensile test pieces.

For dumb-bells it is usual (see ISO 37) to take the width as the width of the die from which the test piece was cut although for accurate work the test piece could be measured by a non-contact method. ISO 37 states that the width of rings should be measured using the standard gauge fitted with suitable curved feet, the nominal distance between the cutters not being precise enough due to distortion of the rubber during cutting. ISO 37 also states that for precise work the ring cross section shall be calculated from its mass, density and mean circumference but no method is given for the precise measurement of circumference.

The circumference of rings is required as this is the initial gauge length for elongation measurements. This measurement is not mentioned in ISO 37, presuming, as will not always be the case, that the nominal circumference is near enough correct. BS 903 suggests that circumference can be measured using a graduated cone and this is indeed a common method.

8.2.3 Apparatus

Sample preparation and measuring apparatus has been discussed and hence this section will deal with the principal item, the 'tensile machine', which as a unit includes grips and extensometer. The 'tensile machine' is in fact very often a universal machine in that, apart from tensile tests, it can also be used for flexural, compression, tear and adhesion tests.

Virtually no description of the test machine is given in ISO 37, ASTM D412 has a few details and BS 903:Part A2 goes a reasonable way towards specifying the important characteristics. One reason why so little detail has been given is that it is rather difficult for rubber technologists to write a specification for a somewhat complex engineering instrument. Very recently an international standard ISO 5893[32] has been published covering such test machinery for rubbers and plastics following a very similar British standard[33] which was produced several years ago. Revisions of tensile standards, and indeed other test methods using a tensile machine, will refer to these documents which specify requirements quite comprehensively.

GRIPS

Rings are held by a pair of pulleys mounted on roller bearings, a mechanism being provided to rotate one or both pulleys automatically during the test. It is not satisfactory to use pegs or fixed pulleys because the rubber does not readily slip over the surfaces and is therefore not uniformly stretched.

Dumb-bells are rather less easy to grip and considerable ingenuity has been devoted to the design of a grip that will hold the end of a dumb-bell with a pressure uniform across its width and adequate to prevent slipping, but without setting up local strains liable to cause failure. The essential design feature is that the grip should close automatically as the tension increases. A widely used and successful design is that due to Gavin shown in Fig. 8.9. The dumb-bell end is held between rollers A, the ends of which pass through slots in the members B and C; the slots in C are horizontal and those in B steeply sloping. By depressing C by hand against the spring D, the rollers are forced apart for the insertion of the dumb-bell; on release the spring pushes C up until the rollers grip the dumb-bell. During the test the tension on the dumb-bell tends to pull the rollers further up and hence, by reason of the inwardly sloping slots, closer together, thus increasing the grip.

Rather surprisingly, lubrication of the dumb-bell ends with talc sometimes improves gripping, presumably by permitting just enough slip to equalise the gripping pressure. In certain extreme cases the plate and roller type of grip with the dumb-bell wrapped around the roller is the most successful. A more modern approach to the problem is the pneumatic or hydraulic grip where flat parallel faces are pushed on to the dumb-bell end under air or fluid pressure. Although expensive, this type of grip can be very convenient and successful.

APPLICATION OF FORCE

The test piece must be stretched smoothly at substantially constant speed and to meet this requirement the drive must have sufficient power to maintain the speed even under maximum force. The standard rate of

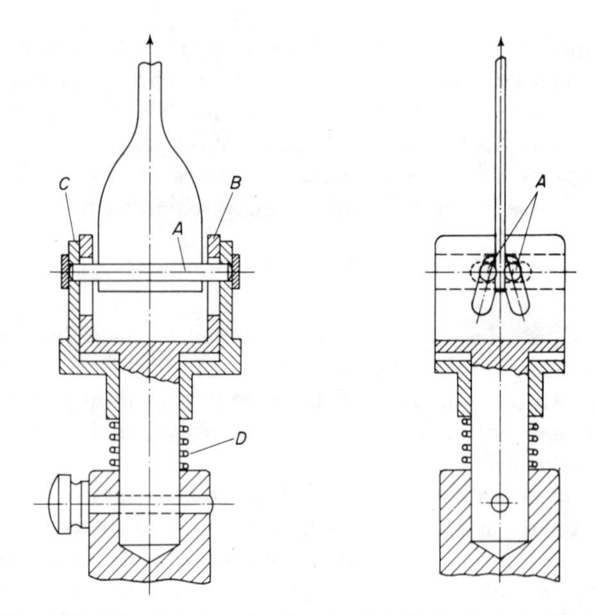

FIG. 8.9. Gavin-type grips for dumb-bell tensile test pieces.

grip separation is 500 ± 50 mm/min but this does not mean that the actual rate of strain in the test piece is being kept constant between equally close limits. The 'fixed' grip may move quite considerably, especially on the later steep part of the stress/strain curve, when a 'soft' pendulum type of machine is used so that the rate of extension is less than the speed of the moving grip. In addition, with a dumb-bell the rate of strain is not constant throughout its length. The strain rate in the centre narrow portion will depend on the free length of test piece between the grips, on the dumb-bell shape (especially the ratio of widths of central and end portions) and on the shape of the stress/strain curve. Hence the strain rate in the centre portion will not always be the same in different tests or even constant during one test.

Generally, speed variations of ± 10% have a negligible effect on the measured tensile strength at the effective strain rates realised in the standard test[34] (i.e. about 650%/min for large rings and 800–1300%/min for type 1 and 2 dumb-bells). Newer materials may be more strain rate dependent and it is sensible to reduce the variation in strain rate by using a 'hard' electrical force measuring system which has other advantages to be discussed later. It would be perfectly feasible to use a constant rate of strain testing machine for rubbers but this complication and expense has never been considered worthwhile.

Tensile properties of rubbers are strain rate dependent and hence any large departure from the arbitrary standard 500 mm/min speed will affect the results. There has been some interest in tests made at very high strain rates and a falling weight driven machine for strain rates between 2.5 and 12.7 m/sec has been described.[35]

FORCE MEASUREMENT

The classical pendulum type of force measuring system was robust, foolproof and relatively cheap—features which are attractive for routine factory use. However, it has the severe disadvantages of being subject to dynamic errors due to inertia and friction as well as being inherently 'soft'.

Electrical transducer systems have many advantages:

(a) The system is inherently very stiff, i.e. there is little movement of the force measuring element.
(b) There is negligible error due to inertia or friction.
(c) Load cells of widely different range and sensitivity can readily be made including those for the measurement of very small forces.
(d) By varying the amplification of the electrical signal a very large range of force, e.g. 1000:1 can be covered with one cell and if required the range can be changed during a test.
(e) The electrical measuring system readily allows recording or further automatic handling of the output data.
(f) Such facilities as offset zero and automatic test piece cross-sectional area compensation are easily provided.

Such systems have very rapidly replaced the pendulum type and improvements in performance are continually being made including a considerable degree of automation of data manipulation.

ISO 37 does not even state the accuracy of the force measurement whilst BS 903 specifies that the error should not exceed 2%. ISO 5893 specifies very fully two grades of accuracy, ±1% and ±2%, respectively, and gives reference to detailed methods of verification. This standard also distinguishes between static and dynamic accuracy, the above tolerances being for the former. Although ISO 5893 considers dynamic calibration to be too difficult to specify at present it does give recommendations to ensure that the recording system used with electrical load cells does not introduce significant inertia errors—it is not often realised that recorders may have very significant dynamic errors and hence lessen the inherent advantage of an 'inertialess' load cell.

ELONGATION MEASUREMENT

Until relatively recently it was common practice to measure elongation at break of dumb-bells with a ruler or piece of string and even 'modulus' results were obtained in the same manner. ISO 37 makes no mention of extensometers and gives no limits for accuracy; BS 903 specifies ±2% where a recording extensometer is used. ASTM D412 suggests that the measurement of elongation in increments of 10% is sufficient.

Again, a much more satisfactory treatment is given in ISO 5893 which recognises the capability of modern extensometers and tensile machines. The grade D^1 extensometer of this standard is the one which would most often be used for rubbers. This has an accuracy of ±2% and a 20 mm gauge length.

In ISO 5893 the same level of precision applies to the measurement of extension by crosshead movement which would be a suitable method for rings when the load cell deflection was insignificant. Generally the elongation of rings is measured by recording the increase in the pulley centre to centre distance from which the extension can be calculated.

The situation with dumb-bells is more difficult in that the distance between the grips does not bear a simple exact relationship to the elongation of the rubber in the central test length and it is necessary to follow the movement of two marks placed on this narrow central part. The gauge length should be central on the narrow portion of the dumb-bell and not more than 25 mm long for a type 1 dumb-bell, 20 mm for a type 2 and 10 mm for the type 3.

Interlaboratory variability can be very high if the old method of marking lines on the test piece and following these with a rule or string is followed, especially for modulus measurements. It is more precise and also very much more convenient to use some form of automatic recording extensometer. ISO 5893 does not give any constructional details of extensometers but notes that they may be of the type attached to the test piece or the optical (non-contact) type. If attached, the extensometer grips must not slip nor affect the test piece in any way.

A number of designs of attached extensometers are in use but they all have certain characteristics. The extensometer grips must be attached with the lightest possible pressure compatible with no slippage, to prevent preferential failure of the test piece at the attachment points. This means that whatever form of support is used for the extensometer grips it must be counterbalanced for its own weight and the grips must be able to move essentially without friction. The movement of the extensometer grips is transmitted to a measuring head by, for example, cords or rods.

The principle of one well tried design originally developed in 1956[36] is shown in Fig. 8.10. The movement of the extensometer grips A and B is transmitted by a cord E to two pulleys D and C. Since the movement by the two grips is not equal and opposite the cord passes around a floating pulley F which, together with weight G, counterbalances the extensometer grips. The movement of the grip causes a differential rotation of the pulleys D and C which can be registered by an electrical or photoelectric system. The electrical output, proportional to elongation, can then be recorded on the same recorder, but using the other axis, as is used to record force, hence giving a force/deflection curve.

It is worth noting that some extensometers record by means of an event marker each 10% elongation and such a simplified system may be quite adequate for routine work. It is also worth noting that slippage of extensometer grips when testing very highly extensible rubbers can be prevented by fitting rubber sleeves over the clamps.

There are two potential advantages in using a non-contact extensometer:

(a) The problems of grip slippage and damage to the test piece are eliminated.
(b) For tests at non-ambient temperatures the extensometer can be placed outside the environmental cabinet.

The principle and use of an optical non-contact extensometer available commercially has been described in some detail[37] (Fig. 8.11). Two photoelectric sensing devices automatically follow, by means of a servo mechanism, contrastingly coloured gauge marks on the test piece. The separation of the auto followers is measured by some form of transducer and the resulting electric signal fed to a recorder. It is apparent that, in addition to the advantages given above, such a system can be used with very weak polymer films and could contribute to increased efficiency and time saving. An unusual approach is to use a television camera.[38] An evaluation of optical extensometers has been made by Hawley.[39]

Because rubbers are virtually incompressible their Poisson's ratio is very near to 0.5. This parameter is rarely required to be measured but can be obtained by means of a dilatometer to detect the reduction of cross-sectional area during a tensile test, at the same time as elongation is measured.[40]

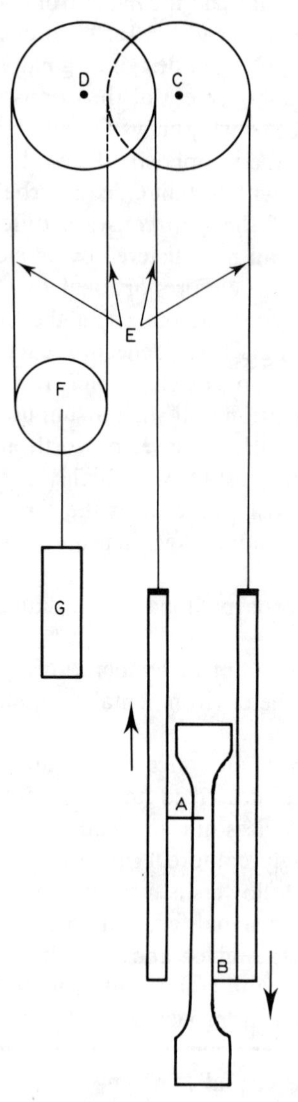

FIG. 8.10. Extensometer for dumb-bell test pieces (not to scale).

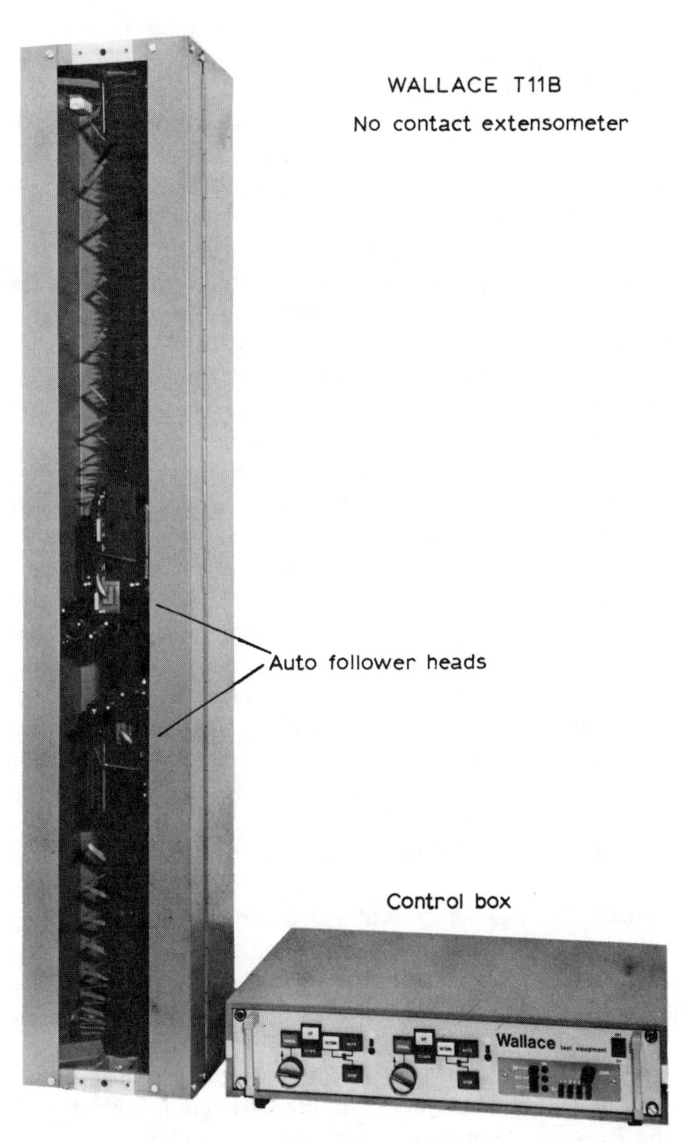

FIG. 8.11. Optical extensometer.

8.2.4 Calculation and Expression of Results

It is usual in rubber testing to calculate tensile stresses, including that at break, on the initial cross-sectional area of the test piece. Strictly, the stress should be the force per unit area of the actual deformed section but this is rather more difficult to calculate and in any case it is the force that a given piece of rubber will withstand which is of interest. The stress calculated on initial cross-section is sometimes called 'nominal stress'.

Extension is always recorded as percentage elongation, i.e. the increase in length as a percentage of original gauge length. It is unusual to find strain units (increase in length divided by initial length) quoted and to avoid confusion the total stretched length should not be given.

With dumb-bells, it is assumed that stress and strain are uniform throughout the gauge length and hence the calculation of stress presents no difficulties. Modulus as such is not normally measured but the stress quoted for a given elongation. It is sometimes debated whether the mean or the minimum cross-sectional area should be used for ultimate stress but whatever the arguments in favour of the minimum, it is rather difficult to measure this and the mean is normally used.

As has been mentioned previously, rings present more of a problem because of non-uniformity of stress and strain. Both ISO 37 and BS 903 calculate strength at break from force divided by twice the cross-sectional area but this is not the true strength (see Section 8.2.1). The elongation at break is calculated on the increase in internal circumference on the assumption that failure starts at the internal, most highly stressed, surface. To be precise, a small correction should be added to the elongation[26] so measured due to the fact that stretching is not uniform round the ring, some parts of which are flat and some curved. However, as the addition varies from about 6% at 100% elongation to 3% at 1000%, it is usually ignored.

To calculate for rings the stress at a given elongation (or the elongation at a given stress), the mean elongation should be used, not that of the internal circumference, since the stress recorded will be the average over the whole cross section. ISO 37 still calls the stress at a given elongation 'modulus' and gives no hints as to which elongation should be taken. BS 903 also wrongly uses the term modulus but specifies an approximation to the mean elongation as follows: 'The internal circumference corresponding to the required elongation (x%) shall be calculated by multiplying the initial mean circumference by $x/100$ and then adding the initial internal circumference.' ASTM D412 calculates the roller separation corresponding to the required elongation using the same approximation to mean circumference as in the British standard but with the calculation expressed in a much more straightforward manner. In a paper comparing data from ring and dumb-bell test pieces, the relationship between internal and mean elongations is given by Scott[41] as:

$$E_m = \frac{2dE_1 - 100(D-d) + 1000(D-d)(100+E_1)^{-1/2}}{D+d}$$

where E_m = mean elongation (%),
E_1 = elongation of internal circumference (%),
d = internal diameter of unstretched ring, and
D = external diameter of unstretched ring.

The calculation used by BS and ASTM is a reasonable approximation to this as it is equivalent to using the simplification $E_m = 2dE_1/(D+d)$.

The magnitude of the error introduced by using the internal circumference elongation (E_1) instead of the mean elongation (E_m) is dependent on the slope of the stress/strain curve but could be as much as 30–40%.[41]

Having omitted to mention elongation when dealing with stress at a given elongation, ISO 37 correctly specifies that the elongation at a given stress is calculated from the mean circumference at this stress. BS 903 calculates the elongation by subtracting the initial internal circumference from the internal circumference at the required stress and expressing the result as a percentage of the initial mean circumference. ASTM D412 has no provision for making this measurement. The BS 903 procedure is an approximation which would not normally lead to very serious errors.

Having waded through the above account of calculation of results for rings together with the detail given in the references, it is not surprising that the majority of people opt for dumb-bell test pieces.

8.2.5 Relaxed Modulus

There are two reasons for using a tensile stress/strain test other than the 'standard' method as typified by ISO 37, BS 903:Part A2, etc. First, it can be sensibly argued that a more useful measure of stiffness is the so-called 'relaxed modulus', i.e. the stress at a given elongation after a fixed time of relaxation; this is essentially a short term stress relaxation test. Secondly, it may be more convenient for quality control purposes to have a simple test in which only one parameter is measured.

In some ways modern tensile testing machines have reduced the need for a separate, particularly simple, routine control test. However, a test which is both simple in the sense of measuring one parameter and provides a relaxed modulus is intrinsically attractive. Such tests in various forms have existed for a long time but do not seem to have attained widespread popularity. A version in which a fixed stress is applied and the elongation after 1 min noted is given in ASTM D1456.[42] ISO and British Standards are currently considering a relaxed modulus test in which the force after 1 min at 100% elongation is taken as the measure of stiffness. This result, as in the normal tensile test, is not really a modulus but the stress at the given elongation. For both these tests specific instruments have been developed which are relatively simple in principle, although in the case of the ASTM method automatic application of the weights results in rather large and complicated apparatus.

8.2.6 Biaxial Extension

There are currently no standard methods for biaxial extension and such measurements are rarely made in industrial laboratories. However, biaxial stressing is of value in the consideration of the theory of elasticity and is involved in certain practical applications of rubber. Recent accounts of biaxial measurements are given in references 43–45.

8.3 COMPRESSION STRESS/STRAIN

A compression stress/strain test is in many ways easier to carry out than a tensile test, and in view of the large number of applications of rubber in compression, should be more often used.

Frequently the 'test piece' may be the complete product and in that case there is no problem as regards shape of the test piece. A compressive force is applied, usually at a constant rate of deformation, and the force and corresponding deformation recorded.

Specially prepared test pieces are usually in the form of a disc or short cylinder, the compressive force being applied to the circular faces. There are in theory two conditions under which the test pieces can be compressed: either with perfect slippage between the rubber and the compressing members or with complete absence of slip. Generally, perfect slippage is impossible to achieve and most applications involve either rubber bonded to metal or compressed between surfaces that virtually eliminate slip.

If there were perfect slippage every element of the test piece would be subjected to the same stress and strain and a cylindrical test piece would remain a true cylinder without any barrelling. Under these conditions stress and strain are approximately related by:

$$\frac{F}{A} = G(\lambda^{-2} - \lambda) = \frac{E}{3}(\lambda^{-2} - \lambda)$$

where: F = compression force,
A = initial cross-sectional area of test piece,
E = Young's modulus,
G = shear modulus, and
λ = ratio of compressed height to initial height.

The strain expressed as a fraction of the original height, $\varepsilon = 1 - \lambda$.

Further approximations to the relation between stress and strains are sometimes seen:

Ignoring second and higher powers of ε:

$$\frac{F}{A} = E\varepsilon$$

Ignoring third and higher powers of ε:

$$\frac{F}{A} = \frac{E\varepsilon}{1 - \varepsilon} = 3G(\lambda^{-1} - 1)$$

The more usual case, both in applications and experiment, is where it is assumed that there is complete absence of slip, stress and strain are not uniform throughout the test piece and 'barrelling' takes place on compression. The pressure distribution over the flat ends of the test piece under these circumstances has been investigated by Hall.[46]

The relationship between stress and strain in a test piece with bonded end pieces is very dependent on the shape factor of the test piece. This is usually defined as the ratio of the loaded cross-sectional area to the total force-free area (Fig. 8.12). The larger the shape factor the more stiff the rubber appears and this property is much exploited in the design of rubber springs and mountings.

The approximate stress/strain relationship incorporating the shape factor has been expressed in several ways but perhaps the most usual is:

$$\frac{F}{A} = \frac{E_c}{3}(\lambda^{-2} - \lambda)$$

where: $E_c = E(1 + 2kS^2)$ (Reference 47).

In the above relationships:

E_c = effective compression modulus,
E = Young's modulus,
F = compression force,
A = initial cross-sectional area,
λ = ratio of compressed height to initial height,
S = shape factor, and
k = numerical factor which varies with modulus.

Values of k have been tabulated by Lindley.[48]

When the shape factor is high, such that E_c/K (where K = bulk modulus) exceeds 0.1, the effective modulus will be below that expected, due to the bulk compression being appreciable. The effective modulus can then be estimated from:[47]

$$\frac{E_c}{1 + \dfrac{E_c}{K}}$$

8.3.1 Standard Test Methods

There is at the time of writing no international standard for compression stress/strain properties of rubbers, which can only be taken as a sad reflection on the order of priorities within the standardisation of rubber testing. There is however a British Standard, BS 903:Part A4,[49] and an ASTM method, D575,[50] and at last a draft is being progressed through ISO TC45.[51]

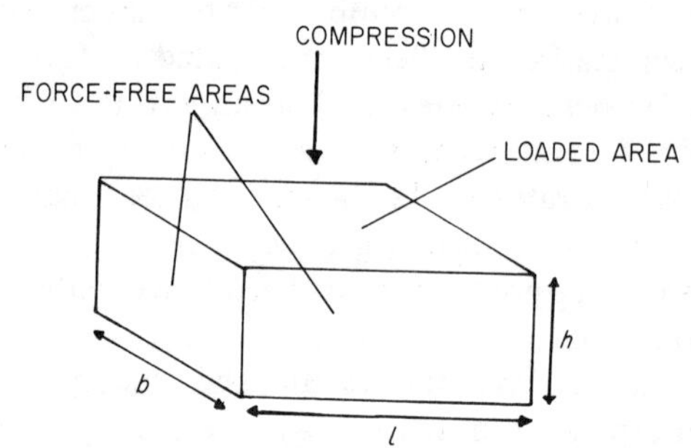

FIG. 8.12. Shape factor: Shape factor $= \dfrac{\text{Loaded area}}{\text{Force-free area}}$

i.e. Shape factor $= \dfrac{lb}{2h(l+b)}$.

The British Standard points out in the scope clause that tests may be made on the complete product but that comparable results will only be obtained on test pieces of the same shape and size and makes no mention of shape factor. The standard specifies cylindrical test pieces 29 ± 0.5 mm diameter and 12.5 ± 0.5 mm thick and states that comparisons should only be made between test pieces whose diameter/thickness ratio differs by less than 0.1. Although cutting of the test piece, as opposed to moulding, is allowed, it is debatable as to whether cut test pieces can be readily produced with sufficient precision.

A brief description of the test machine is given but, as with tensile testing, reference is better made to BS 5214. The usual practice is to use a universal tensile machine in compression mode with autographic recording of force and deflection. If this is the case care must be taken that the machine is sufficiently stiff such that the deflection reading is not significantly affected. BS 903 specifies a rate of compression of 12 ± 2 mm/min, a somewhat curious speed possibly bearing some resemblance to 0.5 in/min. The discussion of strain rate in Section 8.2, 'Tensile Testing', applies also to compression tests. Sandpaper is used between test piece and platens to approximate to the non-slip condition.

The importance of mechanical conditioning has been discussed in Section 5.5 and this British standard is one of the rare examples of where such a procedure is specified. Two conditioning cycles are called for at the standard speed to 5% greater compression than is required on the third, testing, cycle. The standard notes that if for quality control purposes either the stress at a given strain or the strain at a given stress is required the test can be terminated at the given stress or strain on the third cycle. A further variation whereby the strain at a given stress can be determined using dead weights rather than with a power driven machine is presumably intended as a quality control test requiring only very simple equipment.

The ASTM method D575[50] does not primarily recommend the generation of a stress/strain curve but details two methods—force at given deflection and deflection at given force. The test piece is a cylinder 28.6 ± 0.1 mm diameter and 12.5 ± 0.5 mm thick from which all moulded surface layers have been removed. The tight tolerance on diameter is to obviate the need for measurement so that results can be expressed in terms of force per standard test piece. The removal of moulded surfaces is presumably to eliminate skin effects, although such effects, if significant, would be present in a moulded product and one would expect to test in the same condition.

The stress at given deflection method is similar to the British Standard in that two conditioning cycles are applied before the third measuring cycle, all at 12 mm/min. The deflection at given force method specifies only a single cycle using some type of constant force machine, for example dead weights.

The ASTM, and to a lesser extent the BS, methods are simply quality control procedures in which the effect of test piece geometry is recognised but no attempt made to indicate how results on different geometries can be related. The British method recommending the recording of a stress/strain curve readily forms the basis for extension to cover variation in test piece geometry, strain rate and temperature of test.

The ISO draft clearly differentiates between bonded and unbonded test pieces and in an appendix gives the stress/strain relationships, taking account of shape factor. A force/deflection recording is obtained at 10 mm/min and from this the moduli at 10% and 20% strain obtained. Yeoh[52] has described a novel apparatus which enables the modulus at a given strain to be obtained without a relatively expensive universal test machine.

8.3.2 Bulk Compression

Although rubber is generally considered to be incompressible this is not strictly true; it is simply that the bulk modulus is very high, of the order of 10^3 times the Young's or shear moduli. It is certainly not common practice to make measurements of the bulk modulus and not surprisingly no standards are in existence. However, methods of measurement have been reported[53] as has the estimation of bulk modulus from uniaxial compression.[54] In addition, a novel procedure has been described[55] for the determination of both Young's and bulk moduli with a single test piece fitted into an oversize enclosed test cell. Both this and a hydrostatic method have been compared by Stanojevic and Lewis.[56]

8.4 SHEAR STRESS/STRAIN

Shear, like compression, is a more important mode of deformation for engineering applications than tension but despite this even less testing is carried out in shear than in compression, whilst tension remains the most common mode for laboratory stress/strain tests. Testing in shear is certainly no more difficult than testing in tension, the only practical difficulty being the necessity to bond the rubber test piece to rigid members to provide attachments for applying the shearing force.

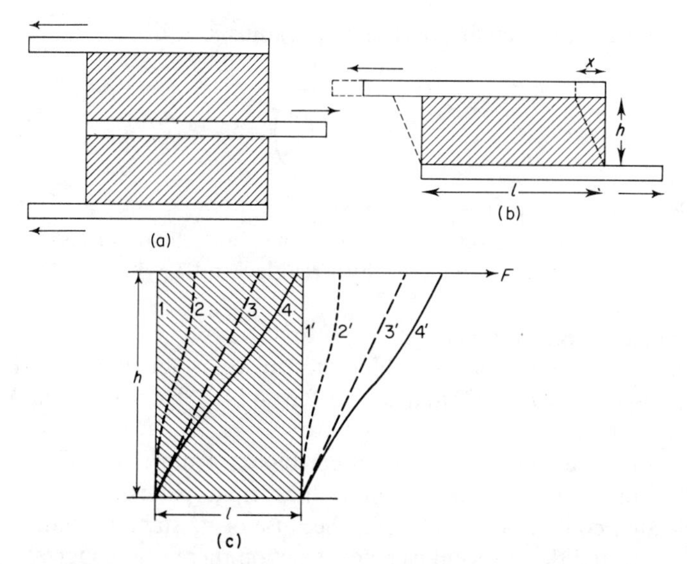

FIG. 8.13. Shear tests. (a) Double sandwich test piece; (b) sandwich test piece showing shear deformation; (c) shearing of rubber block. 1,1′—undeformed shape; 2,2′—bending deformation; 3,3′—true shear deformation; 4,4′—resultant (bending plus shear) deformation.

The stress/strain curve in simple shear is approximately linear up to relatively large strains and can be represented by:

$$\frac{F}{A} = Ge$$

where: F = applied force,
A = cross-sectional area,
G = shear modulus, and
e = shear strain.

With reference to Fig. 8.13 the strain is x/h and area A is $l \times$ the width of the rubber (not shown in the diagram) for a single sandwich and twice this for the double sandwich.

If the ratio of h to l is too great there will be an appreciable bending deformation in addition to the true shear, as shown in Fig. 8.13(c).

This reduces the apparent shear modulus, G_a, which is given by:[2]

$$G_a = \frac{G}{1 + \frac{1}{9}\left(\frac{h}{r}\right)^2}$$

for a test piece of circular cross section, radius r; and

$$G_a = \frac{G}{1 + \frac{1}{3}\left(\frac{h}{r}\right)^2}$$

for a block of square cross section, side r.

Relations between stress and strain for other shear and shear/compression configurations are given by Freakley and Payne.[57]

8.4.1 Standard Shear Tests

A shear test using a quadruple block test piece as shown in Fig. 8.14 is standardised as ISO 1827[58] and BS 903:Part A14.[59] There is not an ASTM equivalent.

The test piece comprises four rubber blocks 4 mm thick, 20 mm wide and 25 mm long, bonded to 5 mm thick rigid plates. This relatively complicated configuration is chosen because of its stability under stress. With single and double sandwich construction there is a tendency for the supporting plates to move out of parallel under load. Creep, dynamic and adhesion tests on rubber are also made in shear and some workers have used circular rubber blocks. The relevant standards committees are currently considering the possibility of a common test piece for all shear tests and circular blocks have been proposed. The rubber is bonded to the metal supports during vulcanisation and with a thickness to length ratio of 0.16 the error due to bending will be negligible.

The test piece assembly is strained in a tensile machine at 25 ± 5 mm/min, at least five conditioning cycles being applied before the measuring cycle. Force at given deflection or deflection at given force is then recorded. No details of apparatus to measure the strain are given but this could be a dial gauge or, with a stiff tensile machine, the crosshead movement.

For the quadruple test piece the shear strain is half the measured deformation divided by the thickness of one rubber block. The shear

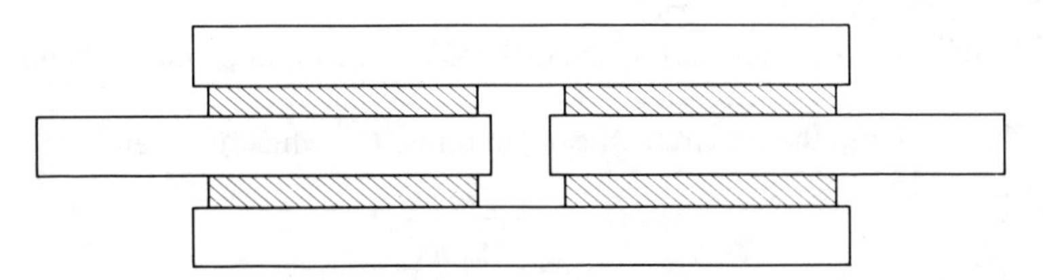

FIG. 8.14. Quadruple shear test piece.

stress is the applied force divided by twice the area of a bonded face of one block.

8.4.2 Torsion

In principle the shear modulus could be measured using test pieces strained in torsion and in engineering practice components such as torsion discs and bushes do operate in this mode. However it is not common practice to test rubber in this manner except as a low temperature test (see Section 15.3.2) when a strip test piece is twisted by means of a torsion wire. The instrument used is not really accurate enough for precise measurement of modulus at room temperature but it would seem reasonable to suppose that an accurate instrument could be devised.

For a strip, force and deflection are related by:

$$\tau = kbt^3 G\theta/l$$

where τ = applied torque,
k = shape factor,
b = width of test piece,
t = thickness of test piece,
G = shear modulus,
θ = angle of twist, and
l = effective length of test piece.

Values of k are quoted in the standards for low temperature stiffness (see Section 15.3.2).

Stress/strain relationships for other torsional configurations can be found in the reference given for shear in general.

8.5 FLEXURAL (BENDING) STRESS/STRAIN

Although rubbers are, by design or accident, deformed by bending in some practical applications it is only very rarely that bending or flexural tests are carried out. This is in contrast to the situation with rigid plastics, including ebonite, where flexural tests are often used and are well standardised.

In most applications where bending apparently takes place the rubber is also deformed in shear, tension or compression for example in a shaped door seal, when the test for stiffness would be a 'compression' test on the actual part. Generally, rubbers are not stiff enough in flexure to support appreciable loads so that there is not much need for flexural tests and at the same time the lack of stiffness makes such tests a little difficult to carry out with precision. There are, however, some cases where stiffness in bend is of interest, for example with thin sheet and coated fabrics as a measure of 'handle'.

Flexural tests for plastics are usually of the three point loading type as in ISO 178[60] where the test piece in the form of a flat strip is supported near its ends and a load applied in the centre. On using such a test for soft rubbers it is immediately apparent that very low forces are realised and that rubbers will deform vastly more than the small strains for which the test is valid. The calculations used in ISO 178 are:

$$\sigma f = \frac{3FL}{2bh^2}$$

and

$$Eb = \frac{L^3 F}{4bh^3 y}$$

where: σf = flexural stress at force F,
L = span length,
b = test piece width,
h = test piece thickness,
Eb = apparent Young's modulus, and
y = deflection at mid span for force F.

These are valid only for very small strains and it would seem sensible that if the force/deformation characteristics of a rubber in flexure are required that the test is made on the particular geometry to be used in practice and no attempt made to calculate stresses or moduli. The theory for large strains (which are not large by normal rubber standards) is extremely complicated.[61] Alternatively, it is valid to make comparisons between materials using the same geometry, this approach having been adopted for films and coated fabrics.[62]

8.6 TEAR TESTS

In a normal tensile test, taken to break, the force to produce failure in a nominally flawless test piece is measured. In a tear test the force is not applied evenly but concentrated on a deliberate flaw or sharp discontinuity and the force to continuously produce a new surface is measured. This force to start or maintain tearing will depend in rather a complex manner on the geometry of the test piece and on the nature of the discontinuity.

Hence, it would be expected that different tear methods, using different geometry will yield different tear strengths. However, there is evidence that for at least a number of rubbers the ranking of compounds is the same regardless of which tear method is used,[63-65] and is the same ranking as found from a tensile test.[63] Dozortsev[66] goes so far as to say that tear measurements are unnecessary for the assessment of the quality of rubbers.

What is certain is that the initiation and propagation of a tear is a real and very important factor in the failure of rubber products, being involved in fatigue and abrasion processes as well as the catastrophic growth of a cut on the application of a stress. There is therefore considerable interest in the tearing resistance of rubbers. What is uncertain is how tear resistance should be measured and the results interpreted.

It is not surprising that, given the importance of tearing and the different levels of result obtained from different geometries, a considerable number of tear tests have been devised, which in part reflect the different stress concentrations found in various products. The arbitrary nature of the geometries means that in general the measured tear strength is not an intrinsic property of the material and it is difficult to directly correlate the results of laboratory tests with the performance of products in service.

Rivlin and Thomas and others made a detailed study of rupture and in particular tearing which has been described in a series of papers.[67-72] They used the concept of 'energy of tearing' which is the energy required to form unit area of new surface by tearing. This energy of tearing is a basic material characteristic and independent of test piece geometry; hence, using this concept and knowing the elastic characteristics of the material, the force needed to tear a given geometry can in theory be predicted. The concept also allows rational analysis of other failure processes in rubber, such as fatigue. Although the concept and importance of tearing energy is now well established, standard methods to date do not make use of it but report the arbitrary tearing force.

8.6.1 Forms of Test Piece

As mentioned above, a large number of different geometries have been used for tear strength measurements. Often they have been chosen with more thought for the convenience of testing than the significance of the results.

Distinction can be made between the force to initiate a tear as distinct from that to propagate a tear. Although it could be argued that once a tear has started the product has failed and the force or energy needed to

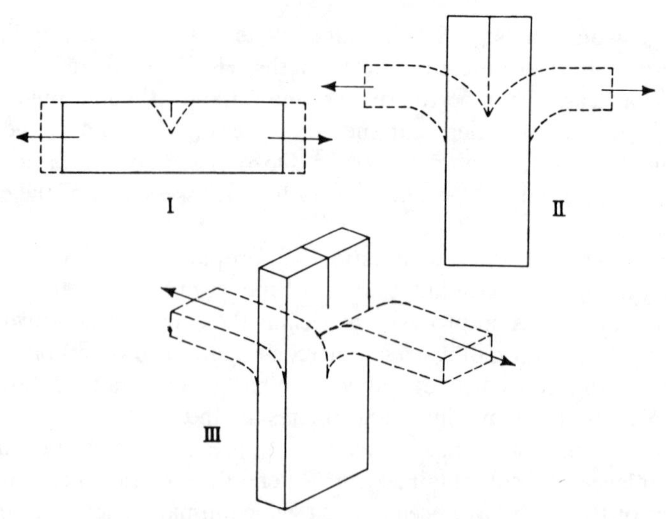

FIG. 8.15. Types of tear test piece. —— Original shape; ----- shape after tearing.

continue the tear is of no interest, in most cases it is the resistance to growth of small flaws or cuts which is important. In most standard tests the maximum force measured during the test is recorded and no distinction is made between initiation and propagation but also most standard tests start with an artificially introduced cut which can be thought of as the initiation.

The discontinuity at which the stress concentration is produced is formed either by a cut, a sharp re-entry angle or both. Only in the case of a sharp angle without a cut will any measure of initiation force be possible.

Test pieces can be categorised into three groups according to the direction of the applied force relative to the plane and long axis of the test piece. With reference to Fig. 8.15, in type I the force is in the plane of the test piece and parallel to its length, in type II it is normal to the plane. In types I and II the stresses in the tip of the tear are essentially tensile, whereas in type III they must include shear stresses.

Traditionally, variants of type I have been most popular but ISO 34[73] now specifies four procedures: trouser (type III), angle (type I) both with and without a cut, and crescent (type I). BS 903:Part A3[74] is identical. ASTM D624[75] has the crescent and angle test pieces and also a modified type of crescent. A further ISO standard, ISO 816[76] specifies the Delft

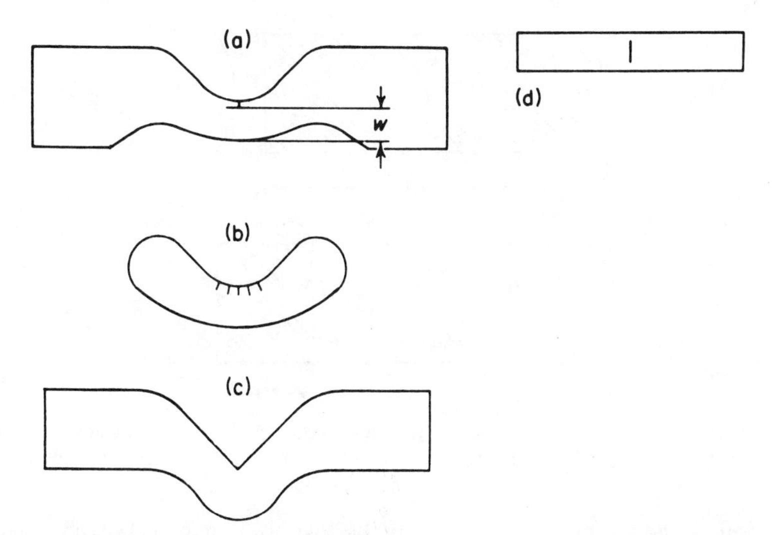

FIG. 8.16. Forms of tear test piece. (a) Crescent with enlarged ends for gripping, w = initial untorn width; (b) simple crescent; (c) angle; (d) Delft.

test piece which is a variant of the type I with an internal slit and is of such a small size as to be more readily cut from finished products. These test pieces are shown in Fig. 8.16.

From studies comparing the various methods[63–66,77,78] a number of general deductions can be made:

The methods all give different absolute values for tearing force but for at least a number of rubbers the ranking order is the same for all methods. A more consistent correlation may be found between certain of the methods, for example trouser and angle as a group compared to crescent and Delft as a second group.

The repeatability of the methods varies, the crescent being generally better than the trouser and angle. It could be argued that greater variability would be expected with the trouser test piece because of the real rubber variability over a long tear path. The variability of the crescent and Delft methods will be increased if the nick is not cut with great care and precision and the angle method requires the angle of the cutter to be accurately maintained.

The trouser test piece is particularly convenient for calculating tearing energy to give more fundamental results and allows the course of tear propagation to be followed as well as being a relatively easy test piece to cut accurately. The Delft has size advantage when cutting test pieces from

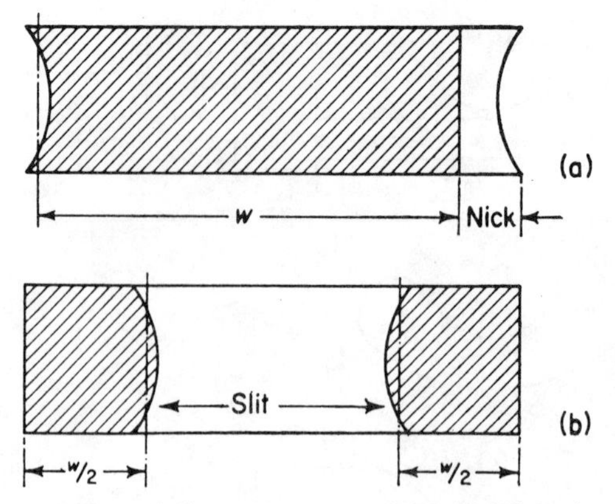

FIG. 8.17. Cross section of die-cut tear test piece. (a) Nicked test piece of type I, Fig. 8.15; (b) Delft test piece. w = effective width of untorn portions.

products and the angle is the only method without an artificially initiated tear.

Clearly it cannot be concluded that one method is universally superior and it is not surprising that there are even more test pieces in use than the generally standardised ones discussed above. The trouser test piece sometimes produces difficulties in practice due to excessive leg extension or deviation of the tear through one leg. Veith[79] has described a modification with fabric reinforced legs designed to overcome these difficulties and Leblanc[80] has approached the same problems by having a thinner centre strip to the test piece. The Pohle method[63] uses a tensile ring with nicks on the internal diameter and the cable test of BS 6899[81] is basically similar to the trouser method.

8.6.2 Preparation of Test Pieces

Tear test pieces are normally stamped from sheet with an appropriate die (see Section 4.2) and then a nick cut if required, although with the Delft test piece both operations are usually done simultaneously.

The depth of nick, or length of cut, is critical for the crescent and Delft test pieces and must be controlled within close limits to obtain consistent results. For the crescent it is necessary to use a special jig to hold the test piece and move a blade across its edge. Such jigs are available commercially. The nick depth can be checked with a travelling or projection microscope, the main difficulty in measuring accurately arising from the concave edges formed by the stamping die (Fig. 8.17). ISO 34 makes no mention of the difficulty in measuring the effective width below the nick but in ISO 816 a procedure for estimating the effective untorn width is given in detail.

For angle test pieces with a nick added the problem is essentially the same as with the crescent test piece. With unnicked angle test pieces the essential requirement is the production of a reproducible re-entrant angle, which is very difficult even with careful die maintenance.

The trouser test piece presents the least difficulty as regards preparation because none of the dimensions, including the nick depth, are as critical as with the other test pieces.

8.6.3 Measurement of Tearing Force

Tear test pieces are gripped and stretched to break in a machine of the same type as used for tensile tests and hence the comments made in Section 8.2.3 also apply generally here. Because of the lower forces involved, gripping is less difficult than with tensile tests. However, small forces mean that a sensitive load measuring device is needed and the rate of change of force can be high and with the trouser test piece the force will rise and fall alternately. Hence, the 'inertialess' type of tester is essential.

The current standards specify a stretching rate of 500 ± 50 mm/min, the same as for tensile tests except for the trouser test which is 100 ± 10 mm/min. These choices are somewhat arbitrary and not surprisingly tear strength can be expected to vary with testing speed. The tearing energy with, for example, natural rubber may be raised or lowered by an increase in speed.[69] The change could be a few per cent for a $\pm 10\%$ speed change and although this would not cause serious discrepancies between results there is perhaps a case for a tighter tolerance on speed than currently standardised.

With crescent, Delft and angle test pieces only the maximum force reached is recorded although the change of force along the short tear path could be obtained with a force measuring and recording system having very low inertia. Over the longer tearing length of the trouser test piece the force may vary very considerably and in an irregular manner, especially in the 'knotty' tearing of some black reinforced rubbers. An accurate trace of force variation can be obtained with suitable measuring and recording equipment.

8.6.4 Expression of Results

The quantity directly measured in a tear test is the force on the test piece during testing. Particularly with the trouser method this force will fluctuate along the tear path. ISO 34 takes the maximum force realised in the case of crescent and angle methods but for the trouser method a median force is determined in accordance with procedures given in ISO 6133[82] 'Analysis of multi-peak traces obtained in determinations of tear strength and adhesion strength'.

ISO 6133 gives three procedures, for traces having less than 5 peaks, 5 to 20 peaks and more than 20 peaks respectively. For less than 5 peaks the median of them all is taken, for 5 to 20 peaks the median of the peaks in the central 80% of the trace is taken and for more than 20 peaks the trace is divided into tenths by nine lines, the peak nearest to each line noted and the median of these taken.

Having obtained the force (maximum or median) the tear strength is given as:

$$T_s = \frac{F}{d}$$

where T_s = tear strength in kilonewtons per metre of thickness,
F = force (maximum or median),
d = thickness.

This method of expression of results could be taken to imply that tear strength is proportional to thickness over a wide range. This would probably not be true and ISO 34 makes it clear that for comparative results the thickness variation of groups of test pieces should be not greater than $\pm 7.5\%$. ASTM D624 used the same calculation.

In ISO 816 the result for the Delft test piece is expressed as the force to tear a test piece of standard width and thickness, corrections for variation within tolerance of width and thickness being given. More emphasis than in ISO 34 is given to measuring the effective test piece width allowing for the curved edges of the slit. It must be noted that this method of expressing the result is not the same as for ISO 34 and hence results are not directly comparable. In fact ISO 34 previously used this method and hence older results which may be found will be in different units to results from the current standard.

The use of the procedure given in ISO 6133 for obtaining a median force can be questioned. It could be argued that the maximum force (highest peak) be used or the lowest force on the basis that these give the best and worst representation of the rubber. ISO 6133 does in fact ask for the range of peak values to be recorded but this is not used in ISO 34 which is an unfortunate omission because with some rubbers the variation can be very large whereas with others a fairly smooth trace is obtained. The commonest quick procedure in the laboratory for tear or adhesion force recordings is to estimate the average level with the aid of a transparent plastic ruler. (See also Section 18.2.1.)

Having obtained a measure of tearing force it is possible with a trouser test piece to readily derive the characteristic energy of tearing which is the most useful measure of tearing performance. The tearing energy is given by:

$$T = \frac{2\lambda F}{t} - wW$$

where T = tearing energy,
λ = extension ratio in legs of test piece,
F = measured force,
w = width of test piece,
t = thickness, and
W = strain energy density.

W can be obtained by extending a test piece without a nick and plotting a stress/strain curve, W being derived from the area under the stress/strain curve up to the extension ratio λ.

If the legs of the test piece are sufficiently stiff so that extension is negligible $\lambda = 1$ and W is zero and the relationship reduces to:

$$T = \frac{2F}{t}$$

This approximation has generally been found adequate for most purposes, even without reinforcing the legs of the test piece to prevent extension.[79]

8.6.5 Cutting Resistance

Although in practice the cutting of rubber by sharp objects is an important way in which damage is made to articles such as tyres there are currently no general standardised test methods. Cutting can take place without any other stress on the rubber, but in the case where the rubber is cut whilst being held under stress the situation might be considered as the sharp object assisting tearing.

The resistance to cutting is contributed to by both the strength properties of the rubber and friction. If the rubber is stressed whilst cutting takes place, one consequence is that friction is much reduced and with it the force needed to cause cutting. Most of the *ad hoc* tests which have been devised to measure cutting or puncture resistance operate under unknown and arbitrary friction conditions and hence do not measure an intrinsic strength property. An example would be a 'stitch tear' test where a thread is looped through two holes in the test piece and the rubber between the holes torn by pulling the thread through it. Quite clearly, any such test will only have relevance to essentially the same stress and geometry conditions in service.

If conditions of service are such that there is a probability of objects being present with sufficient force behind them to cause cutting even in high friction conditions it might be argued that cuts are to be expected and it is resistance to catastrophic propagation of these cuts which is of importance and hence the appropriate test is one for tear strength.

If the cutting resistance of the rubber is required it would best be made under conditions of negligible friction. Lake and Yeoh[83] have described two test geometries which achieve this and also give a comprehensive account of the cutting process using the same fracture mechanics approach as for tearing energy.

REFERENCES

1. Treloar, L.R.G. (1975). *The Physics of Rubber Elasticity*, Oxford University Press.
2. Payne, A.R. and Scott, J.R. (1960). *Engineering Design with Rubber*, Maclaren and Sons.
3. Scott, J.R. (1935). *Trans. IRI*, **11**.
4. Soden, A.L. (1951). *A Practical Manual of Rubber Hardness Testing*, Maclaren and Sons.
5. ISO 48, 1979. Vulcanised Rubbers—Determination of Hardness (Hardness between 30 and 85 IRHD).
6. Steihler, R.D., Decker, G.E. and Bullman, G.W. (1975). Proceedings of the Int. Rubb. Conf., Kuala Lumpur.
7. Yeoh, O.H. (1984). *Plast. Rubb. Process. Appln.*, **4**, (2).
8. ISO 1400, 1975. Vulcanised Rubbers of High Hardness (85 to 100 IRHD)—Determination of Hardness.
9. ISO 1818, 1975. Vulcanised Rubbers of Low Hardness (10 to 35 IRHD)—Determination of Hardness.
10. Scott, J.R. and Soden, A.L., International Rubber Conference, Washington DC. Nov. 1959, Paper 22.
11. BS 903:Part A26, 1969. Determination of Hardness.
12. Brown, R.P. and Hughes, R.C. (July 1972). *RAPRA Bulletin*.
13. ISO DIS 7267. Rubber Covered Rollers—Determination of Apparent Hardness.

14. ASTM D1415–83. International Hardness.
15. ASTM D1414–78. Rubber 'O' Rings.
16. ASTM D531–56. Indentation of Rubber by Means of the Pusey and Jones Plastimeter.
17. Newton, R.G. (1948). *J. Rubber Res.*, **17**.
18. ISO DIS 7619. Determination of Indentation Hardness by Means of Pocket Hardness Meters.
19. ASTM D2240–81. Durometer Hardness.
20. ISO 868, 1978. Determination of Indentation Hardness of Plastics by Means of a Durometer (Shore Hardness).
21. Taylor, Rolla H. (Aug. 1943). *ASTM Bulletin.*
22. BS 2719, 1975. Methods of Use and Calibration of Pocket Type Rubber Hardness Meters.
23. ISO 37, 1977. Rubber, Vulcanised—Determination of Tensile Stress–Strain Properties.
24. BS 903:Part A2, 1971. Determination of Tensile Stress–Strain Properties.
25. ASTM D412–83. Rubber Properties in Tension.
26. Reece, W.H. (1935). *Trans. IRI*, **11**, 312.
27. Scott, J.R. (1949). *J. Rubb. Res.*, **18**.
28. Bartenev, G.M. and Gul, V.E. (1961). *Sov. Plas.*, No. 1, 46.
29. ISO TC45, Interlaboratory Comparison, 1973. Unpublished.
30. Nazeni, D.I. Deutsche Kautschuk—Gesellschaft, Vortragstagung, 4–8th Oct. 1960, West Berlin.
31. Patrikeev, G.A. Thieu, V.V., Kondratov, A.P. and Cherepov, S.B. (1978). *Int. Polym. Sci. Tech.*, **5** (9).
32. ISO 5893, 1985. Rubber and Plastics Test Equipment, Tensile, Flexural or Compression types (Constant Rate of Traverse).
33. BS 5214:Part 1, 1975. Testing Machines for Rubbers and Plastics.
34. Scott, D.C. Jr and Villars, D.S. ASTM Special Publication No. 185, 1956, p.62.
35. Dannis, M.L. (1962). *J. Appl. Polym. Sci.*, **6**, 283.
36. Eagles, A.E. and Payne, A.R. RABRM Res. Memo, No. R405, 1957.
37. Bennett, F.N.B. (1980). *Polym. Test*, **1**, (2).
38. Mitsuhashi, K. and Ito, H. (1980). *Nippon Gomu Kyokaishi*, **53** (5).
39. Hawley, S. (1981). *Polym. Test.*, **2** (1).
40. Lanfer, Z., Diamart, Y., Gill, M. and Fortuna, G. (1978). *Int. J. Polym. Mat.* **6** (3/4).
41. Scott, J.R. (1949). *J. Rubb. Res.* **18**, 30.
42. ASTM D1456–81. Rubber Property—Elongation at Specific Stress.
43. Nottin, J.P., Blanc, D., Evrard, G., Leblanc, F., Prou, J. and Truntzer, G. (1981). Proceedings of 27th Int. Sym. on Macromolecules, Strasbourg, Vol. 2.
44. Kent, R.E. (1978). *Eur. Rubb. J.*, **160** (3).
45. Kawabata, S. and Kawai, H. (1977). *Adv. Polym. Sci.*, No. 24.
46. Hall, M.M. (1971). *J. Strain Analysis*, **6**.
47. Gent, A.N. and Lindley, P.B. (1959). *Proc. Inst. Mech. Engrs*, **173**, 111.
48. Lindley, P.B. (1970). *Nat. Rubb. Tech. Bull.*, No. 8, 3rd edn.
49. BS 903:Part A4, 1973. Determination of Compression Stress–Strain.

50. ASTM D575–69. Rubber Properties in Compression.
51. ISO DP7743. Determination of Compression Stress–Strain Relationship.
52. Yeoh, O.H. (1984). *J. Rubb. Res. Inst. Malaysia*, **32**, Part 3.
53. Ol'khovik, O.E. and Grigoryan, E.S. (Sept. 1974). *Vys. Soed. A*, **16** (9), 2155.
54. Gilmour, I., Trainor, A. and Haward, R.N. (Sept. 1974). *J. Polym. Sci., Polym. Phys. Ed.*, **12** (9), 1939.
55. Worfield, R.W., Cuevas, J.E. and Barnet, F.R. (1968). *J. Appl. Polym. Sci.*, **12**, 1147.
56. Stanojevic, M. and Lewis, G.K. (1983). *Polym. Test.*, **3** (3).
57. Freakley, P.K. and Payne, A.R. (1978). *Theory and Practice of Engineering with Rubber*, Applied Science Publishers.
58. ISO 1827, 1976. Determination of Modulus in Shear—Quadruple Test Method.
59. BS 903:Part A14, 1970. Determination of Modulus in Shear of Rubber.
60. ISO 178, 1975. Determination of Flexural Properties of Rigid Plastics.
61. Heap, R.D. and Norman, R.H. (1969). *Flexural Testing of Plastics*, The Plastics Institute.
62. Brown, R.P. (1975). *RAPRA Members J.* **3** (2).
63. Clamroth, R. and Kempermann, Th. (1986). *Polym. Test.*, **6** (1), 3.
64. Slade, J.L. (1986). *Polym. Test.*, To be published.
65. Brezik, R. (1982). *Int. Polym. Sci. Tech.*, **9** (4).
66. Dozortsev, M.S. (1984). *Int. Polym. Sci. Tech.*,**11** (9).
67. Rivlin, R.S. and Thomas, A.G. (1953). *J. Polym. Sci.*, **10** (3), 291.
68. Thomas, A.G. (1955). *J. Polym. Sci.*, **18**, 177.
69. Greensmith, H.W. and Thomas, A.G. (1955). *J. Polym. Sci.*, **18**, 189.
70. Greensmith, H.W. (1956). *J. Appl. Polym. Sci.*, **21**, 175.
71. Thomas, A.G. (1960). *J. Appl. Polym. Sci.*, **3** (8), 168.
72. Greensmith, H.W. (1960). *J. Appl. Polym. Sci.*, **3** (8), 183.
73. ISO 34, 1979. Determination of Tear Strength (Trouser, Angle and Crescent Test Pieces).
74. BS 903:Part A3, 1982. Determination of Tear Strength (Trouser, Angle and Crescent Test Pieces).
75. ASTM D624–81. Rubber Property—Tear Resistance.
76. ISO 816, 1983. Determination of Tear Strength of Small Test Pieces (Delft Test Pieces).
77. ISO TC45. Interlaboratory Comparison, 1973. Unpublished.
78. Desbrieres, D. (1981). *Caout. Plast.*, No. 612.
79. Veith, A.G. (Nov. 1965). *Rubb. Chem. Tech.*, **38** (4).
80. Leblanc, J.L. (1978). 5th Europ. Plast. Rubb. Conf., Paris, Vol. 2, Paper E22.
81. BS 6899, 1976. Specification for Rubber Insulation and Sheath of Electric Cables.
82. ISO 6133, 1981. Analysis of Multi-Peak Traces Obtained in Determinations of Tear Strength and Adhesion Strength.
83. Lake, G.J. and Yeoh, O.H. (1980). *Rubb. Chem. Tech.*, **53** (1).

Chapter 9

Dynamic Stress and Strain Properties

The term dynamic test is used to describe the type of mechanical test in which the rubber is subjected to a deformation pattern from which the cyclic stress/strain behaviour is calculated. It does not include cyclic tests in which the main objective is to fatigue the rubber, as these are considered in Chapter 12. Dynamic properties are important in a large number of engineering applications of rubber including springs and dampers and are generally much more useful from a design point of view than the results of many of the simpler 'static' tests considered in Chapter 8.

Before considering particular test methods it is useful to survey the principles and terms used in dynamic testing. There are basically two classes of dynamic motion: free vibration in which the test piece is set into oscillation and the amplitude allowed to decay due to damping in the system; and forced vibration in which the oscillation is maintained by external means. These are illustrated in Fig. 9.1 together with a subdivision of forced vibration in which the test piece is subjected to a series of half-cycles. The two classes could be subdivided in a number of ways; for example, forced vibration machines may operate at resonance or away from resonance. Wave propagation (e.g. ultrasonics) is a form of forced vibration method and rebound resilience is a simple dynamic test representative of type (c) in Fig. 9.1 The most common type of free vibration apparatus is the torsion pendulum.

There is an international standard ISO 2856[1] dealing with general requirements for dynamic testing which can be referred to for definitions of terms used and also includes classifications of test machines, preferred conditions, recommended test piece shapes and a bibliography. Other

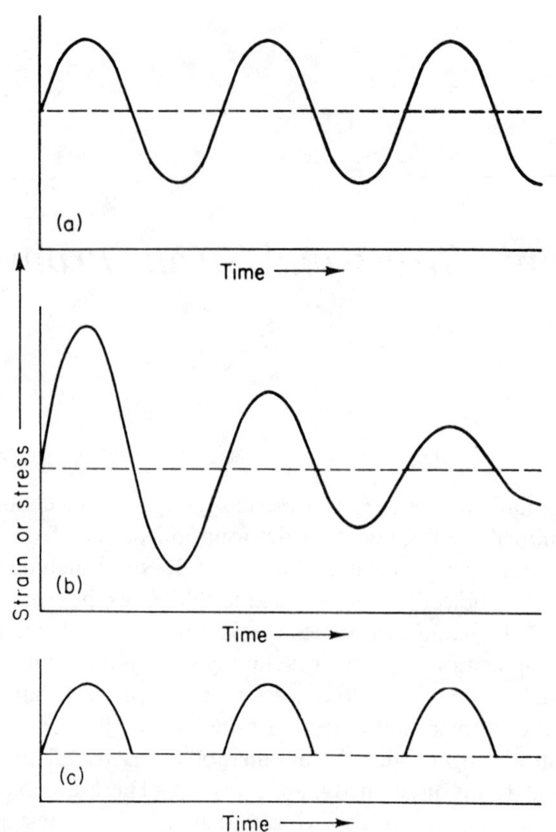

FIG. 9.1. Forms of strain or stress cycles in dynamic tests. (a) Continuous constant amplitude; (b) continuous decaying amplitude; (c) successive half-waves. (The intervals are not necessarily equal to the half-wave duration, as shown here.)

accounts of the basic principles of dynamic testing have been given by, for example, Warner[2] and only a brief outline will be given here.

The static tests considered in Chapter 8 treat the rubber as being essentially an elastic, or rather 'high elastic', material, whereas it is in fact viscoelastic and hence its response to dynamic stressing is a combination of an elastic response and a viscous response and energy is lost in each cycle. A simple model to represent this behaviour is a spring and dashpot in parallel (Maxwell model).

For sinusoidal strain the motion is described by:

$$\gamma = \gamma_0 \sin \omega t$$

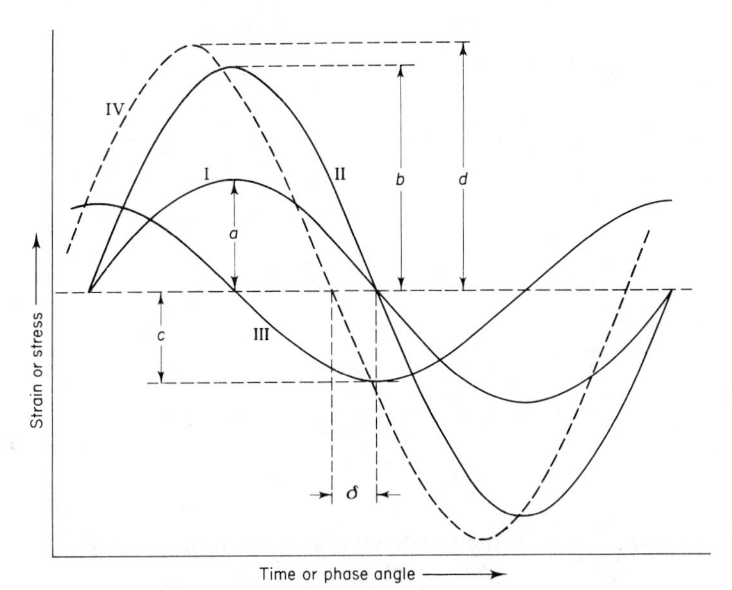

FIG. 9.2. Sinusoidal strain and stress cycles. I—strain, amplitude a; II—in-phase stress, amplitude b; III—out-of-phase stress, amplitude c; IV—total stress (resultant of II and III), amplitude d. δ is the loss angle.

where: γ = strain,
γ_0 = maximum strain amplitude,
ω = angular frequency, and
t = time.

If the rubber were a perfect spring the stress (τ) would be similarly sinusoidal and in phase with the strain. However, because the rubber is viscoelastic the stress will not be in phase with the strain but can be considered to precede it by the phase angle δ so that:

$$\tau = \tau_0 \sin(\omega t + \delta)$$

This is the same as saying that the deformation lags behind the force by the angle δ.

It is convenient to consider the stress as a vector having two components, one in phase with the displacement (τ') and one 90° out of phase (τ'') and to define corresponding in-phase, out-of-phase and the resultant moduli. The sinusoidal motion is illustrated in Fig. 9.2 and the vector in Fig. 9.3.

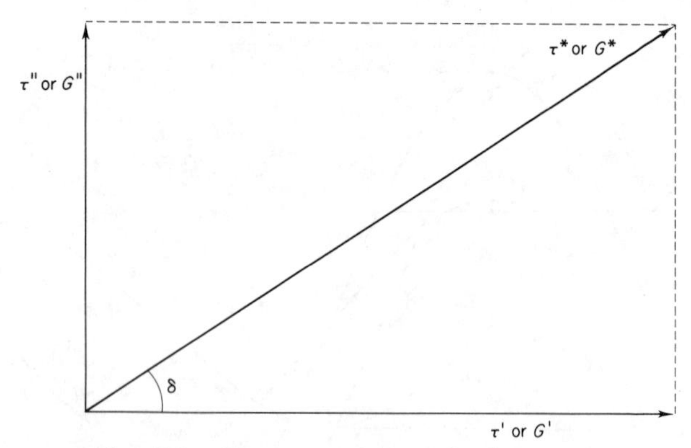

FIG. 9.3. Vector stress or strain diagram.

Considering Fig. 9.3, the vector moduli in shear are defined by:

$$G^* = G' + iG''$$

where G^* = complex (resultant) modulus,
G' = in-phase or storage modulus, and
G'' = out-of-phase or loss modulus.

It can also be shown that:

$$G' = \frac{\tau'}{\gamma_0} = \frac{\tau_0}{\gamma_0} \cos \delta = G^* \cos \delta$$

$$G'' = \frac{\tau''}{\gamma_0} = \frac{\tau_0}{\gamma_0} \sin \delta = G^* \sin \delta$$

$$|G^*| = (G^2 + G^2)^{1/2}$$

$$\text{Tan}\, \delta, \text{the loss factor or loss tangent} = \frac{G''}{G'}$$

[G^*] is the absolute value of the complex modulus but in practical dynamic testing is often written as G^*.

In Fig. 9.2, the in-phase modulus $G' = b/a$ and this is the modulus G assumed to be measured in a static test. The out-of-phase modulus

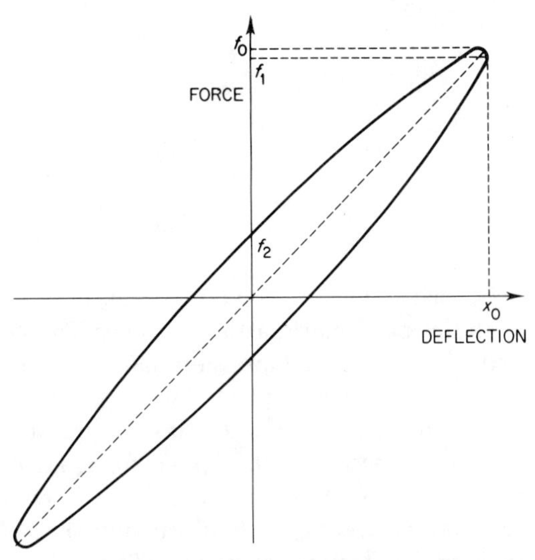

FIG. 9.4. Hystersis loop.

$G'' = c/a$. The magnitude of the complex modulus,

$$|G^*| = \frac{d}{a} = \sqrt{\frac{b^2 + c^2}{a}}$$

The loss tangent, $\tan\,\delta = c/b$.

Similarly, in tension or compression:

$$\text{Young's modulus } |E^*| = (E'^2 + E''^2)^{1/2}, \text{ etc.}$$

If, in a dynamic test with forced sinusoidal oscillation, force is plotted against deflection an hysteresis loop is obtained as shown in Fig. 9.4.

$$\text{Dynamic stiffness } S^* = \frac{\text{force amplitude}}{\text{deformation amplitude}} = \frac{f_0}{x_0}$$

Working in shear:

$$G^* = \frac{f_0 h}{A x_0} = S^* \frac{h}{A}$$

(i.e. stress $\propto$ strain) where h is the test piece thickness and A its effective cross-sectional area.

$$G' = \frac{f_1 h}{A x_0} \qquad G'' = \frac{f_2 h}{A x_0}$$

and

$$\tan \delta = \frac{f_2}{f_1}$$

The viscoelastic behaviour of rubbers is not linear; stress is not proportional to strain, particularly at high strains. The non-linearity is more pronounced in tension or compression than in shear. The result in practice is that dynamic stiffness and moduli are strain dependent and the hysteresis loop will not be a perfect ellipse. If the strain in the test piece is not uniform it is necessary to apply a shape factor in the same manner as for static tests. This is usually the case in compression and even in shear there may be bending in addition to pure shear. Relationships for shear, compression and tension taking these factors into account have been given by Payne[3] and Davey and Payne[4] but because the relationships between dynamic stiffness and the basic moduli may be complex and only approximate it may be preferable for many engineering applications to work in stiffness, particularly if products are tested. ISO 2856 gives in annexes the basic relationships between stress and strain and in the body of the document lists approximate relationships with shape factors for a whole range of test piece shapes and different deformation modes.

The results of dynamic tests are dependent on the test conditions: test piece shape, mode of deformation, strain amplitude, strain history, frequency and temperature. Simple shear is the preferred mode of deformation because stress and strain are linearly related over a greater strain range than in tension or compression and also because in this mode the test piece shape can be chosen so that the strain is homogeneous and a shape factor need not be applied. Measurements at two strain amplitudes would show up any strain dependence. The repetition of strain cycles, as is the normal procedure in a forced oscillation test, may cause progressive change in the dynamic properties for two reasons. First, at the beginning of the test, there may be stress softening as a result of mechanical conditioning (see Section 5.5) and the dynamic properties will, for practical purposes, reach a steady equilibrium level after a few cycles. More troublesome, is a continued change in property level due to the generation of heat within the test piece raising its temperature. This is most likely to occur with materials having a large loss factor tested at high strain

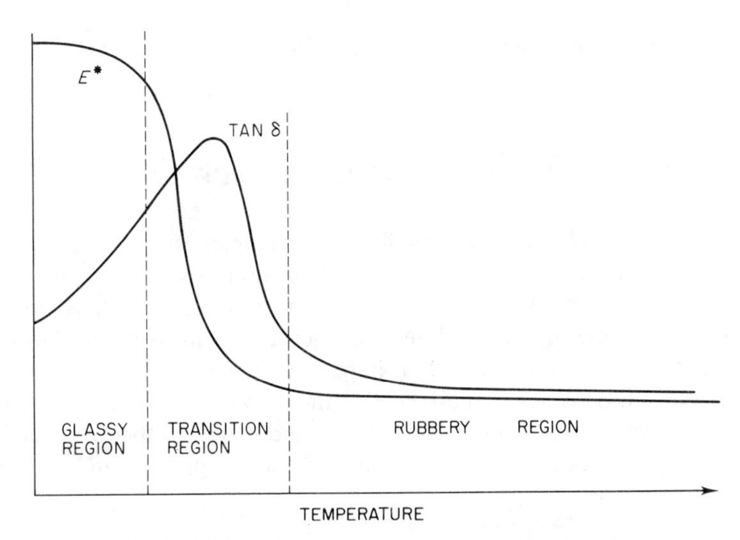

FIG. 9.5. Effect of temperature on dynamic properties.

amplitudes. Hall and Thomas[5] give an example of a high loss material where the time to give a temperature rise of 2°C would be 30 s at 15 Hz.

Dynamic properties are dependent on both frequency and temperature and it is possible to approximately relate the two effects quantitatively. Preferably, results would be obtained over the range of frequencies and temperatures of interest but if it is required to transform modulus results to other temperatures or frequencies use may be made of the so-called Williams, Landel and Ferry (WLF) equation.[6] The general form of the effect of temperature on complex modulus and tan δ is shown in Fig. 9.5. The effect of increasing or decreasing frequency is to shift the curves to the right or left, respectively, along the temperature axis. At room temperature the order of magnitude of the effect of temperature on modulus for a typical rubber is 1% per °C and the effect of frequency of the order of 10% per decade.

Because strain amplitude is controlled, forced vibration rather than resonant or free vibration methods are generally preferred although there are advantages with the other systems in certain circumstances. The preferred waveform is sinusoidal. ISO 2856 states this preference for forced sinusoidal stress or strain in the shear mode, noting that other acceptable forms of stress or strain are free damped oscillations with a logarithmic decrement less than 1, periodic half-sinusoidal cycles and impacts. The oscillations may be superimposed on any form or level of static strain.

9.1 REBOUND RESILIENCE

Rebound resilience is a very basic form of dynamic test in which the test piece is subjected to one half-cycle of deformation only. The strain is applied by impacting the test piece with an indentor which is free to rebound after the impact. Rebound resilience is defined as the ratio of the energy of the indentor after impact to its energy before impact expressed as a percentage and hence, in the case where the indentor falls under gravity, is equal to the ratio of rebound height to the drop height.

Resilience is not an arbitrary parameter but is approximately related to the loss tangent:

$$R = \frac{E_R}{E_I} = \exp(-\pi \tan \delta)$$

$$\frac{E_A}{E_R} = \pi \tan \delta$$

where: E_R = reflected energy,
E_I = incident energy, and
$E_A = E_I - E_R$ = absorbed energy.

The relationship is not particularly accurate because tan δ is strain dependent and in an impact test the form of applied strain is complex and its magnitude not controlled. The value of tan δ is assumed to be that relevant to a frequency of $1/2t$, where t is the dwell time of the indentor.

Despite resilience being such a crude measure of a dynamic property it is an attractive test, especially for quality control purposes, because of its simplicity and the fact that the apparatus required is inexpensive. The two basic forms of resilience apparatus are a ball falling under gravity and an indentor attached to a swinging pendulum; over the years a large number of instruments based on these principles have been devised. Although the falling ball method was probably the earliest in use, the pendulum type is now the most widely used and standardised. It is not particularly clear why this should be the case because the falling ball is rather more simple in concept and is usually a smaller instrument and at the same time robust. It is also free from friction and vibration in the suspension members. However, an advantage of the pendulum is that a greater range of equivalent frequencies can be easily obtained without change of indentor size.

9.1.1 Pendulum Methods

A number of pendulum methods have been standardised over the years but the present ISO standard, ISO 4662[7] is not based on any particular pendulum design but on giving limits for various parameters. A few years back extensive interlaboratory investigations were carried out within TC 45 to determine the effect of pendulum parameters on measured resilience and the conclusions from those investigations have led to the present standard.

The standard states that 'apparent strain energy density' which, with certain assumptions, can be related to impact strain, should be held constant to obtain equivalent results. The body of the standard gives conditions and procedures to obtain 'standard' resilience values and the following parameters are specified:

Indentor diameter $D = 12.45\text{–}15.05$ mm

Test piece thickness $d = 12.5 \pm 0.5$ mm

Impacting mass $m = 0.35\,^{+0}_{-0.1}$ kg

Impact velocity $v = 1.5\,^{+0.6}_{-0}$ m/s

$$\text{Apparent strain energy density} = \frac{mv^2}{Dd^2} = 351\,^{+112}_{-27}\ \text{kJ/m}^3$$

In fact these values correspond to those for the well known Lüpke pendulum and also for a modified version of the Schob pendulum.

It is also necessary for the rigid parts of the pendulum to be sufficiently stiff to avoid spurious vibrations, for the impact to occur at the centre of percussion and for the corrections to be made for friction if necessary. The test piece must either be bonded to a backing plate or very firmly clamped by mechanical means or vacuum.

In an appendix to ISO 4662, apparatus parameters for five different test piece thicknesses are given which it is claimed result in results very near to the 'standard' parameters.

In a second appendix very brief details of three particular penduli are given together with references to full descriptions. The Lüpke takes the form of a horizontal rod with a hemispherical indentor end suspended

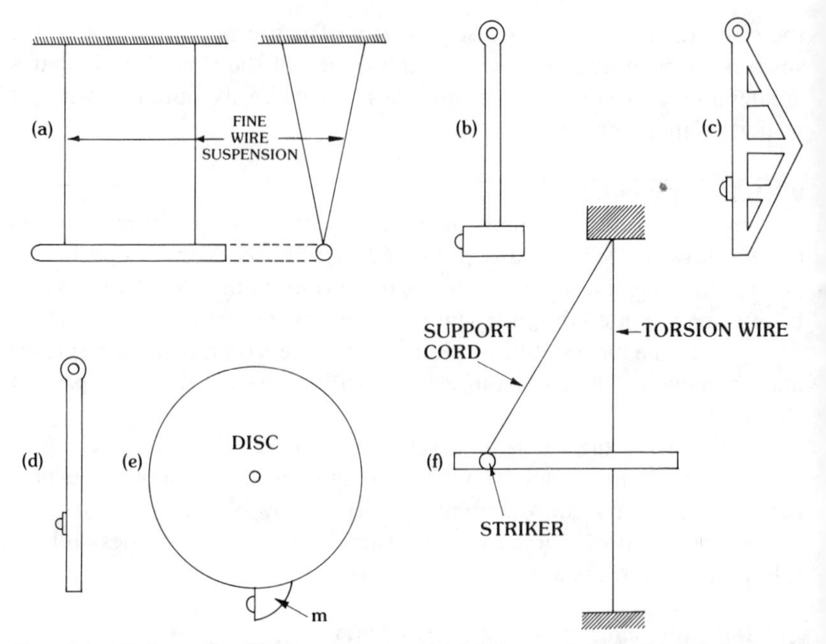

FIG. 9.6. Rebound resilience apparatus. (a) Lüpke pendulum; (b) Schob pendulum; (c) Dunlop pendulum; (d) Goodyear–Healey pendulum; (e) tripsometer (m = off-centre mass); (f) Zerbini torsion pendulum.

by four cords such that it describes an arc of a circle of 2000 mm radius as shown in Fig. 9.6. The scale is normally mounted horizontally and calibrated directly in percentage resilience. The Schob consists of a simple rod pendulum with the majority of the mass located in the 15 mm diameter indentor. The Zerbini pendulum, which exists in various sizes, consists of a rod carrying the indentor and rotating under the action of a torsion wire (see Fig. 9.6).

The British standard, BS 903:Part A8,[8] specifies the Lüpke pendulum and also the Dunlop pendulum and Dunlop tripsometer. The Dunlop pendulum is a compound pendulum shaped to ensure high rigidity, attached by a spindle and ball races to a massive structure. The indentor is 2.5 cm in diameter and the test piece a 50 mm square block, 25 mm thick. The Dunlop tripsometer is an unusual design of pendulum consisting of a 42 cm diameter steel disc mounted on bearings and with an out of balance mass in the form of a bracket carrying the 4 mm diameter indentor attached to its periphery.

ASTM D1054[9] specifies the Goodyear–Healey pendulum which is a simple pendulum consisting of a rod mounted on ball races with an additional mechanism for measuring the depth of penetration of the indentor. The indentor diameter is not given in the standard!

Characteristics of various penduli together with typical results obtained in ISO work are shown in Table 9.1. The contact times are calculated by using the formula given in Chapter 8 for indentation by a ball from which Scott derived that:

$$\text{Contact time (seconds)} = C\left(\frac{W}{E}\right)^{0.426} . V^{-1} . R^{-0.278}$$

where: W = striking energy (kgf cm),
V = striking velocity (cm/s),
E = Young's modulus (kgf/cm^2),
R = striker radius (cm), and
C = constant, approximately 3.2.

The radius R is involved in the energy W and it is found that contact time is proportional to radius.

The calculated times in Table 9.1 are for $E = 25\,\text{kgf/cm}^2$ and the measured (?) times those given by Payne and Scott.[10] (Payne and Scott did not make it absolutely clear that the times in question were measured.) The frequency can be taken as the reciprocal of twice the dwell time. In general, resilience is lower the higher the frequency but the Goodyear–Healey gave surprisingly high resilience figures which could be associated with lack of rigidity. It should be noted that the modified specifications for the Schob given in the draft ISO standard yield results equivalent to the Lüpke.

9.1.2 Falling Weight Methods

One of the oldest and best known falling weight instruments is the Shore Scleroscope in which a hemispherically headed striker is allowed to fall under gravity down a graduated glass tube. This instrument has been used with metals, plastics and rubbers and is sometimes wrongly described as measuring hardness. A falling weight instrument is specified in ASTM D2632,[11] sometimes called the Bashore Resiliometer, in which a shaped plunger falls under gravity with the guidance of a vertical rod. It is obviously essential that care is taken to minimise friction between the plunger and rod.

Most developments of falling weight resilience apparatus following the scleroscope have used a steel ball as the striker. A number of instruments

TABLE 9.1

VARIATION OF RESILIENCE WITH CHARACTERISTICS OF THE APPARATUS

Instrument	*Diameter, cm*	*Contact time, s*		*Resilience %, for vulcanisate of:*			
	velocity, cm/s	*Calculated*	*Payne and Scott*	*Natural*	*SBR*	*Neoprene*	*Butyl*
Schob	0.007 5	0.008 7	—	70	47	42	8.5
Lüpke	0.009 1	0.011	0.003 to 0.005	80	55	52	11
Dunlop	0.012 7	0.019	0.04	85	63	60	20
Tripsometer	0.031	0.070	0.2	87	63	58	23
Goodyear–Healey	0.032	0.028	—	87	67.5	65.5	31.5

have been described including that of Robbins and Weitzel[12] and the ADL tester[13] and claims made[14] that improved falling ball apparatus can be more sensitive than an advanced pendulum such as the tripsometer. A more recently described apparatus[15] illustrates several of the improvements which can be made relative to the scleroscope. A wide tube in relation to ball size is used to eliminate friction and the ball, which is dropped by a magnet, can be varied. Test pieces down to 2 mm thickness can be used, clamped by the dead weight of the dropping tube. Such an apparatus has many advantages such as robust construction and small size and offers possibilities of automation. One disadvantage is that, although frequency is readily varied by change of ball size, the level of frequency is usually much higher than the pendulum apparatus and higher than many applications require. As might be expected, lower values of resilience are recorded than with the Lüpke pendulum.[15]

9.2 FREE VIBRATION METHODS

In free vibration methods the rubber test piece, with or without an added mass, is allowed to oscillate at the natural frequency determined by the dimensions and viscoelastic properties of the rubber and by the total inertia. Due to damping in the rubber the amplitude of oscillations will decay with time and from the rate of decay and the frequency of oscillation the dynamic properties of the test piece can be deduced.

Free vibration methods generally have the advantage that the apparatus is relatively simple compared to forced vibration methods and is convenient to use over a range of temperatures. There are, however, a number of disadvantages; the amplitude of oscillation changes due to damping and to avoid changing conditions because of amplitude dependence of the dynamic properties the method is limited to small amplitudes. The method is also restricted to relatively low frequencies and to change frequency the test piece size and/or auxiliary weights or springs must be changed. Generally, free vibration methods are more appropriate to fundamental material characterisation than to generating engineering data.

The equation of motion of a freely vibrating rubber and mass system can be expressed as:

$$m\frac{\mathrm{d}^2x}{\mathrm{d}t^2} + \frac{S''}{\omega}\cdot\frac{\mathrm{d}x}{\mathrm{d}t} S'x = 0$$

where: m = vibrating mass,
x = displacement,
t = time,
S'' = out-of-phase component of stiffness,
S' = in-phase component of stiffness, and
ω = angular frequency.

The solution of this equation gives:

$$S' = m\omega^2\left(1 + \frac{\Lambda^2}{4\pi^2}\right)$$

$$S'' = \frac{m\omega^2\Lambda}{\pi}$$

and

$$\tan\delta = \frac{\Lambda}{\pi\left(1 + \frac{\Lambda^2}{4\pi^2}\right)}$$

where

$$\Lambda = \text{log decrement}$$

9.2.1 Yerzley Oscillograph

The Yerzley oscillograph is specified in ASTM D945[16] and is shown schematically in Fig. 9.7. It consists of a horizontal beam pivoted so as to oscillate vertically and in so doing deform the test piece mounted between the beam and a fixed support. A pen attached to one end of the beam records the decaying train of oscillations on a revolving drum chart. The dynamic deformation of the test piece can be superimposed on a static strain and the mode of deformation can be either shear or compression. The mass and hence the inertia of the beam can be varied by attached weights.

It would seem reasonable to derive from the data generated values for in-phase and out-of-phase moduli and tan δ. However, D945 specifies the calculation of the in-phase component of modulus and a number of other parameters including set and creep from static loading and the Yerzley resilience and hysteresis. Yerzley resilience is defined as the ratio A_3/A_2 in Fig. 9.7 expressed as a percentage. This quantity approximately equals $\exp(-\frac{1}{2}\pi \tan\delta)$ and must not be confused with rebound resilience. Yerzley hysteresis is defined as (100 − Yerzley resilience).

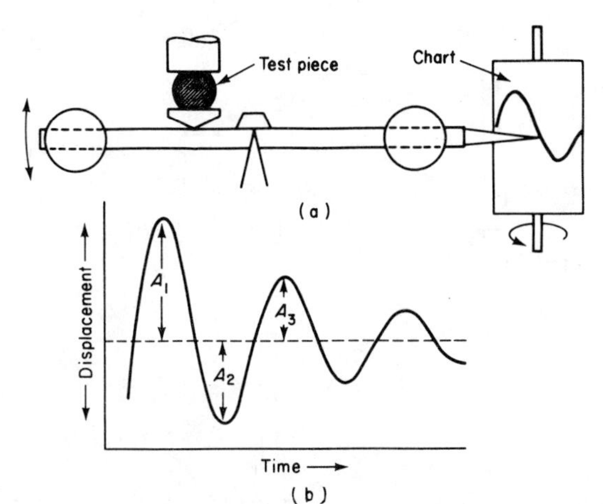

FIG. 9.7. Yerzley oscillograph. (a) Apparatus (diagrammatic); (b) trace of decaying wave train.

The amplitude of deformation with this apparatus must change by a fairly large amount to obtain reasonable precision and consequently it is likely, particularly in compression, that the stress/strain curve will be non-linear over the range measured and hence only an 'effective' modulus measured. The range of frequency obtainable is small at a level of a few hertz.

9.2.2 Torsion Pendulum

The most widely used type of free vibration apparatus is the torsion pendulum which in its simplest form consists of a strip test piece clamped at one end and with a mass to increase inertia at the other. If the strip is twisted and released it will execute a series of decaying torsional oscillations. A number of different designs of torsion penduli have been described of varying complexity and examples are given in references 17–24. A particular form of torsion pendulum, termed torsional braid analysis,[18,22,23] uses a wire or cord coated with the polymer to be tested. The instrument described by Gergen and Keeler[19] is an example of the torsion pendulum moving into the classification of forced oscillation by applying external power to the pendulum to maintain constant amplitude. Several types of torsion pendulum apparatus are now available commercially.

Torsion pendulum methods are standardised in ISO A4663.[24] Three methods are given; in method A the mass of the inertia member is

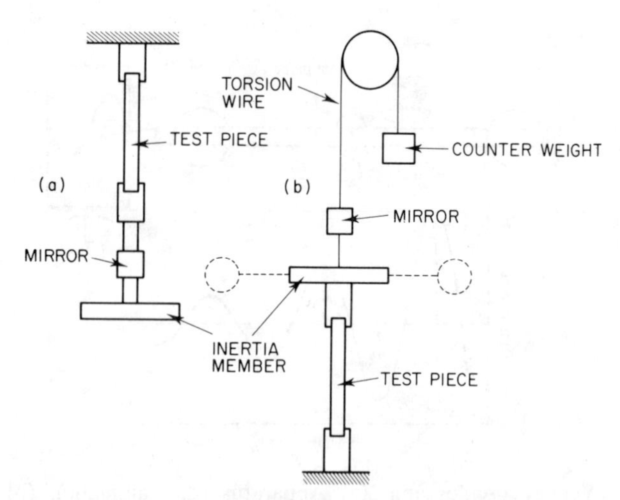

FIG. 9.8. Types of torsion pendulums. (a) Free oscillation apparatus with inertia member supported by the test piece; (b) free oscillation apparatus with inertia member supported by fine wire. In both types of apparatus a lamp and scale is used in conjunction with the mirror to observe the oscillations. The broken lines indicate compensation devices to produce a constant amplitude apparatus.

supported by the test piece; in method B the mass of the inertia member is counterbalanced via a fine suspension wire; and in method C the oscillations are maintained at constant amplitude by supplying energy to the system. Hence method C is not really a free vibration method but it will be convenient to consider it in this section. Schematic diagrams of the apparatus are given in Fig. 9.8. BS 903:Part A31[25] specifies one method which is essentially the same as method B of the ISO standard and notes alternative means of suspending the test piece and of recording the oscillations. ISO 4663 has a bibliography giving reference to original descriptions of the procedures standardised.

The scope of ISO 4663 states that the methods cover the relatively low range of frequencies from 0.1 to 10 Hz at low strains of less than 0.05% in shear. The methods are primarily intended for determining transition temperatures by measuring modulus and damping as a function of temperature and it is suggested that they are not particularly accurate for absolute determination of modulus.

Few details of the apparatus are given in ISO 4663; it is simply stated that means shall be provided to measure frequency to $\pm 1\%$ ($\pm 5\%$ in a transition region), amplitude to $\pm 1\%$ and for method C the supplied energy to $\pm 2\%$. It is suggested that a moment of inertia of about 0.03 gm^2

is suitable for the inertia member which may be a disc or rod. For methods B and C the torsion wire should be of such dimensions that its restoring torque is not more than 25% of the total restoring torque due to the test piece and suspension. BS 903 (equivalent to method B of ISO 4663) suggests that moments of inertia between 50 and 500 g cm^2 are suitable and states that the tensile strain on the test piece should be between 0 and 5%. The British standard also gives methods for determining the moment of inertia of the pendulum.

For rotational oscillations the appropriate relationships are:

$$R' = I\omega^2\left(1 + \frac{\Lambda^2}{4\pi^2}\right)$$

$$R'' = \frac{I\omega^2\Lambda}{\pi}$$

$$\tan\delta = \frac{\Lambda}{\pi\left(1 + \frac{\Lambda^2}{4\pi^2}\right)}$$

where: R' = in-phase component of rotational stiffness (torque/radian),
R'' = out-of-phase component of rotational stiffness,
I = moment of inertia of the vibrating system,
ω = angular frequency, and
Λ = log decrement.

Modulus can be obtained from rotational stiffness by using the formula for static torsion of a strip test piece:

$$G = \frac{Rl}{bh^3k}$$

where: l = free length of test piece,
b = test piece width,
h = test piece thickness, and
k = numerical factor depending on test piece geometry.

Then:

$$G' = 4\pi^2 Ilf^2\left(1 + \frac{\Lambda}{4\pi^2}\right)\cdot\frac{1}{bh^3k}$$

$$G'' = 4\pi Ilf^2\cdot\Lambda\cdot\frac{1}{bh^3k}$$

where: f = frequency in hertz.

For methods B and C of ISO 4663 and the BS method the effective frequency term is $f^2 - f_0$ and the associated log decrements Λ and Λ_0 where:

f = frequency with the test piece,
f_0 = frequency without the test piece,
Λ = log decrement with the test piece, and
Λ_0 = log decrement without the test piece.

The complication of Λ_0 in the term $1 + \Lambda/4\pi^2$ is ignored.

For method B of ISO 4663 and the BS method the log decrement of the rubber,

$$\Lambda_R = \Lambda - \Lambda_0 \cdot \frac{f_0}{f}$$

For method C of ISO 4663:

$$\Lambda_R = \frac{K}{4\pi I f^2 A}$$

where: K = compensating mechanical moment, and
A = amplitude of oscillations.

The above relationships indicate what may be derived from the torsional pendulum measurements. In fact, BS 903 calls for G' and G'' and the log decrement although it does not actually say how to calculate the log decrement for the rubber. BS 903 also allows a circular cross-section test piece, when the term $1/bh^3k$ is replaced by $d^4n/32$ where d is the diameter of the test piece.

ISO 4663 calls for the complex modulus and log decrement. The complex modulus is calculated from:

$$G^* = 12\pi^2 I l f^2 \left(1 + \frac{\Lambda^2}{4\pi^2}\right) \cdot \frac{1}{bh^3k}$$

It is not apparent how this relation was derived but it appears to be the equation for G' with the out-of-phase component being neglected! The factor 12 arises from the factor k in the ISO standard being three times the factor defined in the British Standard.

ISO 4663 gives no advice as to the relative merits of the three methods it specifies. Method C, which is not strictly a free vibration method, removes the difficulties associated with changing amplitude through the course of the test but at the expense of a rather more complex apparatus. When the inertia member is supported by a torsion wire as in method B the tensile strain in the test piece can be controlled to a low level by means of counterweights.

9.3 FORCED VIBRATION METHODS

There are several possible approaches to the measurement of dynamic properties using forced oscillation of the test piece and the methods can be classified in various ways. One classification is to subdivide into:

Forced vibration at or near resonance.
Forced vibration away from resonance.
Transient (as opposed to continuous cyclic) loading.
Wave propagation.

The forced vibration method away from resonance can again be subdivided into those which apply deformation cycles and those which apply force cycles.

The range of frequency covered can be divided into:

Low frequency < 1 Hz
Medium frequency 1 to 100 Hz
High frequency > 100 Hz

Another way of classifying apparatus is according to the means of driving the test piece into oscillation which, other than in wave propagation methods, can be mechanical, electromagnetic or hydraulic. Mechanical activation can be applied in at least three ways. A screw type machine, similar in concept to a 'static' tensile machine can be made to apply force or displacement cycles but is limited to low frequencies, perhaps up to 2 Hz. A rotating eccentric weight will apply force cycles and an eccentric cam can be used to apply displacement cycles. Quite large strain amplitudes can be realised with mechanical activation, being limited only by the force on bearings, etc., but the frequency is restricted, generally to a maximum of about 50 Hz, and fatigue life of the machine may be poor. Generally, mechanically driven machines are today considered outdated.

Electromagnetic vibrators can cover a very wide frequency range, in particular being capable of very high frequencies up to at least 10^4 Hz, although at this level considerable care is needed. Quite high power is obtainable, but electromagnetic drive is more commonly applied to relatively small machines and low strains. It is also used in forced oscillation methods at resonance.

Closed-loop servo hydraulic activation is generally limited to frequencies of up to 100 Hz but in all other respects it is the most versatile method. Either force or strain can be controlled in the same machine and it is possible to use waveforms other than sinusoidal. Relatively large forces sufficient to test products and at large deformations can be realised, usually in the combination of high forces and small deformations or lower forces and large deformations. Multi-axis and mixed mode stresses can be applied to give full characterisation of products. The penalty for this versatility is high cost compared to simple machines and the complexities of operation.

9.3.1 Types of Machine in Use

The number of particular designs of dynamic test machine is virtually legion and there is no question of considering each of them here but reference can be given to a number of types that have been or are used.

Of the mechanically driven machines, the Roelig[26] uses a rotating eccentric weight whereas the RAPRA sinusoidal strain machine[27,28] uses an eccentric cam. The Rotary Power Loss Machine[29] is also in this latter category but the mode of deformation is not simple shear or compression.

The principles of servohydraulic test machines have been outlined by Owens,[30] for example, but for descriptions of machines commercially available reference is best made to the literature of manufacturers of advanced testing systems.

One of the best known examples of an electromagnetic machine is the Rheovibron originally developed by Takayanagi[31,32] which uses small test pieces in tension generally in the frequency range 3.5–110 Hz and over a range of temperatures. This apparatus can also be used in compression[33] and shear[34] and at high elongations.[35]

The Rheovibron could be called a dynamic mechanical analyser (or a dynamic thermomechanical analyser). These terms have come into general use relatively recently and are generally taken as referring to modest sized, bench mounted dynamic test machines which allow the measurement of dynamic properties over a range of frequency and of temperature and which are automated/computerised to a greater or lesser extent. The terms can cover free vibration apparatus as well as forced vibration and an increasing variety of analysers are now on the market, representing numerous different geometries and control systems. As a generalisation, they are machines working with small test pieces which are especially efficient at characterising materials as functions of temperature and frequency to give comparative results but may not give accurate absolute values and are not suitable for characterising products and materials over a range of strains and deformation modes. Perhaps their greatest importance has been in encouraging a much increased amount of dynamic testing because of the efficiency in use. There is of course no fine distinction as to whether an apparatus is a dynamic analyser and the so-called analysers vary considerably in their capability.

The Polymer Laboratories instrument[36] operates in a bending mode and is driven electromagnetically. Also driven electromagnetically is an instrument developed at the NPL[37] which can be used with standard shear test pieces. Rather against the trend, an apparatus described by Yokouchi and Kobayashi[38] uses a mechanical drive in order to obtain higher driving forces. The Dynaliser[39] is rather different to most dynamic test machines, the mode of deformation being indentation and the dynamic characteristics are deduced from a force relaxation curve.

Any rubber test piece with or without added mass has a natural or resonant frequency of vibration determined by the dimensions and viscoelastic properties of the rubber, the total inertia of the system, and the mode of deformation. If constant force amplitude cycles are applied to the rubber and the frequency varied the resulting deformation cycles will have a maximum value when the applied frequency equals the resonant frequency of the test piece system.

At resonance (where the external mass $\gg$ sample mass):

$$f_R = \sqrt{\frac{S'}{m}}$$

and

$$A_R = \frac{F}{S''}$$

where: f_R = resonant frequency,
S'' = in-phase component of stiffness,
m = mass of vibrating system,
A_R = deformation amplitude at resonance,

F = applied force amplitude, and
S'' = out-of-phase component of stiffness.

Fortunately, complications due to inertia effects do not affect these calculations at resonance. From S' and S'', the moduli (working in shear) G', G'' and G^* can be deduced. Usually a mass is added to the rubber test piece to reduce the resonant frequency to levels of practical interest and it is quite feasible to vary the mass at a constant frequency until the system is at resonance.

Commonly electromagnetic actuation is used as in Moyal and Fletcher's machine[40] and the resonant beam apparatus used in the motor industry to test mountings. The latter method involves transmitting the vibration to the test piece via a heavy pivoted beam and the characteristics of the beam significantly affect the results.[41] The vibrating method in which the test piece in the form of a simple cantilever is directly vibrated without added mass has mostly been used for rigid materials but can be applied to rubber. The Dupont DMA[42] and the Brabender–Lonza pendulum are dynamic analysers working at resonance. Tangorra[43] described what is in effect a vibrating hardness tester, the deformation being caused by an electromagnetically driven indentor.

The dynamic properties of rubber can be deduced from the velocity and attenuation of waves passing through it. Quite a variety of systems have been tried and several methods using frequencies up to 200 Hz have been described by Payne and Scott.[10] Sonic or ultrasonic waves can be used, although in the latter case the frequencies are much higher than are usually of practical interest. However, pulsed ultrasonic methods have been successfully used for measurements of moduli of non-isotropic plastics but with rubbers the attenuation is too great to sustain a transverse wave at frequencies greater than 1 MHz.[44]

9.3.2 Forced Vibrations Away from Resonance (Standard Methods)

It is an illustration of development in standardisation to note that the 1964 edition of BS 903:Part A24 described three particular types of apparatus in an appendix, whereas the 1976 edition[45] simply requires that any apparatus can be used as long as it meets the stated performance requirements. The international standard for forced vibration measurements is ISO 4664[46] which also does not specify any particular machine but calls for apparatus giving forced sinusoidal displacement cycles in shear by means of mechanical, servo-hydraulic or electromagnetic actuation, the displacements being determined to an accuracy of $\pm 2.5\%$.

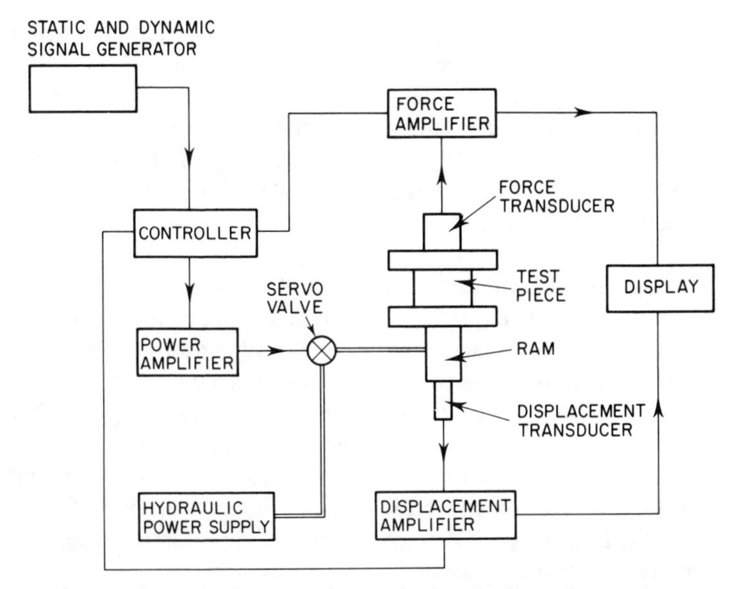

FIG. 9.9. Schematic diagram of servo-hydraulic dynamic test apparatus.

A servo-hydraulic test system is shown diagrammatically in Fig. 9.9. The force measurement should be accurate to $\pm 2.5\%$, tan δ be measurable to $\pm 5\%$ or 0.02 in tan δ and the measurements made at 10 ± 0.5 Hz BS903:Part A24 has a wider scope than ISO 4664 and does not specify any particular frequency, but recommends a series from 15 to 200 Hz, each with a suitable strain amplitude and notes that these may be superimposed on a static strain. The British standard also calls for more accurate measurement asking for ± 0.01 or $\pm 5\%$ in tan δ and a maximum of $\pm 2\%$ in complex modulus $|G^*|$. No figures for strain and force accuracy are given, these being covered by the requirement for $|G^*|$.

Both standards suggest double shear test pieces and do not limit these to any particular size, thus allowing for different machine characteristics, except that the side or diameter to thickness ratio of the rubber elements must be 4:1 (or greater in BS 903) to prevent significant bending. It is of interest that currently discussions are being held to standardise one double shear test piece with circular rubber elements for all shear tests.

Neither standard gives great detail as to carrying out the actual test, this being dependent on the apparatus used. The main difference between the standards is that ISO 4664 is restricted to particular conditions suitable for classification purposes whereas BS 903 is intended to apply more generally. Both specify that complex shear modulus $|G^*|$ and tan δ should be reported, BS 903 recommending that these should be presented as functions of temperature, frequency and strain amplitude. BS 903 illustrates in an appendix how the data can be calculated from a force deflection loop. This is as given at the beginning of this chapter with reference to Fig. 9.3 with the introduction of the factor $\frac{1}{2}$ to account for the double shear test piece.

9.3.3 Comparison of Dynamic Methods

There does not seem to be a fully comprehensive and up-to-date critical account of dynamic test machines and methods for rubbers. Descriptions of several machines available prior to 1960 are given by Payne and Scott[10] and there have been more recent reviews by Hilberry,[41] Brown and Scott[47] and Warner.[2] These and several other of the references given earlier discuss the relative merits of certain techniques and apparatus. The lack of a comprehensive critical review is particularly unfortunate because of the large variety of machines now available and the sometimes confusing claims of manufacturers.

From the standardisation position it is apparent that forced vibration in shear away from resonance, torsion pendulum and rebound resilience are the most commonly specified methods. Resilience measurement remains popular because it is such a simple method for quality control purposes. The more simple torsion penduli allow the determination of limited dynamic data at relatively low apparatus cost. For the generation of engineering design data non-resonant forced vibration methods using versatile and relatively high powered servo-hydraulic apparatus are generally considered the most useful. In one sense the standards are behind practice in that they do not cover the range of automated analysers which perhaps fill the range between simple torsion pendulum and sophisticated servo-hydraulic machines.

Certain of the general characteristics can be summarised: Servo-hydraulic machines can provide relatively large amplitudes of force and deformation with a continuous variation of frequency up to 100 Hz. Electromagnetic actuation is necessary for much higher frequencies but the strain amplitude is more limited. Free vibration methods are not suited to the study of amplitude variation and the frequency range of any apparatus is limited. Forced vibration non-resonant apparatus also has the advantage of being applicable to materials with a large value of tan δ, whereas with free vibration apparatus the vibrations may die out rather too quickly and with resonance methods the amplitude maximum is then not clearly defined. The strain distribution generated by some analysers may be complex and be most suited to comparative work but they are very efficient in operation as a means of dynamically characterising materials. A critical evaluation of analysers would be very useful. Resilience is the most elementary of dynamic tests and its simplicity makes it valuable for quality control purposes. In principle all methods can be used over a range of temperatures.

In principle at least, all the methods should be in agreement if all the effects of frequency, temperature, amplitude and test piece geometry are taken into account but directly comparable results covering several machines do not appear to be readily available. Rebound resilience is so simple and different in concept to the other methods that its correlation with them is often questioned. Results provided by Bulgin are quoted by Payne and Scott[10] showing virtually perfect agreement between measured resilience and that calculated from three types of dynamic testing apparatus. Linear agreement was also obtained by Kainradl[48] using a Schob pendulum except that the measured values were about 20% lower than those calculated. From this it could be deduced that Bulgin's results were obtained with a Dunlop pendulum (see Section 9.1.1)!

It may be somewhat difficult to decide which dynamic test machine should be used in any circumstance and this will depend to a large extent on what the results are needed for and how they will be used. The interpretation and application of measures of dynamic properties is clearly very important and may be complicated. Fairly recent discussion of dynamic data is given in references 49–52.

REFERENCES

1. ISO 2856, 1975. Elastomers—General Requirements for Dynamic Testing.
2. Warner, W.C., ASTM STP553, 1974, pp. 31–55.
3. Payne, A.R. (1959). *Engineer*, **207**, 328.
4. Davey, A.B. and Payne, A.R. (1964). *Rubber in Engineering Practice*, MacLaren and Sons Ltd.
5. Hall, M.M. and Thomas, A.G. (Apr. 1973). *J. IRI*, 65.
6. Ferry, J.D. (1970). *Viscoelastic Properties of Polymers*, J. Wiley and Sons.
7. ISO 4662, 1978. Determination of Rebound Resilience of Vulcanisates.
8. BS 903:Part A8, 1963. Determination of Rebound Resilience.
9. ASTM D1054-79. Resilience Using a Rebound Pendulum.
10. Payne, A.R. and Scott, J.R. (1960). *Engineering Design with Rubber*, MacLaren and Sons Ltd.
11. ASTM D2632-79. Resilience by Vertical Rebound.

12. Robbins, R.F. and Weitzel, D.H. (1969). *Rev. Sci. Instrum.*, **40**, (8), 1014.
13. Raphael, T. and Armeniades, C.D. (Apr. 1964). *SPE Trans.*, 83.
14. Madan, R.C. (1963). *Rubb. Plast. Weekly*, **144**, 679.
15. Brown, R.P. and Boult, B.G. (July/Aug. 1975). *RAPRA Members J.*, 64.
16. ASTM D945-79. Rubber Properties in Compression or Shear (Mechanical Oscillograph).
17. Wheeler, A. (Oct. 1969). *Plastics and Polymers*, 469.
18. Gillham, J.K. and Roller, M.B. (July 1971). *Polym. Eng. Sci.*, **11**(4), 295.
19. Gergen, W.P. and Keeler, T.L. (Sept. 1970). *Rubb. J.*, 81.
20. Bell, C.L.M., Gillham, J.K. and Benci, J.A. Development of a fully automated torsional pendulum with built-in data reduction facilities. Paper presented at SPE 3rd Annual Tech. Conf. Antec 74, California, May 1974.
21. Anzani, I., Xiang, P.Z. and Pritchard, G., (1985). *Polym. Test.*, **5**(4).
22. Gillham, J.K. (1967). *Polym. Eng. Sci.*, **7**, 225.
23. Cowie, J.M. and Fergusson, R. (1983). *Br. Polym. J.*, **15**(1).
24. ISO 4663, 1977. Determination of Dynamic Behaviour at Low Frequencies–Torsion Pendulum Method.
25. BS 903:Part A31, 1976. Determination of the Low Frequency Dynamic Properties of Rubbers by Means of a Torsion Pendulum.
26. Roelig, H. (1945). *Rubb. Chem. Tech.*, **18**, 62.
27. Payne, A.R., RABRM Res. Report 76, 1955.
28. Payne, A.R., RABRM Res. Memo. R411, 1958.
29. Bulgin, D. and Hubbard, D.G. (1958). *Trans. IR*I, **34**, 201.
30. Owens, J.H., Measurement of the Dynamic Properties of Elastomers, SAE/ASTM Symposium, Detroit, Jan. 1973, p.43.
31. Takayanagi, M. (1963). *Memoirs Faculty of Engineering, Kyushu University*, **23**, 41.
32. Massa, D.J. and Flick, J.R. (Apr. 1975). *SEB Petrie, Coat. Plast. Preprints*, **35**(1), 371.
33. Murayama, T. (Sept. 1976). *J. Appl. Polym. Sci.*, **20**(9), 2593.
34. Murayama, T. (Dec. 1975). *J. Appl. Polym. Sci.*, **19**(12), 3221.
35. Voet, A. and Marawski, J.C. ACS Div. of Rubb. Chem., Paper 42, Preprint 012, May, 1974.
36. Wetton, R.E. (1984). *Polym. Test.*, **4**(2–4).
37. Dean, G.O., Duncan, J.C. and Johnson, A.F. (1984). *Polym. Test.*, **4**(2–4).
38. Yokouchi, M. and Kobayashi, Y. (1981). *J. App. Polym. Sci.*, **26**, 4307.
39. de Meersman, C. and Vandoven, P. (1981). *Int. Polym. Sci. Tech.*, **8**(9).
40. Moyal, J.E. and Fletcher, W.P. (1945). *J. Sci. Instrum.*, **22**, 167.
41. Hilberry, B.M., ASTM STP553, 1974, p.142.
42. Gill, P., Lear, J.O. and Leckenby, J.N. (1984). *Polym. Test.*, **4**(2–4).
43. Tangorra, G. (1961). *Rubb. Chem. Tech.*, **34** (1), 347.
44. Brown, R.P., Hughes, R.C. and Norman, R.H. (Sept. 1974). *RAPRA Members J.*, 225.
45. BS 903:Part A24, 1976. Measurement of Dynamic Moduli.
46. ISO 4664, 1978. Determination of Dynamic Properties of Vulcanisates for Classification Purposes (by Forced Sinusoidal Shear Strain).
47. Brown, R.P. and Scott, J.R. (1972). *Prog. Rubb. Tech.*, **36**, 67.
48. Kainradl, P. (1956). *Rubb. Chem Tech.*, **29**, 1082.

49. Howgate, P.G. and Berry, J.P. 125th ACS Rubber Division Meeting, Indianapolis, 1984, Paper 14.
50. Howgate, P.G. 118th ACS Rubber Division Meeting, Detroit, 1980, Paper 35.
51. Gregory, M.J. (1984). *Polym. Test.*, **4**(2–4).
52. Clamroth, R. (1981). *Polym. Test.*, **2**(4).

Chapter 10

Friction and Wear

It is well appreciated that friction and wear are interrelated subjects simply because friction is involved in wear mechanisms. Both friction and wear can be studied and measured in the same experiment and this is done in, for example, the investigation of bearings and sliding joints. In the rubber industry the two topics are more often thought of separately. Friction plays its part in rubber wear mechanisms and is considered when these mechanisms are being studied; furthermore, with a major product, tyres, friction or resistance to slip and wear are two of the most important performance parameters. However, when it comes to laboratory measurements on rubber the two tests are considered as separate subjects and generally are not accredited equal status. Whereas friction tests are rarely standardised and are carried out in relatively few laboratories, wear or abrasion tests appear in many specifications and most rubber laboratories have at least one type of abrasion apparatus. This situation is not unreasonable—far more than simply friction is involved in rubber wear processes and wear is directly a problem in more applications of rubber than is friction.

10.1 FRICTION

It would appear that no account of friction is complete without first stating Leonardo da Vinci's (or Amonton's) laws and Coulomb's law of friction and pointing out that in general polymers do not obey them. The laws are:

(a) The frictional force opposing motion is proportional to the normal force, the constant of proportionality being the coefficient of friction, i.e.

$$F = \mu N$$

where F = frictional force,
μ = coefficient of friction, and
N = normal force.

(b) The coefficient of friction is independent of the apparent area of contact.
(c) The coefficient of friction is independent of the velocity between the two surfaces provided that the velocity is not zero.

In practice rubbers do not normally obey these rules and the coefficient is a 'variable constant', its value depending on the real contact area, normal load, velocity and other factors. In fact friction is sensitive to just about anything, including breathing on the test piece, and any single-point measurement is of very limited use. The factors influencing friction of polymers have been discussed by James[1] who gives a useful bibliography, and only the factors of most importance as regards the test method will be considered briefly here.

The apparent area of contact between two surfaces is much larger than the actual area over which they touch even if the surfaces appear smooth. The frictional force is proportional to the real contact area so that anything that changes the real contact area will change the force measured.

When rubber is brought into contact with another surface it deforms elastically and the real area of contact will increase with increasing normal load and hence the coefficient of friction will decrease with increasing normal load. It is also apparent that the real contact area is dependent on the surface geometry of the test piece. It is hence desirable to measure the friction of rubber over the range of normal forces of interest and to test with the surface geometry to be used in service, which may mean using the product or part of it as the test piece.

The distinction is sometimes made between static and dynamic friction, implying that there is one level of the coefficient of friction just at the point when movement between the surfaces starts and another level when the surfaces are steadily separating. There can of course be no measure of friction without movement so that 'static' friction is actually friction at an extremely low velocity and thereafter the coefficient of friction of rubbers may vary markedly with velocity. Hence it is necessary to measure friction over the range of velocities of interest. Friction is also dependent on temperature, which can lead to inaccuracies at high velocities because of heat build-up at the contacting surfaces.

During a friction test a condition known as 'slip–stick' sometimes occurs in which the relative velocity and the coefficient of friction between the two surfaces both oscillate about a mean value. The essential condition for slip–stick to occur is that in the velocity region being considered the coefficient of friction falls with increasing velocity. An oversimplistic description of the phenomenon is that for a short time the surfaces stick together and the force builds up as in a spring until it exceeds the 'static' coefficient of friction when movement occurs and the friction falls to a lower kinetic value and the spring releases until sticking again takes place. Slip–stick can in fact occur at high velocities when the frequency of vibration can be high enough to cause audible squeals. The amplitude and frequency of the slip–stick vibrations depend on the rigidity and damping of the testing system as well as on the properties of the surfaces. To minimise slip–stick it is necessary to construct the test apparatus, particularly the drive and force measuring elements, to be as stiff as possible. If slip–stick can occur in service its presence can be more important, or rather troublesome, than the actual mean level of friction.

It is fairly obvious that other factors such as lubricants, wear debris, ageing of the surfaces and humidity can also affect friction and once again test conditions must be chosen that resemble those found in service.

10.1.1 Methods of Measuring Friction

The essential requirements for a friction test are two contacting surfaces, a means of creating relative motion between them and a system to indicate the frictional force. A number of different arrangements are then possible, a selection being shown in Fig. 10.1. Although many particular apparatuses have been described in the literature very few friction tests have been standardised. An example of a driven sled apparatus has been described by James;[2] Griffin[3] devised a test for small cylindrical plastic test pieces and Mustafa and Udrea[4] used a rotating steel disc and a stationary plastic test piece. A ball and peg machine is described by Bailey and Cameron,[5] a hemispherically ended pin and flat plate device by West and Senior,[6] a steel pin running on a plastic ring by Jost[7] and a combined inclined and horizontal plane by Wilson and Mahoney.[8] Many more types of apparatus have been described including a comparison of three machines by Robinson and Kopf.[9]

An international standard for plastics has been under consideration for a long time but work has not started for rubber. There is a British standard covering friction in the plastics design data series BS 4618[10] but this does not give definite methods. An inclined plane method is specified for coated fabrics in BS 3424[11] although this edition is formally withdrawn

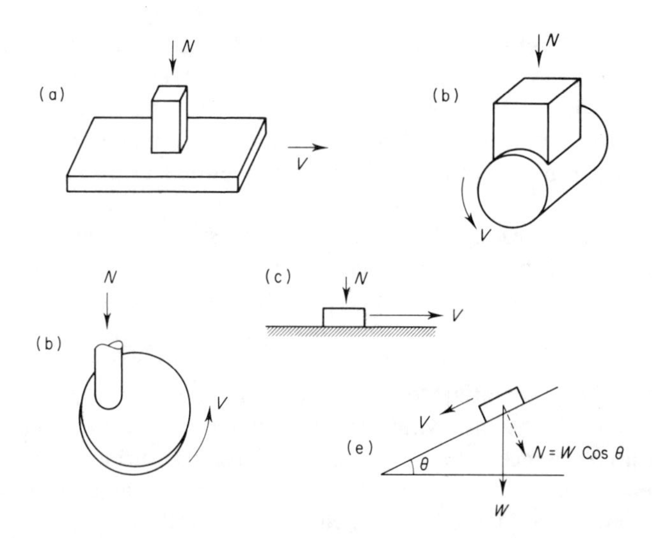

FIG. 10.1. Arrangements for friction tests. (a) Linear track; (b) rotating shaft; (c) towed sled; (d) pin and rotating plate; (e) inclined plane. N = normal force, V = direction of motion, W = weight of the test piece.

and a revision of the friction section has not appeared. The test piece is attached to a sled resting on an inclined plane which is covered with the other surface to be tested. The inclination of the plane is varied until sliding takes place. This is probably the simplest form of friction test but not an accurate one and of course cannot operate at any given velocity.

A towed sled method is specified in BS 2782[12] for plastics film but as described this might be inadequate for more general or very accurate work. The drive linkage is not specified in detail and some apparatus in use would lack stiffness.

An attractive basis for a good quality friction test is a universal tensile testing machine because it can be suitably stiff, gives a very wide range of speeds and has a precise force measuring system. The only difficulty is that such machines operate in a vertical plane and if the normal load on the test piece is applied by a weight acting vertically under gravity the linkage to measure the frictional force must turn through 90° and in doing so there is a danger of introducing friction at a pulley and decreasing the stiffness of the system.

Appreciating these difficulties of precisely measuring the frictional properties of polymers, a novel apparatus operating vertically in a tensile machine was developed at the Rubber and Plastics Research Association and has been described by James and Newell.[13] The advantages of the apparatus (Fig. 10.2) are that in conjunction with a suitable tensile tester, a very stiff system results with very accurate measurement of small forces and a good range of velocity. It can be operated in an environmental chamber, so giving a wide range of temperatures and the test piece geometry can be readily changed, including tests on products or parts of products. Using this apparatus, James and Mohsen[14] have illustrated the importance of specifying sample preparation and test conditions for rubber.

Because of the sensitivity of friction to so many variables it is often desirable, if not essential, to test the actual product in a prototype test rig. This can be the case with tyres, bottle closures and bearings.

The slip resistance of flooring and artificial sports surfaces has attracted considerable interest in recent times. The friction of road surfaces is often measured with a skid tester developed by the Road Research Laboratory[15] and this has also been widely used on other surfaces and floors. It is a pendulum device, the movement of which is arrested by the foot of the pendulum skidding on the surface to be measured. The 'skid resistance' indicated can be approximately related to coefficient of friction by:

$$\text{Skid resistance} \sim \frac{330\mu}{3+\mu}$$

The value of this apparatus is that it can be used to test a very large product in service but it is very questionable as to how realistically it assesses floors and sports surfaces. A number of other instruments have also been developed for this purpose including the Tortus apparatus[16] which is essentially a self-propelled, four wheeled, trolley with a sliding foot. The reasoning behind test methods for slip resistance and descriptions of several apparatus, together with a large bibliography, have been given by James.[17]

Both the skid tester and the Tortus use a block of rubber as the standard slider. Recent work has been directed at evaluating various slider rubbers to enable a composition to be standardised which gives the best correlation with service experience together with good reproducibility between laboratories. Any rubber standardised for this purpose must be produced to a very precise specification. Furthermore, it must be established that the test method gives results which agree with experience. Papers covering recent work on slip resistance are given in references 18–21.

FIG. 10.2. RAPRA friction apparatus.

10.2 WEAR

The terms wear and abrasion are used so loosely that confusion sometimes results. Wear is a very general term covering the loss of material by virtually any means. As wear usually occurs by the rubbing together of two surfaces, abrasion is often used as a general term to mean wear. The mechanisms by which wear occurs when a rubber is in moving contact with any material are somewhat complex, involving principally cutting of the rubber and fatiguing of the rubber. An account of the mechanisms of both friction and wear has been given by Lancaster[22] and one section of a collection of translations of papers[23] describing extensive Russian work on abrasion is devoted to mechanisms. Further discussion is given by Zhang,[24] Rougier *et al.*[25] and Gent and Pulford.[26] It is possible to categorise wear mechanisms of rubber in various ways and one convenient system is to differentiate between three main factors:

(a) Abrasive wear, which is caused by hard asperities cutting the rubber.
(b) Fatigue wear, which is caused by particles of rubber being detached as a result of dynamic stressing on a localised scale.
(c) Adhesive wear, which is the transfer of rubber to another surface as a result of adhesive forces between the two surfaces.

From these definitions it can be seen that the more specific meaning of abrasion is wear by the cutting action of hard asperities. The common practice in the rubber industry of using abrasion as a general term for wear probably results from the fact that most wear tests for rubbers use the action of sharp asperities, for example abrasive paper, to produce wear.

The wear or abrasion of rubber caused by hard asperities is not just simply cutting but involves both plastic and elastic deformation of the rubber.[27] Generally, in any wear process more than one mechanism is involved although one mechanism may predominate. The most important consideration in practice is that the wear process will be complex and critically dependent on the service conditions. It is therefore necessary that any laboratory test must essentially reproduce the service conditions if good correlation is to be obtained. Even a comparison between two rubbers may be invalid if the predominant wear process in the test is different from that in service. It is failure fully to appreciate this which has led to the conclusion that all laboratory abrasion tests are useless except for quality control. In some instances it is virtually impossible to reproduce in the laboratory, with a reasonable degree of acceleration, the complex conditions met in service, although it has been the experience at the Rubber and Plastics Research Association that for many products meaningful results can be obtained by careful modification of standard abrasion tests.

10.2.1 Types of Abrasion Test

A very large number of different abrasion apparatus have been used for testing rubbers and an even larger number of permutations of the various factors would be possible. The first division of test types can be to distinguish between those using a loose abradant and those using a 'solid' abradant. A loose abrasive powder can be used rather in the manner of a shot-blasting machine and this is a logical way to simulate the action of sand or similar abradants[28,29] impinging on the rubber in service, as may be the case with conveyor belts or tank linings. A loose abradant can also be used between two sliding surfaces and this situation occurs in practice through contamination and as a result of the generation of wear debris from a 'solid' abradant. A car tyre is an example of the situation where there is a combination of abrasion against a solid rough abradant, the road, together with a free flowing abradant in the form of grit particles.

'Solid' abradants could consist of almost anything but the most common are: abrasive wheels (vitreous or resilient), abrasive papers or cloth, and metal 'knives'.

The choice of abradant should be made primarily to give the best correlation with service and the usual abrasive wheels and papers really only relate to situations where cutting abrasion predominates. Materials such as textiles and smooth metal plates have been found to give good results for some applications. In practice the abradant is often chosen largely for reasons of convenience and surfaces such as plain steel have the disadvantage of abrading slowly and if the conditions are accelerated give rise to excessive heat build-up. The abrasive wheel is probably the most convenient abradant because of cheapness, mechanical stability and because, by simple refacing, a consistent surface can be maintained. Abrasive papers and cloths are cheap and easy to use but deteriorate in cutting power rather quickly.

The possible geometries by which the test piece and the abradant can be rubbed together are legion and it is not sensible to make any general classification. Some well-known configurations are shown in Fig. 10.3. In type (a) the test piece is reciprocated linearly against a sheet of abradant,

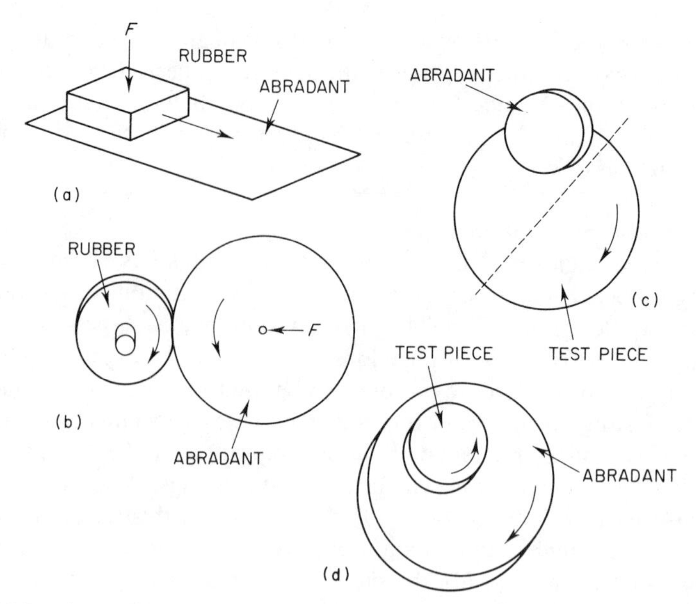

FIG. 10.3. Arrangements for abrasion tests. (a) Reciprocating test piece; (b) Akron type; (c) Taber type (one abrading wheel shown); (d) Schiefer type. F = force pressing rubber and abradant together.

but alternatively a strip of abradant could be moved past a stationary test piece. A further variation is to have the abradant as a rotating disc with the test piece held against its side. Both the abradant and the test piece can be in the form of wheels, type (b), with either being the driven member. In type (c) the abrasive wheel is driven by a rotating flat test piece, and in type (d) both the test piece and abradant are rotated in opposite directions.

10.2.2 Test Conditions

Abrasion occurs when the rubber slips relative to the abradant and the amount of slip is a critical factor in determining the rate of wear. In type (a) of Fig. 10.3 the slip is 100% because the rate of slipping is equal to the rate of movement of the test piece (or abradant). In contrast, with the type (b) arrangement a range of levels can be used by varying the skew angle between the two wheels or in type (c) by varying the distance of the wheel from the centre line of the test piece. Typically, the rate of abrasion with a type (b) apparatus is proportional to something between the square and cube of the slip angle.

An important difference between apparatus of type (a) or (d) and (b) or (c) is that in the former case the test piece is continuously and totally in contact with the abradant and there is no chance for the very considerable heat generated at the contact surface to be dissipated. The actual rate of slip will influence the rate of wear because as the speed is increased heat build-up will rise. Temperature rise during test is one of the important factors in obtaining correlation between laboratory and service.

The contact pressure between the test piece and abradant is another critical factor in determining wear rate. Under some conditions wear rate is more or less proportional to pressure but if with changing pressure the abrasion mechanism changes, perhaps because of a large rise in temperature, then the wear rate may change quite drastically. Again, this is a critical factor in obtaining correlation with service.

Rather than consider contact pressure and degree of slip separately it has been proposed[30] that the power consumed in dragging the rubber over the abradant should be used as a measure of the severity of an abrasion test. The power used will depend on the friction between the surfaces and will determine the rate of heat build-up.

Although temperature has a large effect on wear rate, it is extremely difficult to control the temperature during test, but it is clearly the temperature of the contacting surfaces which is of importance rather than the ambient temperature.

The rate of wear will quite naturally be affected by any change in the nature of the contacting surfaces. Apart from the abradant changing because of its own wear, there can be effects from lubricants, wear debris between the surfaces and clogging of the abradant. Not many commonly used apparatus are suitable for testing in the presence of a liquid lubricant but it is common practice to remove wear debris by continuously brushing the test piece or by the use of air jets, in which case care must be taken to ensure that the air supply is not contaminated with oil or water from the compressor. Clogging or smearing of the abradant is a common problem with abrasive wheels and papers and its occurrence will invalidate the test. It is normally caused by a high temperature at the contact surfaces and although the problem can sometimes be reduced by introducing a powder between the surfaces it should be treated as an indication that the test conditions are not suitable. If high temperatures are to be realised in service a test method in which new abradant is continually used should be chosen. It may be noted that a practical example of a powder influencing abrasion is a car tyre running on a dusty surface. If the abrasion is unidirectional abrasion patterns will develop which can markedly affect abrasion loss.

It should again be emphasised that if correlation between laboratory tests and service is to be obtained the test conditions must be chosen extremely carefully to match those found in the product application.

10.2.3 Standard Rubbers and Expression of Results

Largely because of the critical dependence of the wear rate on the test conditions and particularly because of the difficulty in maintaining a precisely reproducible abradant it is common practice to refer all abrasion results to the results obtained at the same time on a 'standard' rubber. This is an eminently sensible practice as it goes a long way towards eliminating variability due to differences between nominally identical machines and abradants. There is only one drawback, the difficulty of producing an accurately reproducible standard rubber. This produces something of a chicken and egg situation where it is difficult to decide whether it is the abradant or the rubber which is changing.

Within one laboratory the coefficient of variation of abrasion results using different batches of a standard rubber would probably be not much less than 8% and the between laboratory variation would be expected to be greater. Some abradants will certainly be more variable than this but presumably some materials can be reproduced with better precision. It is a fact that standard rubbers are themselves variable but they are of very considerable value, particularly when reference is made only to standards from one batch and where they are used to monitor the change with time of one sample of abradant.

The most common type of standard rubber is based on a tyre tread compound but a shoe sole type material is also widely used. There is no reason at all why, for a particular investigation, an in-house standard representative of the type of material being evaluated should not be used. At the time of writing the standard rubbers given in the British standard for abrasion tests are in the course of revision because the whole standard is being revised and some ingredients specified have fallen into disuse.

In standard abrasion tests it is usually weight loss which is the parameter measured although in certain cases the change in test piece thickness is more convenient. Because it is the amount of material lost which matters it is usual to convert the weight loss to volume loss by dividing by the density. The volume loss can be expressed as the loss per unit distance travelled over the abradant, per 1000 revolutions of the apparatus or whatever. A less usual practice is to express the result as loss per unit energy consumed in causing abrasion. Whatever the loss is related to, it must be remembered that the rate of wear may not be constant because of inhomogeneity of the test piece and gradual change in the nature of the abradant. The experiment should be designed to minimise the latter effect by using standard rubbers, refacing the abradant and running repeat test pieces of a series of materials in reverse order.

Devotees of the use of standard rubbers then finally express the result as an abrasion resistance index defined by:

$$\text{Abrasion resistance index} = \frac{\text{Volume loss of standard rubber}}{\text{Volume loss of rubber under test}} \times 100\%$$

Abrasion resistance is the reciprocal of volume loss.

If the volume loss or abrasion resistance only is quoted it is desirable to have some certification of the abradant used. This is naturally supplied to some extent by specifying a particular grade and source of supply but leaves open to question the variability of that source of supply. Some workers prefer to use a standard rubber to test the abradant. If the abradant is certified in this way by one source only, the procedure has merit but does not then enable corrections to be made for machine variations and ageing of the abradant. It would not seem to be beyond the bounds of ingenuity to find a standard material which is inherently more reproducible than rubber! This could then be used either to certify the abradant or to use in the calculation of abrasion index.

None of the measures of abrasion resistance given above are fundamental properties of the rubber but relate only to the specific conditions of test. One small refinement that can be introduced if the pressure on the rubber during test is constant is to relate the loss to unit surface area giving what is termed a 'specific wear rate'.

10.2.4 Standard Tests

The international method for abrasion testing ISO 4649[31] is essentially based on DIN 53516[32] and the apparatus is commonly known as the DIN abrader. The principle of the machine is illustrated in Fig. 10.4; a disc test piece in a suitable holder is traversed across a rotating drum covered with a sheet of the abradant. In this way there is a relatively large area of abradant, each part of which is passed over in turn by the test piece so that wear of the abradant is uniform and relatively slow. In the standard method there is no provision for changing conditions from those specified but the abradant and the load on the test piece could be changed. The degree of slip is 100% and it would be inconvenient to test

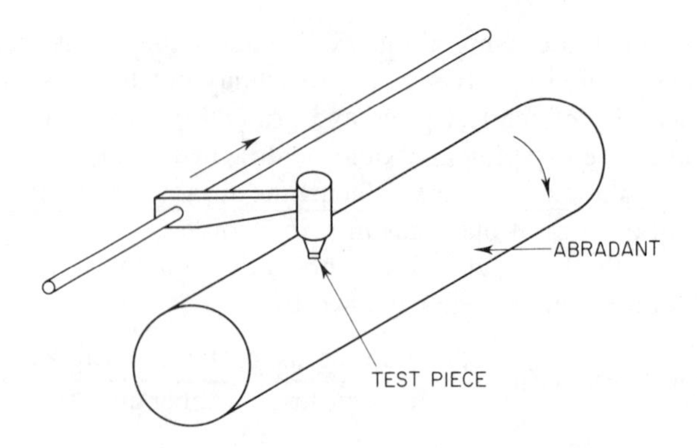

FIG. 10.4. Principle of DIN abrader.

in the presence of a lubricant. Although not versatile, the method is very convenient and rapid and well suited to quality control. The details of procedure and expression of results are something of a compromise, being a compilation of the German approach and the British approach. Results can either be expressed as an abrasion index relative to a standard rubber or as a relative volume loss measured on an abradant standardised with a standard rubber. The standard rubbers in each case are different and both are specified in the standard. This is a fine example of the chicken and egg problem of standard rubber and abradant. To understand the situation it must be appreciated that an accurately specified standard abradant has been available in Germany for many years, but the rubber used to check it has a formulation of no relevance to real products and is very difficult to reproduce in laboratories other than that of its origin.

The current British Standard[33] specifies three types of machine, the Dupont, the Akron and the Dunlop, together with a modification of the Dupont to operate at constant torque, although in fact the constant torque Dupont and the Dunlop have been deleted by amendment. Results are expressed as abrasion resistance index and two standard rubbers, a tyre tread type and a shoe sole type, are specified. Introductory notes to the standard make it clear that different machines are likely to give relatively, as well as absolutely, different values for abrasion loss and states that no close relation between laboratory tests and service are implied, although reference is made to the suitability of all the apparatus for testing tyre treads. The notes state that the tests should not be used in purchasing specifications effectively because of poor interlaboratory reproducibility. For all four methods specified a full test procedure is given including schedules for the times of running-in periods and test runs and for the order of test when comparing several materials.

The Dupont apparatus uses a disc of abrasive paper which rotates whilst a pair of moulded test pieces are continuously pressed against it either with a constant force or with a force adjusted to give a constant torque on the arm holding the test pieces. It is a relatively simple apparatus using an easily replaced abradant but has several disadvantages. The abrasive paper is prone to smearing with soft materials due to heat build-up, few parameters can be varied and the irregular-shaped test piece has to be specially prepared.

The Akron machine is of the form (b) in Fig. 10.3. The test piece is a moulded wheel driven at constant speed and held against the abrasive wheel by a constant force. Its main advantage is that by varying the angle of the test piece relative to the wheel the degree of slip can be varied and hence its effect studied. Any point on the test piece is not continuously in contact with the abradant so that heat build-up is less troublesome than with the DuPont. It is not particularly convenient to change the abradant and the test piece must be specially prepared.

The Dunlop machine also uses a specially prepared test piece in the form of a wheel with a bonded steel centre piece which is driven whilst in contact with the side of an abrasive wheel. The degree of slip is varied by means of an eddy current brake acting on the abrasive wheel. The machine hence has similar advantages and disadvantages to the Akron but is rather more complicated in construction.

A revision of BS 903:Part A9 has been long overdue, largely because of waiting for finalisation of an international procedure. At the time of writing the draft is approaching finalisation and publication can be expected in 1986. It covers four methods, the Dupont, Akron, DIN and Taber abraders. The methods for the Dupont and Akron are revisions of the 1957 edition, but the main essentials of the test are unchanged. The DIN will be identical to ISO 4649.

The 'rotary-platform, double-head' or Taber abrader, unlike those mentioned above, was not developed by the rubber industry but was intended for very general use. It is of the form (c) in Fig. 10.3 but uses a pair of abrasive wheels. Although the degree of slip cannot be varied, the Taber is a very versatile apparatus and uses a simple flat disc as the test piece. In particular, the force on the test piece, and the nature of the abradant are very readily varied and tests can be carried out in the presence of lubricants.

ASTM at present specifies only the Pico abrader[34] for rubber testing although previously a method was given for the Dupont machine. In addition the rotary-platform, double-head abrader is specified for coated fabrics[35] and the NBS abrader for shoe soles and heels.[36]

The Pico abrader uses a pair of tungsten carbide knives which rub the test piece whilst it rotates on a turn-table. The direction of rotation is reversed at intervals throughout a test and a dusting powder is fed to the test piece surface, which doubtless helps to avoid stickiness. The apparatus is calibrated by the use of no less than five standard rubbers and the result also expressed as an abrasion index. Force on the test piece and speed of rotation can be varied and presumably different abradant geometries could be used although the distinctive feature of the Pico is the use of blunt metal knives in the presence of a powder.

The NBS abrader uses rotating drums with abrasive paper wrapped around them onto which the test pieces are pressed by means of levers and weights. There is no provision for traversing the test piece across the abradant as in the DIN machine.

10.2.5 Other Tests

No attempt will be made to give a comprehensive survey of abrasion testers as so many designs have been tried and even some of the standardised types of apparatus are not very commonly used. A review was given by Buist in 1950[37] but many machines have appeared since then. Twenty-one methods are listed in a study of wear of flooring materials[38] and several machines developed in the USSR are described in Section 3 of reference 23. The Lambourne abrader is essentially the Dunlop apparatus given in BS 903, the Conti is used as another name for the DIN apparatus although Buist[37] gives it as a separate machine and the Grasselli is the same as the Dupont.

The Martindale abrader is a four station machine which usually uses cloth as the abradant but coarser and faster acting materials can be substituted. The principal feature of this machine is that the test pieces are rubbed successively in different directions as the motion takes the form of a Lissajous figure.

The Schiefer abrader which is also known in Britain as the WIRA carpet abrader is of the form (d) in Fig. 10.3. Its principal feature is that it produces a constant relative speed between the test piece and abradant at all points on the test piece whilst the direction of relative motion changes steadily around a full circle. It is a versatile machine in that a variety of test piece holders can be fitted and the abradant is readily changed, including the use of serrated metal surfaces.

A number of tests have been developed with particular products in mind. The use of loose abradant to impinge on the rubber[28,29] is particularly suited to surface coatings and a reciprocating test specifically for coatings has also been described.[39] The NBS test[36] is specifically for sole and heel materials and there are a number of other tests used in the footwear industry. A good example of a specialised approach is a machine for 'O' rings[40] where the abradant is a roughened metal disc which is immersed, with the 'O' ring in an abrasive fluid such as drilling mud.

Although it is not really a different method, it should be noted that to obtain correlation with service conditions where wear rate is low, very tiny quantities of material lost have to be measured. These small losses may be difficult to measure by the usual weighing or dimensional methods and radioisotope techniques have been used, as for example in the method of Patel and Deviney.[41]

10.2.6 Comparison of Methods

It is to some extent pointless to compare abrasion testers except in the context of their correlation with a particular product and service condition. If a general comparison is attempted this will inevitably be subjective. However, it can be commented that the DIN abrader is becoming increasingly popular and is probably the most convenient test for routine control use. The Akron is distinctive for its ability to vary slip angle in a simple manner and the Schiefer for giving uniform multi-directional abrasion. Probably the most versatile commonly used apparatus is the Taber because of the very wide range of abradants readily available and its ability to operate with lubricants.

Published accounts of studies of correlation of laboratory abrasion tests with service are not abundant. The study of wear of flooring materials[38] previously mentioned is comprehensive and the same subject has been considered by Gavan.[42] Three abrasion machines were used by Satake *et al.*[43] to study the correlation with tyre wear and an example of tests using the constant power principle versus tyre wear is given by Powell and Gough.[30] Discussion of correlation with tyre wear and of abrasion with other physical properties is contained in Buist's paper.[37] Moakes[44] studied several machines in relation to wear trials on footwear compounds and work at SATRA and CST[45] used PVC and microcellular materials as well as rubber.

Borroff[46] investigated the validity of the NBS test for footwear and Dickerson[47] briefly discusses the conclusions gained from extensive comparative tests on solings and bottoming materials at SATRA. (The results are given in SATRA internal reports.) The reliability of the Taber abrader has been studied by Hill and Nick[48] and its use with polymers discussed by Brown and Crofts.[49] The DIN machine has been compared with the Akron and Taber methods[50] for general use with rubbers. Several papers in an ASTM publication[39] consider the performance of various tests to assess coatings and the use of a metal mesh abradant to assess the wear of conveyor belts is given by Polunin and Gulenko.[51]

REFERENCES

1. James, D.I. (July 1973). *RAPRA Members J.*, 170.
2. James, D.I. (July 1961). *J. Sci. Instrum.*, **38**, 294.
3. Griffin, G.J.L. (1971). *Wear*, **17**, 399.
4. Mustafa, M.R. and Udrea, C. (Dec. 1969). *Plast. Mod. Elast.*, **21**(10), 114.
5. Bailey, M.W. and Cameron, A. (Aug. 1971). *Wear*, **21**(1), 43.
6. West, G.H. and Senior, J.M. (Jan. 1972). *Wear*, **19**(1), 37.
7. Jost, H. (Mar. 1969). *Plaste u. Kaut.*, **16**(3), 175.
8. Wilson, A. and Mahoney, P. (Mar. 1972). *Rubb. World*, **165**(6), 36.
9. Robinson, W.H. and Kopf, R.E. (July 1969). *Mater. Res. Stand.*, **9**(7), 22.
10. BS 4618, Section 5.6, 1975. Guide to Sliding Friction.
11. BS 3424, Method 12, 1973. Determination of Surface Drag.
12. BS 2782, Method 824A, 1984. Determination of Coefficients of Friction of Plastics Film.
13. James, D.I. and Newell, W.G. (1980). *Polym. Test.* **1**(1).
14. James, D.I. and Mohsen, R. (1982). *Polym. Test.*, **3**(2)
15. Giles, C.G., Sabey, B.E. and Cardew, K.H.F., Road Research Technical Paper No. 66, HMSO, 1964.
16. Malkin, F. and Harrison, R. (1980). *J. Physics D*, **13**, L77-9.
17. James, D.I. (1980). *Rubb. Chem. Tech.*, **153**(3).
18. Harrison, R. and Malkin, F. (1983). *Ergonomics*, **26**(1).
19. Perkins, P. and Wilson, M. (1984). *SATRA Bull.*, **21**(3).
20. James, D.I. (1983). *Ergonomics*, **26**(1).
21. James, D.I., RAPRA Members Report No. 94, 1984.
22. Lancaster, J.K. (Dec. 1973). *Plastics and Polymers.*
23. James, D.I. (Ed.) (1967). *Abrasion of Rubber*, Maclaren and Sons.
24. Zhang, S.W. (1984). *Rubb. Chem. Tech.*, **57**(4).
25. Rougier, R., Barquins Mond Courtel, R. (1977). *Wear*, **43**(1).
26. Gent, A.N. and Pulford, C.T.R. (1983). *J. Appl. Polym. Sci.*, **28**(3).
27. Lancaster, J.K. (1969). *Wear*, **14**, 223.
28. Tilly, G.P. (1969). *Wear*, **14**, 241.
29. Egorov, B.N., Poganov, E.V. and Lyashevich, V.V. (1971). *Plast. Massy*, **10**, 65.
30. Powell, E.F. and Gough, S.W., Proc. 3rd Rubb. Tech. Conf., Cambridge, 1954, p.460.

31. ISO 4649, 1985. Determination of Abrasion Resistance Using a Rotating Cylindrical Drum Device.
32. DIN 53516, 1977. Bestimmung des Abriebs.
33. BS 903:Part A9, 1957. Determination of Abrasion Resistance.
34. ASTM D2228-83. Abrasion Resistance (Pico Abrader).
35. ASTM D3389-75. Coated Fabrics—Abrasion Resistance (Rotary Platform, Double Head Abrader).
36. ASTM D1630-83. Abrasion Resistance (NBS Abrader).
37. Buist, J.M. (1950). *Trans. IRI*, **26**, 192.
38. Frick, O.F.V. (1969). *Wear*, **14**(2), 119.
39. Lancaster, J.K. (1981). ASTM Special Publication 769.
40. Burr, B.H. and Marshek, K.M. (1981) *Wear*, **68**(1).
41. Patel, A.C. and Deviney, M.L., ACS Precision treadware measurement at increased mileages by an improved radio-iodine method. Paper presented at Rubber Division Meeting, Detroit, May 1973.
42. Gavan, F.M. (July 1970). *Mater. Res. Stand.*, **10**(7), 24.
43. Satake, K., Sone, T., Hamada, M. and Hayakawa, K. (Oct. 1970). *J. IRI*, **4**(5), 217.
44. Moakes, R.C.W. (Aug. 1971). *J. IRI*, **5**(4), 155.
45. Maddams, J.S. (Dec. 1971). *SATRA Bulletin*.
46. Borroff, J.D. (1977). *Elastomerics*, **109**(12).
47. Dickerson, K. (May 1984). *SATRA Bulletin*.
48. Hill, H.E. and Nick, D.P. (Mar. 1966). *J. Paint Tech.*, **31**, 494, 123.
49. Brown, R.P. and Crofts, D. (Aug. 1971). *RAPRA Bulletin*.
50. Brown, R.P., RAPRA Members Report No. 17, 1978.
51. Polunin, V.T. and Gulenko, G.N. (1972). *Kauch. i. Rezina*, **31**(7), 37.

Chapter 11

Creep, Relaxation and Set

Creep, stress relaxation and set are all methods of investigating the long-term effects of an applied stress or strain. Creep is the measurement of the increase of strain with time under constant force; stress relaxation is the measurement of change of stress with time under constant strain and set is the measurement of recovery after the removal of an applied stress or strain. It is important to appreciate that there are two distinct causes for the phenomena of creep, relaxation and set, the first physical and the second chemical. The physical effect is due to rubbers being viscoelastic, as discussed in Chapter 9, and the response to a stress or strain is not instantaneous but develops with time. The chemical effect is due to 'ageing' of the rubber by oxidative chain scission, further crosslinking or whatever.

In practice it is often rather difficult to distinguish between the two causes and it could be argued that if tests are made under the same conditions as in service it matters relatively little what caused, for example, the creep as long as it can be measured. If any form of accelerated conditions, such as increased temperatures, are used it follows that results could be very misleading. Generally, physical effects are dominant at short times and low temperatures and chemical effects more apparent at longer times and higher temperatures.

Apart from being simply measures of how much a rubber creeps, relaxes or sets under any given conditions, these tests can also be used as measures of ageing characteristics, low temperature resistance or resistance to chemicals. These other applications of the tests are not generally considered in this chapter but it is impossible to make a complete distinction. In particular, set tests commonly used as a quality control tool involve heat ageing, and stress relaxation tests to measure the efficiency of rubbers as seals often involve both heat ageing and exposure to liquids.

11.1 CREEP

A creep test is in essence very simple—a constant force is applied to the rubber and the change in deformation with time monitored—but detailed procedures have rarely been standardised.

At the time of writing there is still no international method, although a draft has reached a fairly advanced stage, and there is no general ASTM procedure. This reflects the relatively small amount of creep testing carried out on rubbers compared to work on thermoplastics, although for particular applications of rubber where creep is important, for example bridge bearings, a considerable amount of data has been generated.

In 1958 there was a British standard which covered both creep and stress relaxation and procedures were given for tests in tension, compression and shear. The subjects have now been separated and the standard for creep, BS 903:Part A15,[1] has methods for shear and compression but not tension, on the basis that engineering components are not generally stressed in this manner.

There has been some controversy over the definition of creep which should be used. Traditionally in the rubber industry creep is defined as the increase in deformation after a specified time interval expressed as a percentage of the test piece deformation at the start of that time interval. In other industries creep is normally defined as the increase in deformation expressed as a percentage of the original unstressed thickness of the test piece. Consequently, care has to be taken when comparing creep values obtained from different sources. If, as is frequently the case with rubbers, the deformation change is linear with log(time) it is convenient to express results as a creep rate, which is the ratio of the creep to the logarithm of the time interval and is quoted as percent per decade.

In the British standard the test pieces for measurements in compression are discs either 13 mm in diameter and 67.3 mm thick or 29 mm in diameter and 12.5 mm thick bonded to end plates, i.e. the measurements are made with no slippage at the compressed surfaces (see Section 8.3). For measurements in shear a double sandwich test piece is used as discussed in Section 8.4, typical dimensions being 25 mm diameter and 5 mm thick.

The principal requirements for the apparatus in compression tests are that one compression plate is fixed and the other is free to move without friction. The force must be applied smoothly and without overshoot and the mechanism must be such that the line of action of the applied force remains coincident with the axis of the test piece as it creeps. The compression of the test piece should be measured to $\pm 0.1\%$ of the test piece thickness. Apparatus for measurements in shear is essentially the same as for compression except for the differences in geometry of the test piece and its mounting. An example of a fairly sophisticated creep apparatus for rubbers which can operate in compression or shear is briefly described by Hall and Wright[2] and illustrated in Fig. 11.1. Although adequate creep data can be obtained with relatively simple apparatus care must be taken in the design to minimise friction, to ensure smooth application of the load which then always acts coincident with the axis of the test piece and to avoid drift in the strain measuring device over long time periods.

The British Standard specifies that a force shall be applied such that a strain of $20 \pm 2\%$ is realised after the time period before the first measurement is made and this first time period is recommended to be 1 min with further measurements after 10, 100, 1000 min, etc. Obviously, the strain and time scale could be adjusted to suit individual applications.

11.2 STRESS RELAXATION

Stress relaxation measurements can be made in compression, shear or tension but in practice a distinction is made as regards the reason for making the test. The most important type of product in which stress relaxation is a critical parameter is a seal or gasket. These usually operate in compression and hence stress relaxation measurements in compression are used to measure sealing efficiency. The measurement of decay of sealing force with time is a most important design consideration and, although this has been long appreciated, the widespread use of such tests has been inhibited by the inherent instrumental difficulties in holding a constant deflection and at the same time monitoring the force exerted by the test piece. Relatively recently considerable advances have been made in apparatus and the standardisation of test procedures.

Stress relaxation measurements can also be used as a general guide to ageing, and it is particularly relaxation due to chemical effects which is then studied. Such measurements are normally made in tension and will be considered in Chapter 15 as an ageing test. Hence, in this section,

FIG. 11.1 RAPRA creep apparatus for rubber.

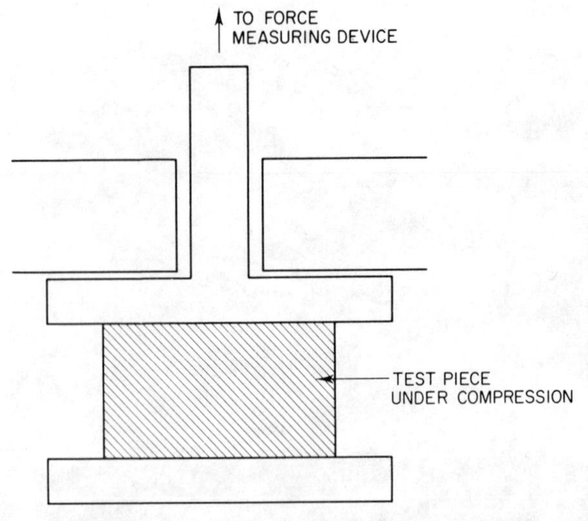

FIG. 11.2. Principle of stress relaxation in compression jig. The stress exerted by the test piece is measured when the top platen and the body of the jig are just separated by application of a small additional compression.

only relaxation tests in compression will be discussed as this mode of deformation is the only one commonly used and standardised to directly estimate the relaxation of rubbers in service. For an application in tension the methods described in Chapter 15 could of course be adapted. It must be appreciated that the methods in compression do not only measure relaxation due to physical effects, especially when elevated temperatures and liquid environments are used.

11.2.1 Standard Methods in Compression

ISO has standardised two methods for relaxation in compression, a general method using disc test pieces and a method using ring test pieces mainly for tests in liquids. The use of rings reflects the importance of 'O' ring seals and allows the maximum surface area of test piece to be exposed to a test liquid.

The ISO methods together with British and ASTM equivalents are all basically similar. The test piece is compressed between platens to a constant strain, and the force exerted by the test piece is measured at intervals by applying a very small additional strain. A common way of doing this is shown in Fig. 11.2 where a small additional compression results in the top platen being just separated from the body of the jig and the force is transmitted via the central rod to a force measuring device. Alternatively, the force measuring device can be continuously balancing the force exerted by the test piece.

In ISO 3384,[3] which is the general method, either continuous measurement or intermittent measurement by addition of a small strain or force is allowed, the specified accuracy being $\pm 1\%$. If a small extra strain is added this must be less than 0.05 mm. The test piece is a disc either 13 mm in diameter and 6.3 mm thick or 29 mm in diameter and 12.5 mm thick, the same as specified for creep and also for compression set. Unlike the usual creep methods, the test pieces are not bonded to end pieces but compressed with lubrication of the ends with a fluorosilicone fluid.

ISO 3384 gives two test procedures but in both cases the applied compression is either 25% or 15%. In procedure A, the test piece is compressed at the test temperature and all force measurements are made at that temperature, whilst in procedure B compression and force measurements are made at 23°C, the test piece being subjected for intervals to the test temperature. With procedure A, there may be difficulty with some designs of apparatus in loading hot and with procedure B, an apparatus with a large thermal capacity may take a long time to cool. In both methods the initial force measurement F_0 is made 30 min after compressing the test piece and the standard length of test is 168 h.

The force measurements are normalised to the initial force measurement and expressed as a percentage:

$$R_t = \frac{F_0 - F_t}{F_0} \times 100$$

where R_t is the stress relaxation after time t.

With this method of expressing results the force units used are of no importance and the result is not in general critically dependent on the degree of applied strain. The standard notes that it is also common practice to plot the ratios F_t/F_0 as a function of time on a logarithmic scale.

The British standard BS 903:Part A42[4] is based on ISO 3384 and the differences are relatively minor. The ASTM method for stress relaxation in compression is D1390[5] which specifies a compression jig with intermittent measurement of force by application of an additional strain and uses only the larger sized disc with the moulded faces buffed off of the test piece.

Details of timing differ from the ISO method and only the procedure making measurements at ambient temperature is given. In contrast to the increasing interest in stress relaxation measurements in Europe, apparently the ASTM method is little used and has been threatened with withdrawal.

ISO had intended to produce a method for use at low temperatures but in fact the procedure is not always satisfactory under these conditions because as the rubber stiffens the additional overcompression can generate a relatively very large force.

Although stress relaxation measurements in liquids are essentially the same as those in air it is normal for the former to use 'O' or square section ring test pieces, so simulating the geometry of practical seals and giving a relatively large surface area to volume ratio so that equilibrium swelling is reached reasonably quickly. The rings can be tested in air if required.

The ISO method for rings is ISO 6056.[6] The rings specified are either 2 mm square section and 15 mm internal diameter or 'O' rings of 2.65 mm cord diameter and 14 mm internal diameter. The apparatus is essentially similar to that specified in ISO 3384 with, additionally, provision for circulation of fluid within the volume enclosed by the ring. Three different procedures are given. Procedures A and B are equivalent to A and B of ISO 3384, measurements being made at the test temperature and ambient temperature respectively. In procedure C the compression is applied at room temperature but the measurements of counterforce are made at the test temperature. Practical applications arise in which any of the three testing procedures is the most relevant and the procedure should be chosen to suit the application. However, it will be found that some designs of apparatus make certain of the procedures very difficult.

The British standard, BS 903:Part A34[7] is essentially the same as ISO 6056 and will probably be totally aligned eventually. The 'O' ring preferred is of 2.4 mm cross section and internal diameter 14.6 mm, but this difference should not be very significant. The force measurement is required to be accurate to 1% rather than 2% and there are many editorial differences. BS 903:Part A34 does allow the use of sub-ambient temperatures but, as pointed out earlier, the method may not be satisfactory under these conditions.

It is clear that at future revisions there is scope to align details of ISO 6056 and ISO 3384 as well as to eliminate differences in the British methods. There is still relatively little experience with these standard methods and refinement is doubtless desirable. Birley *et al.*[8] have recently studied a number of factors and make recommendations for change.

11.2.2 Apparatus

It is perfectly feasible to have a force measuring element permanently attached to each test piece holding jig but although this eliminates certain instrumental problems is likely to be expensive and not often done unless springs are used as the force measuring element.[9] It is more usual to have individual jigs which are placed under a single force measuring head in turn.

Although the principle of a jig such as that illustrated schematically in Fig. 11.2 is fairly simple in practice, there are many difficulties. The two essential problems are to provide an efficient and reproducible detection system for the point at which the extra compression is applied and to prevent the platens tilting whilst not introducing appreciable friction.

The force measuring head, together with provision for applying the extra compression can be a beam balance, as used in the well-known Lucas apparatus, a universal tensile machine or a specially designed electronic load cell unit.[10,11] The point at which the small amount of extra compression has been applied can be detected by breaking an electrical circuit. An early apparatus used a load cell attached to an arbor press. The operator manually lowered the press until the break in the electrical circuit was indicated visually by the extinction of a light. The Lucas apparatus has a similar detection system, the balance weights are adjusted manually until the force exerted by the beam just overcomes the force exerted by the test piece. Both of these approaches involve a somewhat delicate operation. The use of a tensile machine affords some reduction in experimental difficulty but a very slow speed must be used to avoid overshoot as the increase in compression is very small, less than 0.05 mm in ISO 3384.

An apparatus developed at RAPRA[12] has been designed to reduce as far as possible operator dependence. A load cell is driven onto the jig by a pneumatic ram; at the moment when a very small additional compression on the test piece is detected electrically the ram is automatically stopped, the force reading digitally recorded and the ram reversed (Fig. 11.3). This measuring head can be used with a variety of jig designs including Lucas and ASTM.

The jigs illustrated in ASTM D1390 utilise ball bushings for the load applicator to slide in and the electrical contact is made through a circular plate. In principle this arrangement gives good lateral stability but introduces friction which may be excessive and the large electrical contact area is not conducive to a clean break.

The Lucas jigs use a ball contact to make and break the electrical

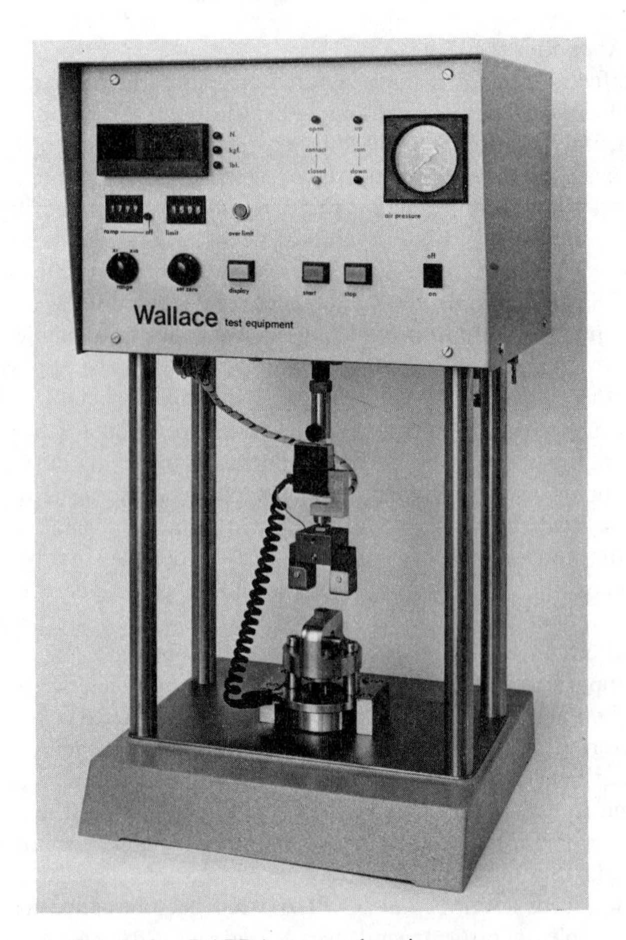

FIG. 11.3. RAPRA stress relaxation apparatus.

circuit but as they have no lateral constraint, are prone to tilting and can hence only be used with rings having a large diameter to height ratio.

The RAPRA jigs use a ball electrical contact but provide lateral support through circular leaf springs with high lateral but low vertical stiffness which eliminates friction. A small correction is made to the measured force for the vertical spring stiffness. These jigs utilise a ceramic insulator element and corrosion resistant steel which has allowed several years stable use in water, and are of relatively massive construction to accommodate stiff button test pieces.

An alternative approach to detecting slight overcompression with an electrical circuit is to arrange for a preset amount of overcompression. Simple jigs of this type relying on a mechanical stop will probably produce excessive overcompression and be somewhat variable in use.

Fernando *et al.*[13] have introduced the concept of an ideal loading curve which is the condition when a completely smooth and instantaneous uptake of load occurs on compression. They show that this does not occur with a fixed overcompression jig and hence errors are likely. They do not present data for electrical contact jigs saying that although they are ideal in theory they have other disadvantages. The one disadvantage the best ones certainly have is complexity and cost and probably this was an important reason why they developed a non-electrical contact jig to have as near a perfect loading curve as possible. Their jig is constructed with particular attention to parallelism of the important surfaces to give a smooth and stable uptake of load which is deduced from a recording of the stress/strain curve.

With any of the designs of apparatus considerable care must be taken to ensure accurate alignment of components and to standardise procedures if reproducible results are to be obtained. James and Peppiatt[14] have made a detailed study of the Lucas jig and suggest improved test procedures.

If, as is usually the case, it is required to make measurements at non-ambient temperatures, the detailed mechanical design and thermal capacity are important.

The Lucas apparatus has lightweight jigs, and an 'oven' built into the measuring head. It is hence fairly convenient for procedures B and C of the British method for rings but is very difficult to load hot and therefore not convenient for procedure A. (This apparatus is not suitable for ISO D3384 using disc test pieces.) The ASTM and the RAPRA designs of apparatus have relatively high thermal capacity and are not convenient for procedure C unless a longer heating time than standard is used.

The difficulties and expense of compression stress relaxometers led Bassi and Zerbini[15] to consider an alternative approach which they term pressure relaxation. Compressed air is applied through a small hole in the centre of the one platen and the pressure when air leaks past the test piece recorded. This pressure being nominally that to equalise the pressure exerted by the compressed rubber is related to the compression force.

11.2.3 Use of Stress Relaxation Data

The standard methods for stress relaxation are generally restricted to periods of about a week, whereas in practice the performance of seals is required to be known over periods of years. In a large-scale study conducted at RAPRA on behalf of industry (but not as yet published) the stress relaxation of a range of rubbers for pipe seals was measured in both air and water for several years. A generalised conclusion was that in most cases a simple extrapolation of short-term data was not valid, particularly in liquids. Meier *et al.*[16] made measurements for up to 17 years in the dry and obtained some success in correlating results from short-term tests by the method of reduced variables, but noted that at longer times the dominance of chemical relaxation could cause errors.

Because compression set measurements have been used traditionally as an indication of sealing performance and because they are relatively simple tests to perform, it is of interest to know what correlation exists between compression set and compression stress relaxation. Ebbul and Southern[17] have investigated the correlation for a number of compounds and conclude that although a reasonably good general correlation existed it was not good enough to predict one property from the other. The RAPRA work mentioned above also included compression set measurements and did not find good correlation with stress relaxation. In theory set can be related to the difference between continuous and intermittent stress relaxation but this is generally of little use in the case of seals when only continuous relaxation is of interest. However, Sprey[18] has demonstrated that predictions of compression set can be made from intermittent and continuous tension/relaxation measurements.

11.3 SET

The rubber industry has traditionally paid more attention to measuring the recovery after removal of an applied stress or strain, i.e. set, than creep or stress relaxation. This is partly because relatively simple apparatus is required but also because it appears at first sight that set is the important parameter when judging sealing efficiency. Set correlates with relaxation only generally and it is actually the force exerted by a seal that usually matters, rather than the amount it would recover if released.

Set tests are made in either tension or compression and for their prime use, quality control, the choice of mode can be made according to the convenience of the test piece available. If intended to simulate service conditions, e.g. indentation of flooring, the most relevant mode of deformation would be used. Tests can be carried out in which the test piece is subjected to either constant stress or constant strain but as the

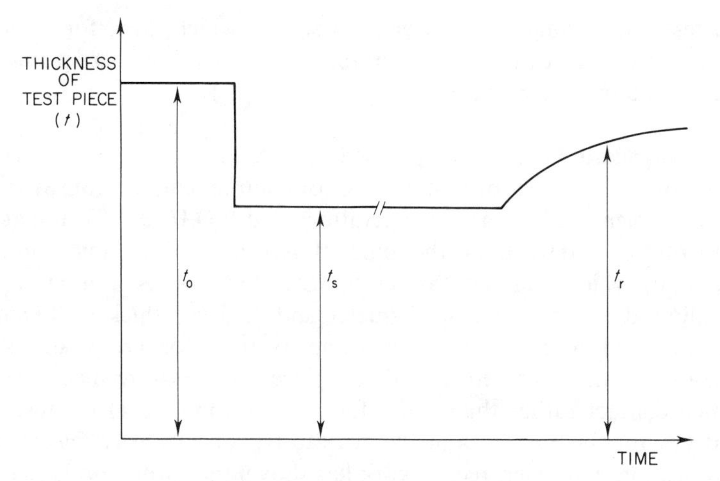

FIG. 11.4. Illustration of compression set: t_0 = initial thickness; t_s = compressed thickness; t_r = recovered thickness.

latter is by far the more widely used the illustration of set measurement given in Fig. 11.4 is based on constant strain in the compression mode.

Referring to Fig. 11.4, the test piece is more or less instantly compressed and held at that compression for a fixed length of time. The test piece is released and its recovered height measured. It is common practice to measure the recovered height 30 min after release of the test piece but this is an abitrary time. The term permanent set is sometimes used but if this has any meaning it would be referring to the set remaining after an infinite recovery time. Set is normally expressed as a percentage of the applied deformation, i.e.

$$\text{set} = \frac{t_0 - t_r}{t_0 - t_s} \times 100\%$$

but can be expressed as a percentage of the original thickness.

The measurement of set is a very effective quality control test as it is a relatively simple test and the results are sensitive to state of cure. Because of the widespread use of set measurements as a guide to seal performance it is worth pointing out that the usual short-term set measurements do not necessarily correlate well with long-term performance.[19] If set is to be used as a measure of performance it is necessary to largely disregard the arbitrary conditions specified in standard methods and test under conditions relevant to service which may, for example, involve recovery at the test temperature and measuring set as a function of time and/or temperature.

11.3.1 Standard Tests in Compression

There are two international standards for compression set, ISO 815[20] for use at ambient and elevated temperatures and ISO 1653[21] for use at low temperatures although at the time of writing a draft revision is in circulation which combines the two. In ISO 815 two sizes of disc test piece are allowed either 29 mm in diameter and 12.5 mm thick or 13 mm in diameter and 6.3 mm thick, the same as used for creep and stress relaxation. The measurements of test piece thickness are made using a domed contact rather than a flat foot (see Section 7.2.1) because of a tendency for the end faces of the discs to be concave after release from compression. However, recent work has shown that ordinary flat feet are perfectly satisfactory especially with lubricated test pieces and the draft revision makes this change.

Only a constant strain method is specified with a standard strain of 25% and the compression is made between very smooth platens which may, optionally, be lubricated. Hence, the compression is made either with some attempt at perfect slippage or in no-man's land between this and bonded end pieces. Fairly obviously, the degree of slip and the test piece shape factor can affect the measured values of set. At one time it was standard to use glass-paper between the test piece and the platens to prevent slip but this produces greater concavity of the ends after release. The draft revision has gone to the opposite extreme and specifies lubrication only.

The test piece is compressed at 23°C and then held at the test temperature, commonly 70°C or 100°C, for a set time, commonly 24 h, and then the test piece immediately removed and allowed to recover at 23°C for 30 min. This is not a particularly logical sequence as it would seem more reasonable to either allow recovery at the test temperature or to cool in the compressed condition, rather than allow recovery at an undefined cooling rate which must differ between the two sizes of test piece. Recovery is speeded up at an elevated temperature and slowed down if cooling takes place under compression.

ISO 1653 is essentially the same as ISO 815. The apparatus is required to be fitted with a quick release device to relieve the compression and a low temperature cabinet is specified with the near impossible requirement of control to ±0.5°C. The test piece is compressed at 23°C and, after the required exposure time at a low temperature, released and a series of measurements of recovery made at that temperature over the period from 10 s to 2 h. The standard suggests that a plot of test piece thickness against log(recovery time) may produce a straight line and recommends the reporting of set after 10 s and 1800 s. This more sensible treatment of recovery compared to ISO 815 is essential because set at low temperatures is totally due to physical mechanisms and the degree of set will be critically dependent on the recovery temperature.

The British standard for compression set, BS 903:Part A6,[22] is very similar to ISO 815. Originally it also contained a procedure for compression set under constant stress but this was deleted in 1974 because of lack of use. Recently there have been suggestions that it should be reinstated because of its resemblance to the service conditions of certain products such as flooring and footwear. The truth one suspects is that there is no need for this test for quality control purposes but a constant stress method with conditions chosen to simulate a particular application may be useful. The standard method used a calibrated spring to apply the compression force which ranged from 1100 N to 7300 N depending on test piece and test temperature and set was expressed as a percentage of original thickness. BS 903:Part A39[23] is identical to ISO 1653.

There are no less than four methods for compression set of solid rubber listed in ASTM. D945 refers to set measured when dynamically testing with the Yerzley oscillograph (see Chapter 9), D1424 is a collection of test methods for 'O' rings, the normal standard D395[24] and D1229[25] for use at low temperatures.

D395 specifies a constant strain and a constant stress method for use in air. The constant strain procedure is similar to the ISO and British methods, using the same test pieces, recovery procedure and expression of results. A second constant strain method describes a jig specifically for use in liquids, but also specifies recovery at the test temperature. The constant stress method uses a calibrated spring to apply a force of 1.8 kN on the larger compression set disc and expresses the result as a percentage of the original thickness. D1229 is a constant strain method, essentially the same as D395, with the measurement of recovery being made, as in the ISO equivalent, at the test temperature.

11.3.2 Standard Tests in Tension

The international method for tension set, ISO 2285,[26] uses a constant strain procedure with either strip, dumb-bell or ring test pieces. The actual length of a strip test piece is not specified but the preferred reference length is 50 mm and this is also the preferred distance between the square ends of the dumb-bell. The advantage of the square-ended dumb-bell is that it can be simply clipped into a slotted bar and does not need grips as such. Similarly, rings are relatively easily fitted over pulleys. It is doubtful whether the difference in size between the various test pieces would significantly affect the results as long as the necessary precision in measurement was maintained.

The test apparatus is simply a rod or other suitable guide fitted with a pair of grips or pulleys, one of which is movable, and a measuring device accurate to 0.1 mm. It should be noted that, although simple, the straining and measuring devices need to be carefully constructed as the tolerances on measurement are quite small. More precision is required than in earlier standards when a dumb-bell with a reference length of 100 mm was used. It is usual to measure the reference length of ring test pieces along a straightened portion of the ring, in which case a rigid channel is required to straighten the test piece.

A choice of strains is given with 100% the preferred value, the most usual test time is 24 h and the test temperature is commonly 70°C or 100°C. The test piece is strained at between 2 and 10 mm/s and the reference length measured between 10 and 20 min later. If it does not fall within given tolerances about the nominal strain, the test piece is rejected. Presumably, if appreciable relaxation occurs in this period, it may be necessary to overstrain initially with test pieces clamped well outside of the reference length. For tests at elevated temperatures, the test piece is placed in the oven between 20 and 30 min after applying the strain. At the end of the test, the test piece is allowed to cool in the strained condition (note the difference from compression set), released and the measurement of final length made after a further 30 min at room temperature. Set is expressed as a percentage of the applied extension. It is apparent that this standard procedure is a little more involved than the corresponding procedure for compression set and also that the procedures differ in the details of timing and the conditions under which recovery takes place. This will affect, as will the large difference in the bulk of the test pieces, the correlation between the two types of set measurement, all of which illustrates the arbitrary nature of these standard methods.

The equivalent British standard, BS 903:Part A5[27] is virtually identical to ISO 2285 except that ring test pieces are not included. Apart from a procedure in the standard D1414 for 'O' rings, the only ASTM method for tension set is that given in ASTM D412[28] for tensile properties. Two very simple procedures are given: either a test piece is strained, held for

10 min, released and the reference length measured after a further 10 min, all at room temperature; or the set at break is found by fitting the broken pieces back together and measuring the reference length. ASTM does not give tension set the same status or attention as compression set.

REFERENCES

1. BS 903:Part A15, 1982. Determination of Creep.
2. Hall, M.M. and Wright, D.C. (Sept./Oct. 1976). *RAPRA Members J.*, No. 5.
3. ISO 3384, 1979. Determination of Stress Relaxation in Compression at Normal and High Temperatures.
4. BS 903:Part A42, 1983. Determination of Stress Relaxation.
5. ASTM D1390–76. Stress Relaxation in Compression.
6. ISO 6056, 1980. Determination of Compression Stress Relaxation (Rings).
7. BS 903:Part A34, 1978. Determination of Stress Relaxation of Rings in Compression.
8. Birley, A.W., Fernando, K.P. and Takir, M. (1986). *Polym. Test.*, **6**(2).
9. Stenberg, B. and Jansson, J.F. (Dec. 1973). *Rubb. Chem. Tech.*, **46**, 1316.
10. Aston, M.W., Fletcher, W. and Morrell, S.H, 4th Int. Conf. on Fluid Sealing, Philadelphia, 1969.
11. Solid Rubber Seals and Joints, RAPRA Technical Review, No. 53, Nov. 1970.
12. Brown, R.P. and Bennett, F.N.B. (1981). *Polym. Test.*, **2**(2).
13. Fernando, K.P., Birley, A.W., Hepburn, C. and Wright, N. (1983). *Polym. Test.*, **3**(3).
14. James, D.I. and Peppiatt, A. (1981). *Polym. Test.*, **2**(4).
15. Bassi, A.C. and Zerbini , V. (1986). *Polym. Test.*, **6**(1).
16. Meier, U., Kuster, J. and Mandell, J.F. (1984). *Rubb. Chem. Tech.*, **57**(2).
17. Ebbul, M.D. and Southern, F. (1985). *Plast. Rubber Process. and Appln.*, **5**(1).
18. Sprey, R. (1984). *Rubb. World*, **191**(1)
19. Moakes, R.C. (Sept./Oct. 1975). *RAPRA Members J.*, 77.
20. ISO 815, 1972. Determination of Compression Set under Constant Deflection at Normal and High Temperatures.
21. ISO 1653, 1975. Determination of Compression Set under Constant Deflection at Low Temperatures.
22. BS 903:Part A6, 1969. Determination of Compression Set under Constant Strain.
23. BS 903:Part A39, 1980. Determination of Compression Set under Constant Deflection at Low Temperatures.
24. ASTM D395–82. Rubber Property—Compression Set.
25. ASTM D1229–79. Rubber Property—Compression Set at Low Temperatures.
26. ISO 2285, 1981. Determination of Tension Set at Normal and High Temperatures.
27. BS 903:Part A5, 1974. Determination of Tension Set.
28. ASTM D412–83. Rubber Properties in Tension.

Chapter 12

Fatigue

Fatigue could be defined as any change in the properties of a material caused by prolonged action of stress or strain, but this general definition would then include creep and stress relaxation. Here, fatigue will be taken to cover only changes resulting from repeated cyclic deformation which means, in effect, long-term dynamic testing.

Subjecting a rubber to repeated deformation cycles results in a change in stiffness and a loss of mechanical strength. In some products, even a relatively small change in stiffness can be important, but this measure of fatigue is relatively little used, certainly not in standard tests. It would be relatively straightforward, although perhaps expensive in machine time, to continue a dynamic test as discussed in Chapter 9 over a very long period and monitor the change in modulus. Alternatively, modulus could be measured at intervals after dynamic cycling on a separate apparatus. In many products, notably tyres, it is the loss in strength shown by cracking and/or complete rupture which is considered to be the important aspect of fatigue and this is the measure of fatigue which is normally used in laboratory tests on rubbers.

The manner of breakdown will vary according to the geometry of the component, the type of stressing and the environmental conditions. The mechanisms which may contribute to the breakdown include thermal degradation, oxidation and attack by ozone as well as the basic propagation of cracks by tearing. In rubber testing it is normal to distinguish between two types of fatigue test; tests in which the aim is to induce and/or propagate cracks without subjecting the test piece to large increases in temperature, and tests in which the prime aim is to cause heating of the specimen by the stressing process. The former type is generally referred to as flex-cracking or cut-growth tests and the latter as heat build-up. This division leaves out specialised tests for particular products which may have characteristics of both types. For example, tyres fail by fatigue in which heat build-up is important and also suffer from groove and sidewall cracking.

12.1 FLEX-CRACKING AND CUT-GROWTH TESTS

The vast majority of flex-cracking tests strain the test piece in flexure, representing the mode of deformation experienced in service by such important products as tyres, belting and footwear. Unfortunately, despite the obvious logic of this approach, there are disadvantages. The principal problem is that it is difficult to control the degree of bending, which may, for example, vary with the modulus of the rubber and, because the fatigue life of rubber will be sensitive to the magnitude of the applied strain, misleading results may be obtained. The more nearly the deformation produced in the laboratory test reproduces that experienced in service the better should be the hope of correlation. It is hardly necessary to add that most products are subjected to a most complex pattern of straining. The alternative approach is to use a simple but reproducible mode of deformation such as pure tension.

A variety of flex tests have been used, many intended for particular products such as belting, footwear and coated fabrics, but a number of them, although once well known, are not now in such common use. A useful review was given by Buist and Williams in 1951.[1]

Four types of machine in which bending is produced in different ways are shown schematically in Fig. 12.1. The most widely known and standardised apparatus, the De Mattia, has the action shown in Fig. 12.1(a). The test piece, which is a strip with a transverse groove, is fixed in two clamps which move towards each other to bend the strip into a loop. The maximum surface strain at the critical point X is somewhat indeterminate. The 'flipper' or Torrens machine (Fig. 12.1(b)) is now little used. The strip test piece, fixed in a slot in the periphery of a rotating wheel is bent against a fixed, but freely rotatable, roller. Again, the radius of curvature and hence the maximum surface strain is not precisely controlled. In the Dupont machine (Fig. 12.1(c)) the test pieces are connected together to form an endless belt and run over a series of pulleys of specified diameters. Although the overall radius of bending is controlled, the surface strain in the test piece is complicated by it having several transverse 'V' grooves. The Ross machine (Fig. 12.1(d)) bends the test

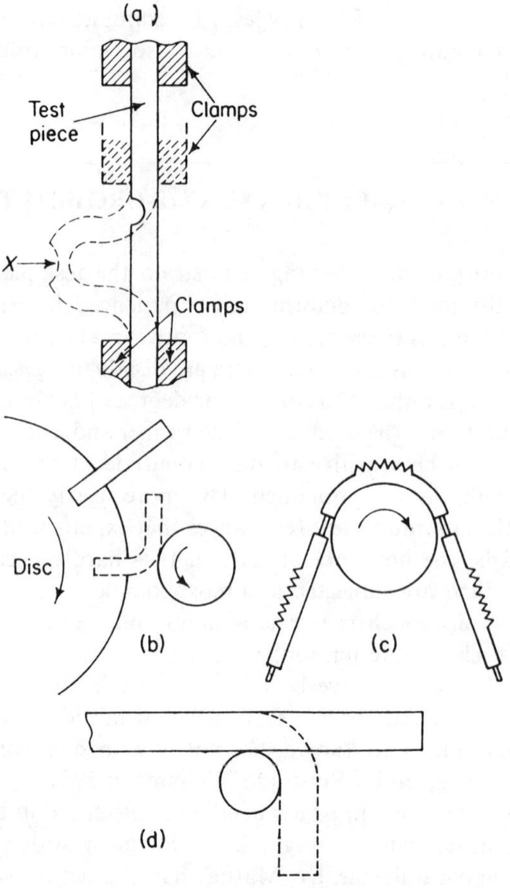

FIG. 12.1. Flexing tests. (a) De Mattia machine; (b) 'flipper' (Torrens) machine; (c) Dupont machine; (d) Ross machine. In (a), (b) and (d) the flexed form is shown by broken lines.

piece through 90° over a rod and the maximum surface strain is rather more controlled than in the other machines described.

Even more important than the control of maximum strain in a flexing cycle is the control of minimum strain because cracking is particularly severe if this is zero. In all the methods described above the strain is deliberately intended to be zero but only in the Ross apparatus is this achieved precisely and in a reproducible manner. In all bending methods the maximum strain depends on the thickness of the test piece and hence this must be closely controlled.

12.1.1 Standard Methods

There are two international standards for flex testing, ISO 132[2] and ISO 133[3] which illustrate the essential distinction between crack initiation and cut growth. The former standard is for crack initiation and the latter for cut growth and the different meaning of the two tests is illustrated by the fact that natural rubber fairly quickly develops fine cracks in a flex-cracking test but is relatively resistant to the further growth of these cracks or of a purposely made cut, whereas SBR shows just the opposite behaviour. Both standards use the De Mattia apparatus and the same test conditions, the essential difference being that in ISO 133 a cut is made through the bottom of the groove in the test piece before flexing is started.

In flex-cracking tests, one of the most difficult problems is how to assess the degree of cracking. Visual examination is the only really feasible procedure and inevitably the assessment is subjective and operator dependent. Unfortunately, the pattern of cracking in a De Mattia test varies with the type of rubber and is likely to start at the edges of the test piece, although this can be virtually eliminated by radiusing the edges. Alternative grading systems were discussed by Boss and Greensmith[4] and the 'modified code' they suggested is now essentially the procedure specified in ISO 132. It is based primarily on the length of the largest crack present at any stage and the depth of the crack is ignored. Any more complicated process involving the measurement of length, depth and number of cracks is generally unacceptable and in any case any precision gained is usually masked by between test piece variability. Judging against a standard set of photographs is only of any use if the rubber under test follows the same pattern as that illustrated. Judging on the time to the first appearance of cracks gives only a single point measurement and is liable to be more variable than taking a series of grades of increasing severity.

In the cut-growth method, ISO 133, a 2 mm cut is made through the rubber with a tool of specified geometry. The length of this cut is measured at intervals and the number of cycles for it to increase by 2 mm, 6 mm and 10 mm is deduced.

The equivalent British Standards BS 903:Parts A10 and 11[5,6] are identical to the ISO methods. The ASTM standard for flex cracking is D430[7] and this specifies three machine including the De Mattia. The Scott Flexer is included specifically to test for ply separation of composites such as belting or tyres and is not used for rubber alone. The De Mattia method is not the same as the ISO and BS methods. In addition to using the machine to repeatedly bend the test piece a procedure is given in which a dumb-bell is cycled in pure tension. In both cases the degree of cracking is judged by visual observation but the grading is made against a standard set of specimens or photographs as mentioned above. The use of the direct tension mode of straining will be considered in Section 12.1.2. The third machine specified is the Dupont Flexer, briefly described in Section 12.1.

ASTM D813[8] specifies the De Mattia apparatus for cut-growth measurements and suggests that it should be used for materials which do not readily initiate cracks when tested by the methods given in ASTM D430. The fact that some materials are difficult to initially crack but will readily propagate tears is obviously of great practical importance but it is very debatable whether the two forms of test should be considered on an either/or basis. The procedure and expression of results are not identical with ISO 133.

A second method for cut growth using the Ross Flexer is given in ASTM D1052[9] but there appears to be no cross reference between this and D813. As discussed in Section 12.1, the Ross has the particular advantage of controlling the maximum and minimum strains rather more precisely than in other bending tests. Because of this, it is a little surprising that the method is not more widely used. In Britain, although not generally standardised, the Ross is used to test soling materials for footwear. A slower speed than in the ASTM standard is used to prevent heat build-up in the test piece and it is common practice to test at 0°C or −5°C.

12.1.2 Tests in Tension

All of the bending methods are to some extent arbitrary as to the degree of strain used and in most tests neither maximum nor minimum strains are well defined. By cycling in simple tension strains can be reproduced more easily and a range of strains and prestrains can be readily realised with one apparatus. A standard procedure for fatigue in tension has now been adopted by ISO and has arisen out of the MRPRA work on the concept of tearing energy (see Section 8.6). BSI and ASTM are expected to publish identical or similar methods.

The MRPRA work[10,11] leads to the following expression for fatigue life of a strip in tension:

$$N = \frac{G}{(2KW)^n} \cdot \frac{1}{C_0}$$

where: N = fatigue life in cycles to failure,
G = cut growth constant,

K = function of the extension ratio,
W = strain energy per unit volume,
C_0 = initial depth of cut (or intrinsic flaw), and
n = strain exponent dependent upon the nature of the polymer.

Hence, at a constant value of K a plot of $\log(N)$ against $\log(W)$ will have a slope n. The value of n has been found to be about 1.5 for a natural rubber tyre tread and 3 for an SBR tread. If no artificial cut is introduced then C_0 is the effective size of a naturally occurring flaw. The strain energy density, W, can be found from the area under the stress/strain curve for the test piece and is strain dependent. The fatigue life is independent of the specimen geometry when expressed in these terms. At low strains the equation does not adequately describe the fatigue behaviour and there is a fatigue limit corresponding to tearing energy below which there is virtually no cut growth and fatigue life becomes very long.

Test can be made at a number of extensions and compounds can be compared in terms of fatigue life at the same strain or at the same strain energy. In the latter case absolute comparisons can be made on compounds of different modulus. When comparing different rubbers, it is necessary to test at a number of strains or to define the severity of conditions which will occur in service, because with the number of variables (G, K, W, n and C_0) the ranking order may vary with the maximum strain employed.

Fatigue life is influenced by the environmental conditions under which the test is carried out, in particular temperature, oxygen and ozone. The effects of these have been discussed by Derham *et al.*[11] and Clapson and Dove,[12] the latter authors also giving examples of the application of the tensile form of fatigue testing to practical applications.

ISO 6943[13] for fatigue in tension specifies two different types of test pieces, rings and dumb-bells, which corresponds to the geometries used on commercially available apparatus. A ring apparatus is shown in Fig. 12.2. There is in principle little difference between the two forms of test piece but dumb-bells are necessary for studying directional effects. They are also easier than rings to cut from sheet but normally a specially moulded sheet is required such that the dumb-bells have a raised bar across each tab end to aid location and gripping. There are no gripping problems with rings and the roller separation is a direct measure of strain.

The dumb-bells specified are the same as those for tensile stress/strain tests except the preferred thickness is 1.5 mm. The ring is also the same as the tensile ring which means that the bulk of the two types of test piece are different.

FIG. 12.2. A ring apparatus.

The range of frequency specified is between 1 and 5 Hz and the standard only covers strain cycles passing through zero, although prestrains could be applied. It is suggested that at least five test pieces should be tested at each strain and that usually it is desirable to test at a number of maximum strains. The strain on ring test pieces is calculated on the internal diameter (see Section 8.2.1). The test is continued until complete failure of the test piece occurs and then the number of cycles recorded.

During the course of a test the stress/strain relationship of the test piece will change and there will also be a degree of set. It is recommended that both these parameters are measured at intervals and the results reported as well as the fatigue life. The results can be presented in graphical form as log(fatigue life) against strain, log(stress energy density) or log(stress). An annex gives explanatory notes including a section on interpretation of results which introduces the concept of a fatigue limit.

Now that this standard is published it is hoped that more workers will apply the fracture mechanics approach which underlies it to the prediction of fatigue in rubber products. The concept can be extended to shear and compression situations and examples of the latter have been given by Stevenson.[14]

12.2 HEAT BUILD-UP

It is rather confusing that the 'heat build-up' type of fatigue test is carried out on an apparatus generally called a flexometer, which brings to mind flex-cracking and cut-growth tests. The term heat build-up is not in fact a particularly good one as rupture of the test piece, set and changes in stiffness can also be measured, but it serves to distinguish the tests from those where only surface cracking is of interest and the test piece geometry is such that temperature rise is minimised.

Flexometers or heat build-up fatigue apparatus operate in compression, shear or a combination of the two and various designs have been in use and standardised, particularly by ASTM, for many years. The test piece geometry and deformation cycle used are inevitably somewhat arbitrary and this has perhaps contributed to the lack of any international or British standard method. There is now an international standard, ISO 4666,[15] which has the title 'Determination of Temperature Rise and Resistance to Fatigue in Flexometer Testing', and is split into three parts; basic principles, the rotary flexometer and the compression flexometer. The ASTM standard is D623[16] which specifies two types of apparatus, the Goodrich flexometer and the Firestone flexometer. The compression flexometer of the ISO standard is essentially the same as the Goodrich and operates by superimposing a cyclic compression strain on to the static deformation caused by a constant force. The ISO rotary flexometer and the Firestone both operate by superimposing a cyclic shear deformation on to a static compressive deformation but the cyclic action of the two machines is not the same and the ISO apparatus is derived from the St Joe flexometer which at one time was included in the ASTM standard (up to 1962).

For the Goodrich flexometer test given in ASTM D623, the 17.8 mm diameter by 25 mm high test piece is cycled at 1800 cycles/min with a stroke of 4.45 mm for 25 min and the temperature rise recorded. A choice of three static loads is given, alternative strokes suggested and two ambient temperatures, 50°C and 100°C, recommended. Apart from measuring temperature rise, the static deflection of the test piece, its dynamic deflection, compression set, and indentation hardness are recorded.

The Firestone flexometer method in D623 is not very specific. The standard test pieces are in the shape of a frustum of a rectangular pyramid but the use of any suitable shape is permitted when cut from products. The apparatus operates at 800 cycles/min and a range of compression loads and amplitudes of oscillation are possible but no particular conditions are specified. The test piece is fatigued until a definite but unspecified decrease in the height of the test piece is reached which is supposed to represent the onset of internal porosity. Parameters such as temperature rise and changes in compression are reported.

The specification of the ISO rotary flexometer is not much better. It uses cylindrical test pieces and operates at 14.6 or 25 Hz. The axial compression can be either constant stress or constant strain and loading conditions are suggested for both measurements of temperature rise and resistance to fatigue breakdown. Breakdown is not precisely defined.

BS903:Parts A49 and A50 are identical to ISO 4666:Parts 1 and 3. There is no British equivalent to the rotary flexometer, simply because such an arbitrary apparatus was not considered worthy of standardisation and it is not used in the UK.

The vagueness of the ASTM and ISO methods for rotary flexometers reflects the arbitrary nature of these tests. The first part of ISO 4666 attempts to describe the basic principles of fatigue testing to give guidance on the interpretation of results using particular apparatus and test conditions. This information would seem to be very necessary because results obtained under any particular conditions are quite arbitrary and have no significance apart from the conditions used.

It is commonly found that the relationship between 'fatigue life' and applied stress or strain is of the form shown in Fig. 12.3. The important feature of this so-called 'Wöhler curve' is that on reducing the stress or strain towards a particular value the fatigue life increases virtually to infinity giving rise to the concept of a limiting fatigue life (see also Section 12.1.2), and an ultimate fatigue strength or deformability.

Most fatigue tests apply a fixed pre-stress or strain partly because without bonding of the test piece it is necessary to hold the rubber in

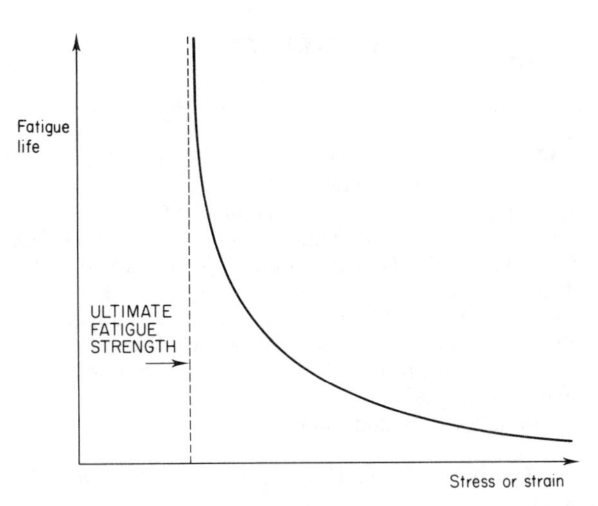

FIG. 12.3. Relationship between fatigue life and applied stress and strain (Wöhler curve).

place. The amount of pre-stress or strain will affect the fatigue life; in particular the fatigue life is appreciably shortened if the cyclic deformation passes near to or through zero strain.

A fatigue test can be made with cycles of either constant strain amplitude or constant force amplitude. With constant strain the resultant stresses are greater for higher stiffness rubbers, so that these are stressed more highly and, other things being equal, will develop more heat. On the other hand, with constant stress a stiff rubber will deform less and consequently tend to give a better fatigue performance. Consequently, to avoid conflicting results it is necessary to choose conditions which correspond with those met in service. It is also possible to use cycles of constant energy which is quite commonly the situation met by such products as dampers and shock absorbers, but this is more difficult from the apparatus point of view. The 'pre-stress' can in fact also be a constant stress or a constant strain.

ISO 4666:Part 1 remarks on the care necessary in measuring temperature rise and the fact that the result depends on where the temperature is measured and on the test piece geometry.[18] It recommends testing at a series of strain or stress levels because a comparison of rubbers at one level only can be misleading. The draft also mentions the measurement of creep and set in the test piece after periods of dynamic cycling.

REFERENCES

1. Buist, J.M. and Williams, G.E. (1951). *India Rubber World*, **127**, 320, 447 and 567.
2. ISO 132, 1983. Determination of Flex Cracking (De Mattia).
3. ISO 133, 1981. Determination of Crack Growth (De Mattia).
4. Boss, J.D. and Greensmith, H.W. (May/June 1967). *J. IRI*, 165.
5. BS903:Part A10, 1984. Determination of Flex Cracking (De Mattia).
6. BS 903:Part A11, 1976. Determination of Resistance to Cut Growth (De Mattia Type Machine).
7. ASTM D430–73. Rubber Deterioration—Dynamic Fatigue.
8. ASTM D813–59. Rubber Deterioration—Crack Growth.
9. ASTM D1052–55. Measuring Rubber Deterioration—Cut Growth using Ross Flexing Apparatus.
10. Gent, A.N., Lindley, P.B. and Thomas, A.G. (1964). *J. Appl. Polym. Sci.*, **8**, 455.
11. Derham, C.J., Lake, G.J. and Thomas, A.G. (1964). *J. Appl. Polym. Sci.*, **8**, 455.
12. Clapson, B.E. and Dove, R.A. (Dec. 1972). *Rubb. J.*, No. 12, 41.
13. ISO 6943, 1984. Determination of Tension Fatigue.
14. Stevenson, A. (1984). *Polym. Test.*, **4**(2–4).
15. ISO 4666, 1982. Determination of Temperature Rise and Resistance to Fatigue in Flexometer Testing.
16. ASTM D623–78. Heat Generation and Flexing Fatigue in Compression.
17. BS 903:Parts A49 and A50, 1984. Determination of Temperature Rise and Resistance to Fatigue in Flexometer Testing.
18. Buckley, D.J., ACS Division of Rubber Chemistry Meeting, Toronto, May 1974, Paper 34.

Chapter 13

Electrical Tests

Rubbers are usually electrically insulating and this property is widely exploited in cables and various components in electrical appliances. They can also be made anti-static and even conducting by the addition of suitable carbon blacks. In all cases it is the combination of the electrical properties and the inherent flexibility of rubbers which makes them attractive. In principle, the measurement of the electrical properties of rubber is the same as for any other material yielding results of the same order of magnitude, but particular precautions have to be taken because of distortion of this relatively low stiffness material by applied electrodes and the very high contact resistances which may exist between the electrodes and the rubber surface. In addition, the properties are often very sensitive to the past history of the rubber, including mechanical stress, which can lead to a large degree of variability. It is not intended to dwell here on the aspects of electrical measurement which are common to all materials but to emphasise the aspects which are more particular to rubbers. Electrical testing is a specialised subject and as much of the apparatus used is complex the work is normally undertaken by those having suitable training.

The tests commonly made on rubbers can be classified as:

Resistance or resistivity
Power factor and permittivity
Electric strength

Rather less often tests are made for tracking resistance and surface charge. For anti-static and conductive rubbers only resistivity, and sometimes surface charge, measurements are useful because such rubbers would not be used in situations requiring low dielectric loss or involving high voltages.

13.1 RESISTANCE AND RESISTIVITY

Because the surface of rubbers may conduct electricity more easily than the bulk of the material, it is usual to distinguish between volume resistivity and surface resistivity. Volume resistivity is defined as the electrical resistance between opposite faces of a unit cube, whereas surface resistivity is defined as the resistance between opposite sides of a square on the surface. Resistivity is occasionally called specific resistance. Insulation resistance is the resistance measured between any two particular electrodes on or in the rubber and hence is a function of both surface and volume resistivities and of the test piece geometry. Conductance and conductivity are simply the reciprocals of resistance and resistivity, respectively.

In practice, resistivity is calculated from the resistance measured with a known, fairly uncomplicated geometry. If a distinction is being made between volume and surface resistivity then the test arrangement is chosen to minimise the effect of the unwanted component of resistivity. With insulating rubbers the surface is frequently more conductive than the bulk of the material because of adsorbed moisture or contamination and a sensible distinction can be made between surface and volume resistivity. With most anti-static and conducting rubbers the surface layer is no more conductive than the bulk and then, whatever the geometry chosen, the current will largely take the 'easy' route through the bulk of the material and surface resistivity has no real meaning. Hence, with these lower resistance rubbers it is usual to assume that resistivity consists of the volume component alone and to refer to the measured value as simply, resistivity. Similarly, if the surface of an insulating rubber is no more conductive than the bulk a surface measurement will have no meaning.

It is not possible to make a clear cut distinction between insulating, anti-static and conducting rubbers. The definitions should be made with respect to the resistance between two relevant points on a product rather than to the resistivity of the rubber because if you took a long enough length of a fairly low resistivity rubber, the total resistance from end to end would effectively make it an insulator. Generally, resistances of up to $10^4\,\Omega$ are considered conductive, between 10^4 and $10^8\,\Omega$ anti-static and above this insulating, although various definitions have been given by different workers or according to the hazard associated with the product. The test methods for insulating and anti-static/conducting rubbers differ considerably and the two will be discussed separately.

In all cases there is, in addition to the resistance of the rubber, a contact resistance at the electrode/rubber interface. The magnitude of this contact resistance is a complex function of the electrode system used, the rubber under test and the applied voltage, and the mechanisms which produce it have not been fully elucidated. When testing insulating rubbers the contact resistance is ignored because, although it may be high, there is generally no way of measuring it. Tests are, however, made with well-defined electrode systems. Contact resistance is a very troublesome complication when testing conductive and anti-static rubbers and will be discussed in Section 13.1.2.

13.1.1 Tests on Insulating Rubbers

The usual standard methods for volume and surface resistivity both use the same test piece and electrode geometry and essentially the same measuring circuit. There are no ISO methods, but national standards for rubbers are usually adaptations of the IEC Publication 93[1] for insulating materials in general. The relevant British standards are BS 903:Part C1[2] for surface resistivity and Part C2[3] for volume resistivity .

Sections through circular electrode systems are shown in Fig. 13.1. If the arrangement shown in Fig. 13.1(a) is used for volume resistivity measurement, current will pass over the surface as well as through the volume of the rubber and, likewise, if the arrangement in Fig. 13(b) is used for surface resistivity some current will pass through the volume of the rubber. Also, in arrangement (a) the current in the volume of the rubber near the electrode edges flows in a curved path giving a 'fringing' effect and increases the effective electrode area. As mentioned previously, the surface resistivity of insulating rubbers is generally lower than their volume resistivity so that the effect of current flow in the bulk of the rubber in arrangement (b) for surface resistivity is not serious.

The following means, as shown in Fig. 13.1(c) and (d), are used in the BS 903 methods to minimise or eliminate the problems:

(a) The thickness of the test piece at 1 to 3.2 mm is much less than the electrode diameter of either 15 or 5 cm—this reduces the effect of fringing.

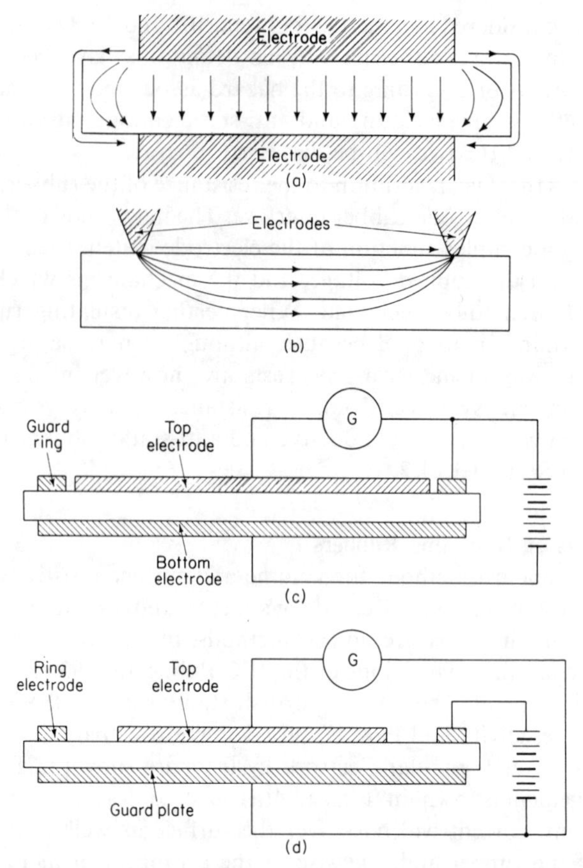

FIG. 13.1. Electrode systems (diagrammatic). (a) Showing 'fringing' effect and surface conduction in volume resistance measurement; (b) showing volume conduction in surface resistance measurement; (c) electrodes, guard ring and circuit for volume resistance measurement; (d) electrodes, guard plate and circuit for surface resistance measurement. (G = galvanometer or other current measuring device.)

(b) A 'guard ring' or annular electrode surrounding the top electrode (Fig. 13.1(c)) is used for volume resistivity and a 'guard plate' electrode (bottom electrode in Fig. 13.1(d)) for surface resistivity measurements. In arrangement (c) the guard ring is connected so that any current passing over the test piece surface is not included in the current measured. In BS 903 the gap between the centre and guard electrode is 1 cm and the effect of fringing would be reduced if this was made smaller but such a refinement is not really warranted.

The British standards specify that the electrodes shall be formed by painting on a colloidal graphite suspension in water, by vacuum deposition of a metallic film or using conducting silver paint. In all cases the electrodes are backed by rigid metal (usually brass) plates. The graphite electrodes are the type most widely used. Other types of conductive paint or tin foil backed on to wet graphite are probably equally good but in any case it is essential that the electrodes are prepared and applied with very great care. Mercury electrodes are not necessarily better and being a health hazard are no longer specified.

The overall dimensions of the electrode system are not really critical but two sizes are specified. The large of the two should be used when the resistivity of the test material is very high.

In Fig. 13.1 the voltage source is shown simply as a battery and the current measuring device as a galvanometer, but in practice the circuits needed are more complex. Some guidance on construction is given in the standards, including two short appendices, and reference made to BS 6233 (identical to IEC 93) which gives more details. The measurement of very high resistivities is beset with practical difficulties and hidden errors. It is essential that apparatus is selected and operated with great care and expertise to achieve success. Helpful advice is given by Norman[4] who has also discussed at length the same measurements for plastics.[5]

Resistivity is sensitive to temperature and humidity and tests are usually made after conditioning in the standard atmosphere of $23 \pm 2°C$ and $50 \pm 5\%$RH. Surface resistivity is particularly sensitive to humidity and the standard humidity should be maintained during the test. Where results are to be used as design or performance data it would be advisable to test over a range of humidities and (perhaps) temperatures. The resistance of the test piece is measured after the test voltage has been applied for a set period, usually 1 min, although this will very often not be an adequate time for the current, and hence the measured resistivity, to reach an equilibrium value. If it is suspected that this is the case, resistivity can be monitored as a function of time of electrification.

The quantity directly measured is the resistance of the test piece which is then converted into resistivity by means of the appropriate formula involving the dimensions of the electrode system:

$$\text{Volume resistivity } (\Omega\,\text{cm}) = \frac{AR_v}{h}$$

$$\text{which with circular electrodes} = \frac{\pi}{40}(D_1 + 5)^2 \frac{R_v}{t}$$

Surface resistivity (Ω, or sometimes Ω per square)

$$\text{with strip electrodes} = \frac{lR_s}{d}$$

$$\text{with circular electrodes} = \frac{\pi(D_2 + D_1)}{(D_2 - D_1)} R_s$$

where: R_v = volume resistance of test piece (Ω),
R_s = surface resistance of test piece (Ω),
A = area of guarded electrode,
t = thickness of test piece (mm),
D_1 = diameter of top inner electrode (mm),
D_2 = internal diameter of ring (top outer) electrode (mm),
l = length of electrodes (mm), and
d = distance between electrodes (mm).

Note that the unit for surface resistivity does not involve length as does the unit for volume resistivity. Neither involve the unit of volume.

Strictly there should be a correction to A and l for the effect of fringing but this is not normally considered significant. It is not usually possible to obtain very great precision in measurements of high resistivity and results are never quoted beyond two significant figures. Often, between sample variability is such that two materials would be considered really significantly different only if their resistivities differed by a factor of 10.

The ASTM method for resistivity is D257[6] which covers all insulating materials and is hence not specific to rubbers and is more equivalent to IEC 93. It is a fairly comprehensive document and contains a bibliography.

The discussion given in BS 4618, Sections 2.3 and 2.4[7] on design data for plastics applies in principle to rubbers, and stresses the advantage of measuring resistivity as a function of temperature, humidity, electric stress and time of electrification.

Measurements of resistance are not normally made using an applied AC voltage and if they were, account would need to be made of the fact that the total impedance would be comprised of conductance and capacitance terms. Results of both DC and AC measurements have been given by Buller *et al.*[8]

There is an international method, ISO 2951[9] for the determination of insulation resistance. The test pieces specified are either flat sheets or tube or rod and the electrodes either conductive paint or metal bars. It is apparent by the reference to rigid materials that the wording has been 'lifted' from a general document for insulating materials, presumably IEC 167.[10] There is no mention of metal backing plates for the paint electrodes and to obtain consistent results with rubbers the electrode system would need to be defined more precisely.

The resistance is measured using a 500 V supply, in the same manner as for resistivity, after 1 min electrification. Paint electrodes are spaced 10 mm apart and should be 100 mm long. The measured resistance is normalised to these conditions if any other length of electrode is used, as would presumably be the case for rod or tube. Bar electrodes are longer than the test piece width and the measured resistance is normalised to what is termed a 25 mm electrode length, meaning a 25 mm wide test piece, with the electrodes 25 mm apart. It would be reasonable to suggest that this standard would have far greater value if it were written around products, which is where resistance tests as opposed to resistivity tests are required.

The British equivalent BS 903:Part C5[11] covers plastics as well as rubbers and includes methods using taper pins which are not applicable to rubber. However, the methods for tube, and rod and sheet are very similar to ISO 2951 but metal electrode attachments are illustrated and a more realistic tolerance of $\pm$ 50 on the applied voltage is specified.

13.1.2 Tests on Conducting and Anti-static Rubbers

A most comprehensive and detailed account of conductive polymers and associated test methods is given by Norman[12] in what has become the standard textbook on the subject and this should be referred to if an in-depth understanding is required. The account given here will be restricted to the standard methods.

The international standard for measurement of resistivity is ISO 1853[13] which details one procedure only, the potentiometric or four electrode method. The principle of the method is shown in Fig. 13.2; the strip test piece has metal current electrodes clamped at each end and is connected in series with a voltage source and a means of measuring the current flowing. The 'potentiometric' electrodes are placed on the test piece

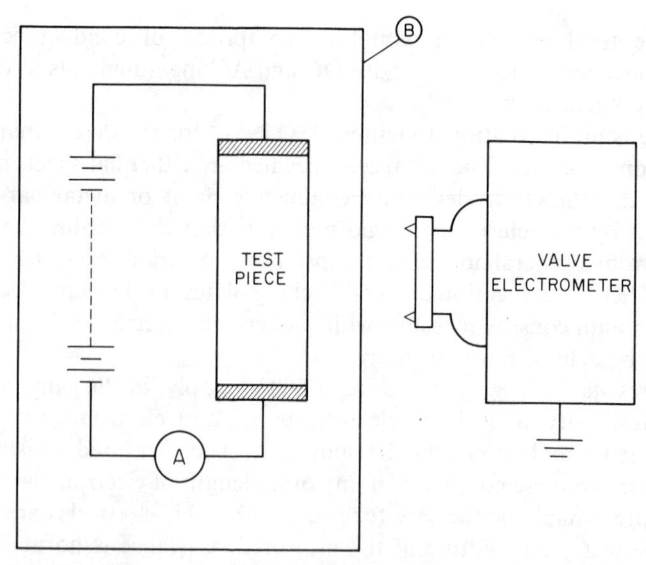

FIG. 13.2. Circuit for potentiometric (four electrode) method. All components within the rectangle B should be insulated (better than 10^{12} ohm) from earth.

between the two current electrodes and connected to an electrometer such that the voltage drop between them can be measured. The advantage of this procedure is that contact resistances, which are the biggest problem in measurements on anti-static and conducting rubbers, are virtually eliminated. The contact resistance at the current electrodes does not matter and those at the potentiometric electrodes do not affect the measurement if no current is taken by the very high impedance electrometer. The resistivity is calculated from the measured current and voltage drop and the cross section and length of the test piece between the potentiometric electrodes.

Mechanical conditioning can greatly affect the measured resistivity but the effect of deformation is not entirely permanent and recovery can be accelerated by heating. For this reason ISO 1853 specifies conditioning at 70°C, followed by conditioning at 23°C and 50% RH without disturbance of the test piece. The potentiometric electrodes are shown in detail and a suitable electrometer described in an annexe. The potential of the current source is not specified but this is commonly a 1.5 V dry cell. For low resistivities a multi-ammeter can be used for the current measurement but for low currents an electrometer would be used to measure the voltage drop across a known standard resistor. The method is generally restricted to resistivities of less than $10^6\,\Omega$ cm and much of the apparatus must be very highly insulated from earth.

The alternative approach to overcoming the contact problems is to use electrodes which have a very low contact resistance. In the equivalent British standard BS 2044[14] three procedures are given, with Method 2 being essentially the same as ISO 1853. Method 1 uses a strip test piece with brass electrodes at either end bonded in during vulcanisation. This gives a very efficient electrode system with negligible contact resistance and is hence the preferred method, although it is only suitable when laboratory prepared test pieces are available and the rubber will bond to brass. Any suitable resistance measuring instrument can be used as long as it does not dissipate more than 0.25 W in the test piece, so avoiding any heating. The third method uses less efficient electrodes in the form of tin foil wrapped on to wet colloidal graphite painted on to the test piece. It is hence similar to method 1 but less accurate and should only be used when methods 1 and 2 are not possible.

It is worth noting that, as regards testing, a resistivity of about $10^7\,\Omega$ cm is taken as the dividing line between anti-static and conducting rubbers and insulating rubbers, the methods of Section 13.1.1 being used above this level. In practice this can be very annoying when unknown rubbers around 10^6–$10^9\,\Omega$ cm are to be measured, although some overlap is possible, because of the great difference in test piece geometry between the methods.

On anti-static and conducting products it is usual to measure the resistance between specified points. More or less efficient electrode systems are used, contact resistance included and the resistance usually measured with a commercial 'insulation tester'. The relevant international method is ISO 2878[15] and the British method BS 2050.[16] The ASTM method, D991,[17] appears from the title to cover products but is actually a resistivity method on strip test pieces using a four electrode system. It is hence similar to ISO 1853 and BS 2040, method 2 but the electrode construction is different and care would need to be taken to avoid leakage between the current and potentiometric electrodes.

13.2 SURFACE CHARGE

The mechanism of electrostatic charging of polymer surfaces is complicated and as yet not fully understood.[18] The fact remains that an insulating rubber can become charged by direct application of a voltage or by friction against another material and the main purpose of anti-static rubbers is to allow the rapid decay of such a charge in circumstances where it would be a hazard or a nuisance.

The likely efficiency of a material to dissipate charge rapidly can be estimated from a measurement of resistivity and the rate of charge decay for a surface with its edges held at zero potential has been shown to be linearly dependent on resistivity.[19,20] The international and British standards previously mentioned for anti-static and conducting rubber products use resistivity as the means of classification. A direct measure of the charge and its rate of decay is desirable, but for rubbers such measurements are rarely carried out and there are no ISO or British methods. The various methods which have been used for measuring charge are discussed in some detail by Norman[12] and the procedures fairly commonly used for plastics film are standardised in BS 2782.[21]

The voltage due to the charge on an insulator is influenced by any measuring instrument brought near to it and the charge can only be measured by Faraday ice-pail types of experiment where the sample is put completely inside the screened, insulated electrode of an electrometer. All that can usually be measured is the field strength above the sample with a given geometry of metal or whatever around the measuring area. This is the procedure adopted in method 250A of BS 2782 where the sample is charged, earthed and the decay in field strength monitored with a rotating vane electrometer. The field strength is proportional to charge for the given test geometry and the result is expressed as the time to decay to 50% of the initial meter reading, which is far less critically dependent on geometry than is the absolute meter reading.

13.3 ELECTRIC STRENGTH

Electric strength is usually taken as the nominal voltage gradient (applied voltage divided by test piece thickness) at which breakdown occurs under specified conditions of test. The specified conditions of test are important as the measured electric strength is not an intrinsic property of the material but depends on test piece thickness, time of electrification and the electrode geometry, as well as on conditioning of the material.

There is no ISO standard for electric strength but there is a British method BS 903:Part C4[22] which takes into account the general method for insulating materials, IEC 243.[23] Two procedures are specified in BS 903, both using AC voltage at 50 Hz and covering both rubbers and plastics. The normal test piece for rubbers is a disc of minimum diameter 100 mm or a square of side not less than 100 mm, with thickness of 1.25 ± 0.2 mm. Sheet can be tested at other thicknesses and there is also provision for tube and tape.

In the first procedure the voltage is applied at a uniform rate from zero such that breakdown occurs on an average of between 10 and 20 s. In the second method a step by step procedure is used whereby a voltage is applied for 20 s and if failure does not occur successively higher voltages are applied for 20 s until breakdown does occur. Results can be expressed as the breakdown voltage or the electric strength taking account of thickness. With so many permutations this standard is editorially somewhat complex and because results will vary with procedure and test piece thickness it is necessary to check carefully exactly which procedure has been used. Previously BS 903 included a method whereby the breakdown voltage corresponding to 1 min electrification was deduced from the times to breakdown for a series of applied voltages but this has been dropped. For product specifications it is usual to simply apply a proof voltage for a specified period and the result is either pass or fail. For design purposes it may be desirable to measure breakdown strength as a function of test piece thickness, time of electrification and of electrode geometry.

The ASTM method, D149,[24] is not specific to rubbers but covers all insulating materials and allows several electrode configurations. Three procedures are listed, a rapid rate of rise until breakdown, a slow rate of rise until breakdown and a step-by-step method. The details of the procedures are left to the material specification and again it should be emphasised that results are only comparable if exactly the same procedure and conditions are used.

Equipment for breakdown tests is often constructed by the user from bought-in components. Some information is given in the British and ASTM standards but this would only be sufficient for the electrical expert. General requirements for equipment at power frequencies is given in BS 2918,[25] and BS 923[26] is a comprehensive guide in four parts on high voltage testing techniques.

13.4 TRACKING RESISTANCE

The term tracking relates to the development of conductive paths or tracks on a test piece surface between electrodes under discharge conditions. There are, however, two types of tracking tests, low voltage and high voltage. The low voltage tests (up to 1000 V) rely on a contaminant

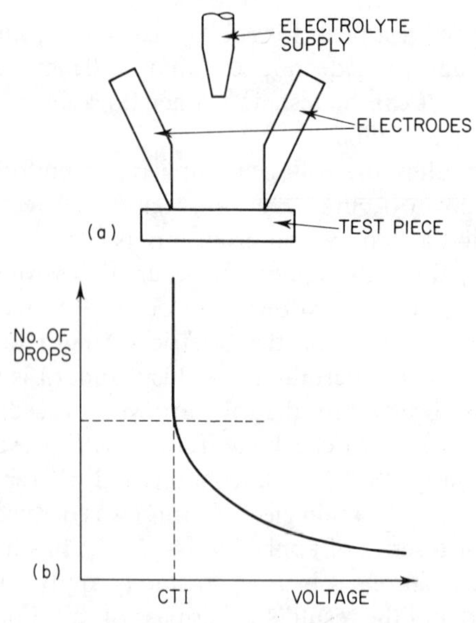

FIG. 13.3. Tracking test. (a) Electrode system; (b) effect of voltage on number of electrolyte drops causing 'failure'. CTI is the comparative tracking index.

applied between the electrodes to induce initial flash over whilst the high voltage tests may or may not use a contaminant and use equipment similar to that for breakdown tests with specific electrode systems. Neither type of test is very commonly applied to rubber but the low voltage test is generally the more important of the two.

The best known test is that given in IEC recommendation 112[27] and the identical procedure in BS 5901.[28] The basis of the test is to position electrodes as shown in Fig. 13.3 and to allow drops of an electrolyte to fall at fixed intervals on to the test piece surface between the electrodes whilst an AC voltage is applied to them. The number of drops of electrolyte to cause tracking, as indicated by a 'continuous' current passing between the electrodes, is noted. The procedure is repeated at different voltages and a curve constructed as shown in Fig. 13.3. An investigation into the effect of variables in this test on the result has been given by Yarsley *et al.*[29]

13.5 PERMITTIVITY AND POWER FACTOR

These properties are sometimes grouped as the dielectric properties but this is not entirely logical as dielectric simply means insulating. Relative permittivity of a material can, for practical purposes, be defined as the ratio of the capacitance of a condenser having the material as the dielectric, to the capacitance of a similar condenser having air, or more precisely a vacuum, as the dielectric. The word relative is usually dropped and the property simply called permittivity and is the same thing as used to be called dielectric constant.

The power factor of a material may be described loosely as the fraction of the electrical energy stored by the condenser in each cycle which is lost as heat. This arises because the phase difference between voltage and current deviates from 90° (which it would be for a perfect dielectric, e.g. vacuum) by the loss angle, δ.

The actual power loss in a dielectric in an AC field is proportional to the loss factor, i.e. the product of permittivity and power factor, so that to achieve the minimum loss both of these parameters must be small. Power loss is also proportional to frequency so that at high frequencies, for example, in telecommunications applications, it may be especially necessary to use low permittivity and low loss materials. For making capacitors a high permittivity is desirable because then the physical size of the component for a given capacitance can be as small as possible. A high loss is not normally wanted in a capacitor so that the ideal dielectric would have a high permittivity and a low power factor. A high loss factor may be desirable where it is required to absorb energy, for example, in radio frequency or microwave heating.

13.5.1 Standard Methods

For most materials, permittivity and power factor are not constant over the very wide range of frequencies of industrial interest and, because the same apparatus and method cannot be used at all frequencies, standard methods are usually subdivided into procedures for different frequency ranges.

The general international standard for insulating materials is IEC Publication 250[30] but there are no ISO methods for rubbers in particular. Similarly the ASTM standard D150[31] is applicable generally to solid insulating materials. The presentation for use as design data of permittivity and loss tangent values of plastics is covered in BS 4618[32] and the brief information given is relevant to rubbers also.

BS 903:Part C3[33] specifically covers the measurement of loss tangent and permittivity of rubbers at power and audio frequencies (45 Hz–10 kHz). Test pieces are either flat sheet on round tube, electrodes for the former being similar to those for resistivity of insulating material but with a smaller guard gap. The measuring equipment is either a Schering bridge with Wagner earth or a transformer ratio arm bridge.

Previously, BS 903 also included a method for radio frequencies which referred to BS 2067[34] for the Hartshorn and Ward apparatus used. BS 2067 is rarely used and was declared obsolete in 1985.

REFERENCES

1. IEC Publication 93, 1980. Methods of Test for Volume Resistivity and Surface Resistivity of Electrical Insulating Materials.
2. BS 903:Part C1, 1982. Determination of Surface Resistivity.
3. BS 903:Part C2, 1982. Determination of Volume Resistivity.
4. Brown, R.P. (Ed.) (1979). *RAPRA Guide to Rubber and Plastics Test Equipment*, RAPRA.
5. Brown, R.P. (Ed.) (1981). *Handbook of Plastics Test Methods*. George Godwin.
6. ASTM D257–78. DC Resistance or Conductance of Insulating Materials.
7. BS 4618, 1975. The Presentation of Plastics Design Data. Sections 2.3 and 2.4. Volume and Surface Resistivity.
8. Buller, M., Mayne, J.E.O. and Mills, D.J. (Oct. 1976). *J. Oil Col. Chem. Assoc.*, **59**(10), 351.
9. ISO 2951, 1974. Vulcanised Rubber—Determination of Insulating Resistance.
10. IEC Publication 167. Methods of Test for the Determination of the Insulation Resistance of Solid Insulating Materials.
11. BS 903:Part C5 (1983). Determination of Insulation Resistance.
12. Norman, R.H. (1970). *Conductive Rubbers and Plastics*, Elsevier.
13. ISO 1853, 1975. Conducting and Antistatic Rubbers—Measurement of Resistivity.
14. BS 2044, 1984. Determination of Resistivity of Conductive and Antistatic Plastics and Rubbers. (Laboratory methods.)
15. ISO 2878, 1978. Anti-static and Conductive Rubber Products—Determination of Electrical Resistance.
16. BS 2050, 1978. Electrical Resistance of Conductive and Antistatic Products Made from Flexible Polymeric Material.
17. ASTM D991–83. Volume Resistivity of Electrically Conductive and Antistatic Products.
18. Morris, W.T., RAPRA Technical Review No. 47, July 1969.
19. Boyd, J. and Bulgin, D. (1957). *J. Test. Inst.*, **48**, 66.
20. Beesley, J.H. and Norman, R.H., RAPRA Members Report No. 5, 1977.
21. BS 2782, Methods 250 A-C, 1983. Anti-static Behaviour of Film.
22. BS 903:Part C4, 1983. Determination of Electric Strength.
23. IEC Publication 243. Recommended Methods of Test for Electric Strength of Solid Insulating Materials at Power Frequencies.
24. ASTM D149–81. Dielectric Breakdown Voltage and Dielectric Strength of Electrical Insulating Materials at Commercial Power Frequencies.
25. BS 2918, 1957. Electric Strength of Solid Insulating Materials at Power Frequencies.

26. BS 923, 1980. Guide on High-Voltage Testing Techniques.
27. IEC Publication 112, 1979. Method for Determining the Comparative and Proof Tracking Indices of Solid Insulating Materials under Moist Conditions.
28. BS 5901, 1980. Determining the Comparative and Proof Tracking Indices of Solid Insulating Materials under Moist Conditions.
29. Yarsley, V.E., Grant, W.J. and Ives, G.C., ERA Report A/T 136, 1980.
30. IEC Publication 250. Recommended Methods for the Determination of the Permittivity and Dielectric Dissipation Factor of Electrical Insulating Materials at Power, Audio and Radio Frequencies Including Metre Wavelengths.
31. ASTM D150–81. AC Loss Characteristics and Permittivity (Dielectric Constant) of Solid Electrical Insulating Materials.
32. BS 4618, 1970. The Presentation of Plastics Design Data. Sections 2.1 and 2.2. Permittivity and Loss Tangent.
33. BS 903:Part C3, 1982. Determination of Loss Tangent and Permittivity at Power and Audio Frequencies.
34. BS 2067, 1953. Determination of Power Factor and Permittivity of Insulating Materials (Hartshorn and Ward Method).

Chapter 14

Thermal Properties

What is covered by the term 'thermal properties' can mean different things to different people. When talking loosely, heat ageing, low temperature tests and fire resistance are sometimes included but these are more properly dealt with, as in this volume, under Effect of Temperature (Chapter 15).

The thermal properties of rubber are of very great importance, particularly in the processing stages, but there is a remarkable dearth of reliable data. Traditionally, the approach to heating and cooling problems has been empirical rather than by careful analysis. The data needed for such analysis has not been available largely because of the undoubted experimental difficulties to be overcome but, even with data, somewhat complicated calculation is required.

In the last few years the value of an analytical approach to heat transfer problems has been increasingly realised and considerable effort has been devoted to developing measuring techniques and convenient methods of calculation. However, the measurement of thermal properties remains a very specialised subject, there is little evidence of standard procedures and the tests are carried out in relatively few laboratories. As was said of another specialised area of rubber testing, electrical testing, the measurement of the thermal properties of rubber is in principle the same as for other materials although there are particular difficulties, especially when testing unvulcanised rubbers at processing temperatures.

14.1 THERMAL ANALYSIS

Probably the largest growth area in testing in recent times has been thermal analysis. Essentially, thermal analysis is the study of one or more properties of a material as a function of temperature. A thermal analyser is hence an apparatus which allows the automatic monitoring of the chosen property with temperature change. In principle any property can be measured (and most have been); for example dynamic thermal mechanical analysis (see Chapter 9), thermogravimetry to measure weight changes and thermodilatometry to follow dimensional change.

It becomes clear that although thermal analysis is often treated as one subject the information gained and the use to which it is put are very varied. Thermodilatometry is essentially a dimensional measurement which can measure thermal expansion or possibly dimensional change resulting from a chemical reaction. Dynamic thermal mechanical analysis essentially measures damping and dynamic modulus but the change of these with temperature is one way of detecting the glass transition point. Thermogravimetry is primarily a chemical analysis method. Reference has been made to relevant techniques in the appropriate chapters, e.g. Chapter 9 (Dynamic Stress and Strain Properties) and Chapter 15 (Effect of Temperature).

As regards thermal properties, the techniques of interest are differential thermal analysis (DTA) and its variant differential scanning calorimetry (DSC). In these techniques heat losses to the surrounding medium are allowed but assumed to be dependent on temperature only. The heat input and temperature rise for the material under test are compared with those for a standard material. In DTA the two test pieces are heated simultaneously under the same conditions and the difference in temperature between the two is monitored, whereas in DSC the difference in heat input to maintain both test pieces at the same temperature is recorded.

There have been many papers describing applications of thermal analysis to polymers including comprehensive reviews by Chia[1] and, more recently, Brazier.[2] Richardson[3] has described applications of DSC.

14.2 SPECIFIC HEAT

Specific heat is the quantity of heat required to raise the unit mass of the material through 1 °C, i.e. the heat capacity of unit mass.

The principal specific heats are those at constant volume and constant pressure but the specific heat at constant pressure is the quantity normally measured. The specific heat at constant volume, which is virtually impossible to measure, can be calculated from:

$$C_p - C_v = \frac{TB\beta^2}{\rho}$$

where: C_p = specific heat at constant pressure,
C_v = specific heat at constant volume,
T = absolute temperature,
B = bulk modulus,
β = coefficient of volume expansion, and
ρ = density.

The difference between the two specific heats is usually small enough to be ignored.

Specific heat is measured by supplying heat to a calorimeter containing the test piece and measuring the resulting temperature rise. An adiabatic calorimeter is one in which no exchange of heat between the calorimeter and its surround is allowed and this is achieved by surrounding the calorimeter by a jacket which is heated to follow the temperature change of the calorimeter itself. A variety of adiabatic calorimeters have been described and a number have been referenced by Hands[4] in a review of specific heat of polymers. Although simple in principle, the design of these calorimeters will be complex to obtain the highest levels of precision. A variation, which results in a more simple apparatus, is the drop calorimeter. The test piece is heated (or cooled) externally, dropped into the calorimeter and the resultant change in temperature monitored. For the simplest measurements, the calorimeter need not be surrounded by an adiabatic jacket but in that case corrections for the heat exchange with the surroundings must be applied.

Except where the very highest precision is required, when an adiabatic calorimeter would be used, it is usual nowadays to measure specific heat by a comparative method using differential thermal analysis (DTA) or its variant, differential scanning calorimetry (DSC), the latter being the preferred and usual technique. The experimental procedure has been outlined, for example, by Richardson.[5]

14.3 THERMAL CONDUCTIVITY

Thermal conductivity may be defined as the quantity of heat passing per unit time normally through unit area of a material of unit thickness for unit temperature difference between the faces. In the steady state, i.e. when the temperature at any point in the material is constant with time, conductivity is the parameter which controls heat transfer. It is then related to the heat flow and temperature gradient by:

$$q = -KA\frac{d\theta}{dx}$$

where: q = rate of heat flow,
K = thermal conductivity,
A = test piece area,
θ = temperature, and
x = test piece thickness.

Thermal conductivity is of importance in the design of products which will have a thermal insulation function and also in the design of rubber processing equipment.

In a review of thermal transport properties of polymers, Hands[6] discusses theory and methods of measurement, which can be divided into steady state methods and transient methods, and sources of error. Steady state methods are most widely used as these are mathematically more simple, but because, particularly for materials of low conductivity, they can be very time consuming and involve expensive apparatus, non-steady state or transient methods have been developed. These can have experimental advantages once the much more difficult mathematical treatment has been worked out. Parrot and Stuckes[7] have written a comprehensive volume on the thermal conductivity of solids and there are standard methods for materials in general in BS 874[8] which also defines a wide range of thermal insulating terms.

The thermal conductivity of solid rubbers is of the order of 1–2×1^{-10} W/ m K which is in the region of fairly low conductivity where experimental errors due to heat loss will be greatest. A heated disc method as given in BS 874, Clause 4.2.3 is satisfactory for many purposes but for the lowest conductivity materials a guarded hot plate will give more precise results. A guarded hot plate procedure is given in BS 874 Clause 4.2.1, in BS 4370[9] for foams and in ASTM C177.[10] References are given in BS 874 to papers giving more details of the various apparatus than are given in the standard.

Additional methods for conductivity measurement are given in appendices to BS 874. These are transient methods or they use a heat flow meter in the steady state with the intention of giving more rapid measurement than with the standard methods, although the precision will not generally be as great. A heat flow meter method suitable for rubbers is also given in ASTM C518[11] and a semi-automatic heat meter apparatus is described by Howard *et al.*[12] A 'system identification' procedure is given by Peter and Tarvin[13] and a procedure involving the velocity of heat wave across the test piece by Sourour and Kamal;[14] it is also possible to measure conductivity with DSC apparatus.[15]

An enclosed method of measuring conductivity described by Hands and Horsfall[16] is of particular importance because apart from eliminating heat losses from exposed edges of the test piece by enclosing the heat source and test pieces inside the heat sink, the apparatus allows measurements to be made through both the solid and liquid phases, which is of great value when studying polymers at processing temperatures.

A cheaper alternative in the form of a one-sided unguarded hot plate which is more accurate but will only operate with materials in the solid state and at ambient temperature was used by Hands[17] to study the effect of orientation on conductivity. This form of apparatus has also been modified to allow measurements at elevated temperature.

14.4 THERMAL DIFFUSIVITY

Thermal diffusivity is the parameter which determines the temperature distribution through a material in non-steady state conditions, i.e. when the material is being heated or cooled. It is a function of the thermal conductivity, specific heat and density:

$$\alpha = \frac{K}{\rho C_p}$$

where: α = thermal diffusivity,
K = thermal conductivity,
ρ = density, and
C_p = specific heat at constant pressure.

Thermal diffusivity is of little interest in many thermal insulation applications, for example civil engineering, where approximately steady state conditions normally exist. However, in rubber processing when temperatures are changing rapidly it is of more value than conductivity.

Conductivity, density, specific heat and diffusivity are all temperature dependent, but if the values of the first three parameters are known at the temperature in question then diffusivity can be calculated. Diffusivity is actually rather easier to measure experimentally than conductivity because it is only necessary to measure the change in temperature with time at three points in the material but the mathematical treatment required for such non-steady state measurements is relatively complicated.

Despite the importance of diffusivity in transient conditions it received less attention than conductivity until recent times, and there are no standard methods for polymers. Methods of measurement have been reviewed by Hands[6] and apparatus developed at RAPRA described by Hands and Horsfall.[18] For data used in the study and prediction of flow and processing properties, diffusivity is in fact now more often measured than conductivity and the Hands and Horsfall apparatus has been further developed. Very recently Smith[19] produced an updated review which covers conductivity as well as diffusivity.

14.5 SURFACE HEAT TRANSFER COEFFICIENT

The surface heat transfer coefficient can be defined as the quantity of heat flowing per unit time normal to the surface across unit area of the interface between two materials with unit temperature difference across the interface. If there is no resistance to heat flow between the surfaces then the transfer coefficient is infinite.

It is immediately apparent that in many processes involving rubber heat flow across the interface between two surfaces has to be considered. This is true in mixing, moulding, cooling after processing and conditioning of test pieces, but nevertheless very little attention has been paid to the measurement of the coefficient. The effect of the heat transfer coefficient on net heat flow is greatest with thin articles and where one of the materials is a gas. It is probably reasonable to assume a value of infinity for the transfer coefficient when rubber is pressed into intimate contact with a metal but in other cases it will be finite. Very little data are available on the measurement of heat transfer coefficient. Hands[6] mentions the empirical nature of the coefficient and the numerous factors which will affect its value, particularly between rubber and a fluid. Griffiths and Norman[20] calculated the heat transfer coefficients for rubbers in air and water.

REFERENCES

1. Chia, J. (1974). *J. Macromol. Sci.*, **A8**(1), 3.
2. Brazier, D.W. (1980). *Rubb. Chem. Techn.*, **153**(3).
3. Richardson, M.J. (1984). *Polym. Test.*, **4**(2–4).
4. Hands, D., RAPRA Technical Review, No. 51, February 1970.
5. Richardson, M.J. (1976). *Plast. Rubb. Mats. Appl.*, **1**(3–4).
6. Hands, D. (1977). *Rubb. Chem. Tech.*, **50**, 480.
7. Parrot, J.E. and Stuckes, Audrey D. (1975). *Thermal Conductivity of Solids*, Pion Ltd.
8. BS 874, 1973. Determining Thermal Insulating Properties with Definitions of Thermal Insulating Terms.
9. BS 4370:Part 2, 1973. Methods of Test for Rigid Cellular Materials.
10. ASTM C177–76. Steady State Thermal Transmission Properties by Means of the Guarded Hotplate.
11. ASTM C518–76. Steady State Thermal Transmission Properties by Means of the Heat Flow Meter.
12. Howard, J.F., Coumou, K.G. and Tye, R.P. (Sept./Oct. 1973). *J. Cell. Plast.*, **9**(5), 226.
13. Peter, J.W. and Tarvin, R.L. (Apr. 1976). *Polym. Engng. Sci.*, **16**(4), 229.
14. Sourour, S. and Kamal, M.R. (July 1976). *Polym. Engng. Sci.*, **16**(7), 480.
15. Sircar, A.K. and Wells, J.L. (1982). *Rubb. Chem. Tech.*, **55**(1).
16. Hands, D. and Horsfall, F. (1975). *J. Phys. E. Sci. Instrum.*, **8**, 687.
17. Hands, D. (1980). *Rubb. Chem. Tech.*, **53**(1).
18. Hands, D. and Horsfall, F. (1977). *Rubb. Chem. Tech.*, **50**, 253.
19. Smith, D. (1986). PhD Thesis, Bradford University.
20. Griffiths, M.D. and Norman, R.H. (Jul./Aug. 1976). *RAPRA Members Journal*, 87.

Chapter 15

Effect of Temperature

Distinction should be made between the short-term and long-term effects of temperature In general, short-term effects are physical and reversible when the temperature is returned to ambient, whilst the long-term effects are mostly chemical and not reversible. The long-term chemical effects are usually referred to as the results of ageing.

All physical properties of rubber vary in the short-term when the temperature is changed, some to a greater extent than others. Throughout this book the need has been emphasised to measure properties over a range of temperatures in order to fully characterise the material. In principle, almost any property could be used to monitor the general temperature sensitivity of a rubber but obviously some properties are more satisfactory than others. In practice, the simpler mechanical tests are frequently chosen to give experimental simplicity, but these are not necessarily the most sensitive. It is most sensible, where possible, to monitor the properties which are most relevant to the service applications, and this principle also applies when measuring the ageing effects of long-term exposure. Particular types of test for the short-term effect of temperature which require individual comments are thermal expansion, the detection of glass transition point, and low temperature tests, the last subject also including longer term tests for measurement of crystallisation.

15.1 THERMAL EXPANSION

The coefficients of linear and volume expansion are defined respectively as:

$$\alpha = \frac{1}{l}\cdot\frac{\partial l}{\partial T} \qquad \beta = \frac{1}{V}\cdot\frac{\partial V}{\partial T} \quad \text{at constant pressure}$$

where: l = length, V = volume, and T = temperature.

In practice an average value of the expansion coefficient over a given temperature range is often taken. This is the case in crude measures of mould shrinkage (see Section 7.2.5) where expansion, or rather contraction, of the rubber is the main contribution to the property measured. For an isotropic and homogeneous material the coefficients are related by $\beta = 3\alpha$.

The coefficient of linear expansion can be measured as an average over tens of °C to reasonable precision using a precision cathetometer. Most contact methods of measuring length are not suitable because of the low stiffness of rubber; and for the same reason it is often necessary to support the test piece in a horizontal position. A convenient procedure is to lay a test piece on a smooth surface in a glass-sided bath, but it must be emphasised that only moderate precision can normally be obtained, unless an extremely sensitive optical measuring device is employed.

The classical method for volume expansion is to use a liquid-in-glass dilatometer. The test piece is placed in a chamber and covered with a known mass of liquid. As the temperature is raised the increase in volume is detected by the rise of the liquid up a graduated capillary and the expansion of the rubber can be deduced after making corrections for the expansion of the liquid and container, etc. Such a method is specified for plastics in ASTM D864[1] and a detailed account of the dilatometer technique has been given by Bekkedahl.[2] Although the dilatometer method is essentially simple and can yield very precise results, in practice great care has to be taken in calibrating and operating the apparatus. The procedure is somewhat lengthy and, if a cathetometer is used to follow the level in the capillary, tiring for the operator. Automatic dilatometers and thermodilatometers have been devised using various 'transducers' to measure the capillary height.

A useful review of methods of measuring the thermal expansion of polymers has been given by Griffiths.[3] This, naturally, covers dilatometers and various methods for detection of length changes but, in addition, methods measuring pressure change rather than length, the use of capillary rheometers and the use of thermomechanical analysers are discussed.

15.2 TRANSITION TEMPERATURE

A crystalline solid will melt when heated and this change of state is known as a first order transition, and is accompanied by a discontinuous (step) change in density and heat content. A second order transition is one in

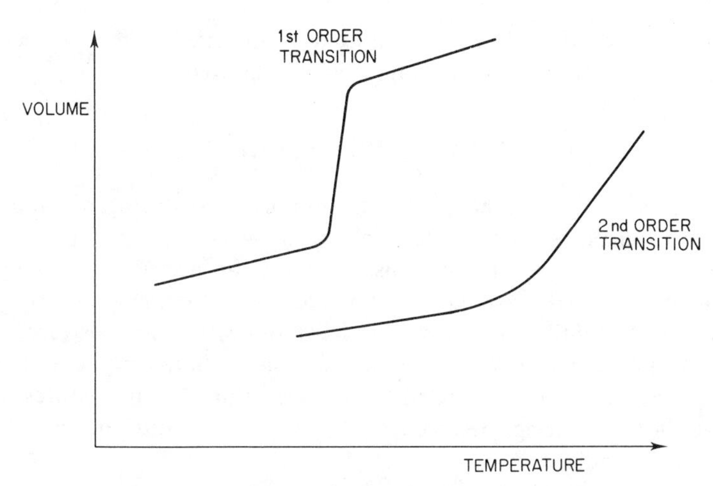

FIG. 15.1. First and second order transitions.

which these 'primary' properties do not show a step change but their rate of change with temperature alters abruptly. This is illustrated diagrammatically in Fig. 15.1. Rubbers show such a transition temperature when changing from the glassy state to the rubbery state.

The glass transition (T_g) is also marked by a large change in modulus; as the temperature is decreased the material loses its rubber-like characteristics, passes through a leathery state near the transition temperature and becomes a hard brittle glass. Other physical properties, mechanical, thermal and electrical, change by various degrees at the glass transition and hence there are potentially a number of different tests which can be used to measure the transition temperature. In Chapter 9 the interdependence of temperature and frequency was briefly discussed; increasing frequency is equivalent to decreasing temperature. Consequently the measured T_g will depend on the frequency of the test used, a 'fast' test yielding a higher T_g than a 'slow' one. Furthermore, the measured T_g will depend on the heating or cooling rate if temperature is changed continuously and in any case a gradual transition will often be seen rather than the idealised sharp transition shown in Fig. 15.1.

There are no standard methods for measuring T_g, but traditionally change in expansion (or density) by dilatometry is the most often used. Thermal properties, for example specific heat, are also widely used, particularly the methods of differential thermal analysis[4] (see also Section 14.1). Measuring T_g by mechanical methods is usually done with the dynamic methods as discussed in Chapter 9 but estimates could be made from the standardised low temperature tests discussed in the next section. Occasionally electrical methods have also been used.

15.3 LOW TEMPERATURE TESTS

Any physical test can be made at sub-normal temperatures and for particular purposes it will be desirable to follow changes in, for example, tensile strength, dynamic modulus, resilience, or electrical resistivity as the temperature is lowered. With reduction of temperature, rubbers become stiffer until finally becoming hard and brittle, and also recovery from an applied deformation becomes more sluggish. Largely for practical convenience a number of specific low temperature test procedures have evolved for measuring these general trends in behaviour and have been widely standardised.

These low temperature tests can be grouped as follows:

Rate of recovery (set and retraction)
Change in stiffness
Brittleness point

In addition some rubbers, for example, natural rubber and polychloroprene, stiffen at low temperatures by partial crystallisation. This is a gradual process continuing over many days or weeks and is most rapid at a particular temperature characteristic of each polymer, for example −25°C for natural rubber. Hence, tests intended to measure the effect of crystallisation must detect changes in stiffness or recovery after periods of 'ageing' at a low temperature.

15.3.1 Recovery Tests

The most straightforward way to measure the effect of low temperatures on recovery is by means of a compression set or tension set test. Tests in compression are favoured and a method has been standardised internationally. The procedure is essentially the same as set measurements at normal or elevated temperatures and has been discussed in Section 11.3.1.

As the recovery of the rubber becomes more sluggish with reduction of temperature the dynamic loss tangent becomes larger and the resilience lower (see Chapter 9), and these parameters are sensitive measures of the effects of low temperatures. Procedures have not been standardised but rebound resilience tests are inherently simple and quite commonly carried out as a function of temperature. It is found that resilience becomes a

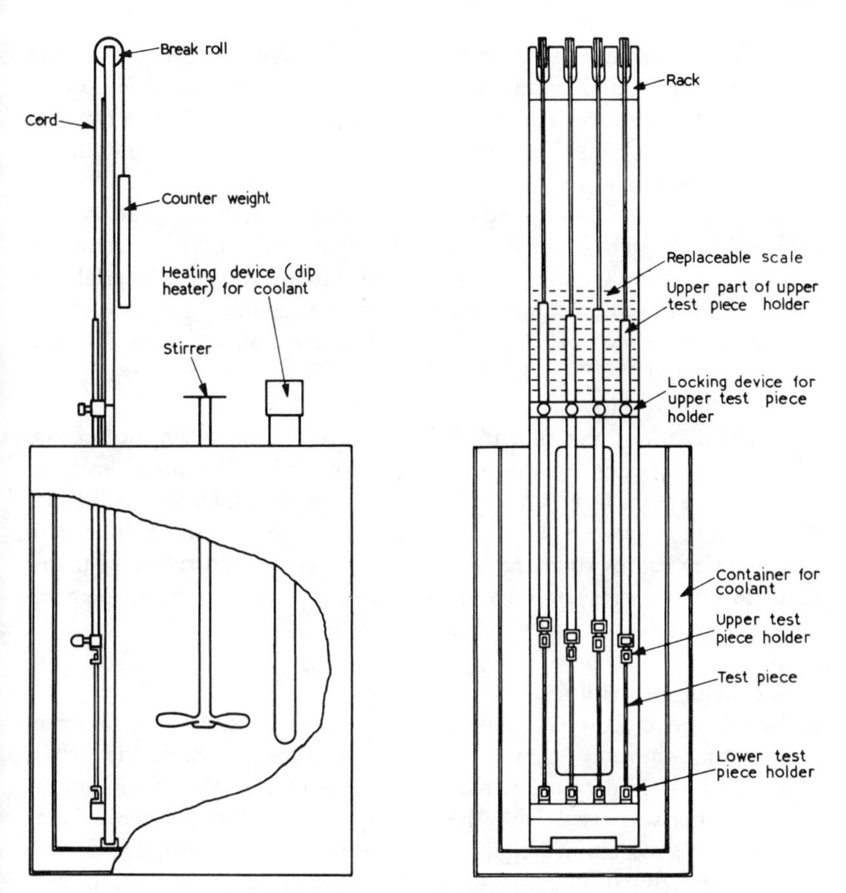

FIG. 15.2. Retraction (TR) apparatus.

minimum when the rubber is in its most leathery state and rises again as the rubber becomes hard and brittle.

A particular form of recovery test developed as a measure of low temperature behaviour is the so-called temperature–retraction test (TR test) which is standardised internationally as ISO 2921.[5] The test consists of stretching a dumb-bell test piece, placing it in the stretched condition in a bath at −70°C and allowing it to retract freely whilst the temperature is raised at 1°C per minute. The temperatures are noted at which the test piece has retracted by 10%, 30%, 50% and 70% of the applied elongation and these temperatures designated TR10,TR30, etc.

A suitable form of apparatus is shown in Fig. 15.2. The upper test piece clamp is counterbalanced to give a small stress of between 10 and 20 kPa on the test piece, and it is essential that the cord and pulley system is virtually friction free. The standard makes no mention of any automatic heater control to raise the temperature at 1 °C per minute but this is desirable if not essential.

The same test is also standardised in BS 903:Part A29[6] and ASTM D1329.[7] With such an *ad hoc* method it is essential that the details of procedure given in the standard are followed to achieve good interlaboratory agreement. The ISO and BS methods are identical and the ASTM appears to have no really significant differences but all allow different elongations and ISO 2921/BS903 note that different elongations may not give the same results.

ASTM does not recommend the method for specification use because of insufficient evidence as to reproducibility, and in Britain at least the test does not seem to be very widely used. It is also stated in ASTM that the difference between TR10 and TR70 increases as the tendency to crystallise increases, that TR70 has been found to correlate with low temperature compression set and TR10 to correlate with brittle temperature, but no evidence or references are given.

15.3.2 Change in Stiffness

Although torsional tests are little used to measure stiffness at ambient temperature, they have proved very convenient for measuring the change of stiffness as temperature is reduced. Originally a Clash and Berg type of apparatus was used with the torque to twist the test piece being provided by a system of weights, cord and pulleys, but current standards favour the Gehman apparatus with torque provided by a torsion wire. The main advantage of this type of test over, for example, a tensile test, is that the apparatus is relatively simple and cheap with an integral, liquid filled, low temperature bath and can readily accommodate several test pieces at once.

The Gehman test is standardised in ISO 1432[8] and the apparatus is shown diagrammatically in Fig. 15.3. A strip test piece is held in two clamps, the lower one fixed and the upper one being capable of being attached to a stud at the bottom of a torsion wire. The top of the torsion wire is fixed to a torsion head which can be turned through 180°. A pointer attached to the bottom of the torsion wire moves over a protractor to indicate the degree of twist. In most apparatus the test pieces and clamps are contained in an insulated bath with a liquid heat transfer medium cooled by solid carbon dioxide but the standard also allows a gaseous transfer medium. A liquid gives a quicker approach to temperature

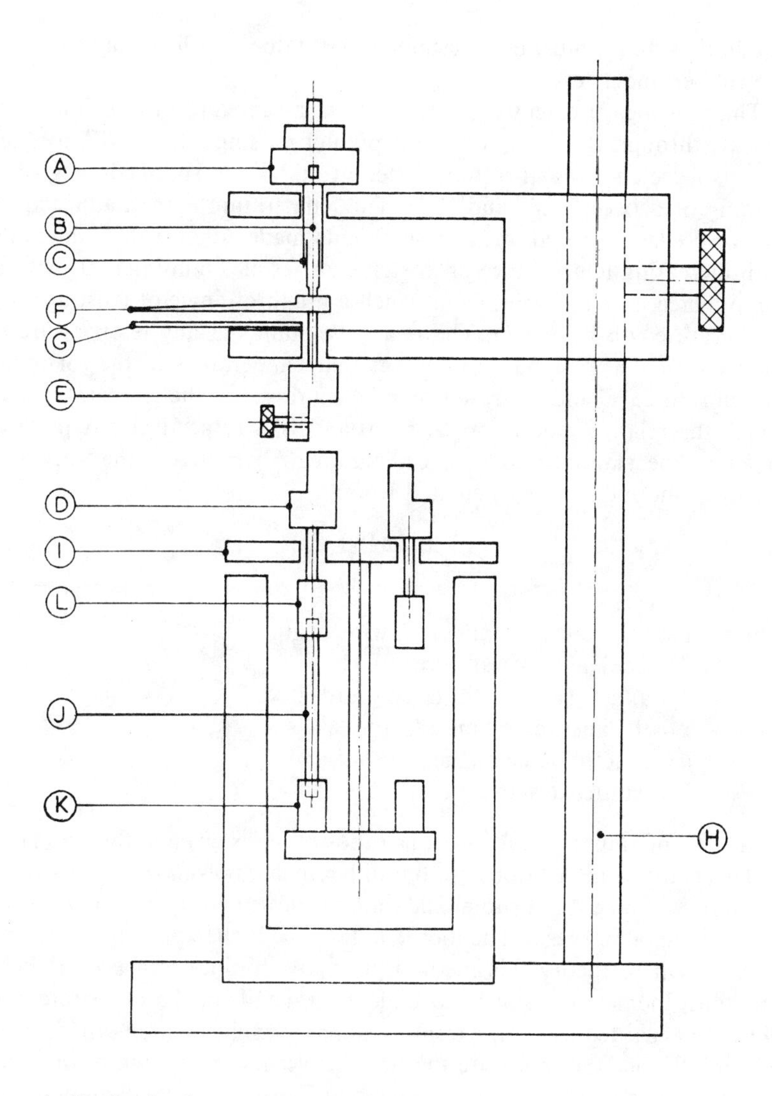

A. Torsion head
B. Torsion wire
C. Sleeve
D. Clamp stud
E. Screw connector
F. Pointer
G. Movable protractor
H. Supporting stand
I. Rack
J. Test piece
K. Bottom clamp
L. Top clamp

FIG. 15.3. Gehman apparatus.

equilibrium but it must be ascertained that it does not have any effect on the rubber under test.

The first measurement is made at 23°C by moving the torsion head quickly through 180° and noting the pointer reading after 10 s. A torsion wire is selected from the three different stiffnesses specified to give a reading of between 120° and 170°. The temperature is then adjusted to the lowest level desired and measurements made either at 5°C intervals with increasing temperature or in step changes at 5 min intervals with a ramp increase of 1°C/min, from which a graph of angle of twist against temperature can be drawn. The relative modulus at any temperature is the ratio of the torsional modulus at that temperature to the torsional modulus at 23°C and the results can be expressed as the temperatures at which the relative moduli are 2, 5, 10 and 100, respectively. A table is given in the standard to save calculation. Alternatively the apparent torsional moduli can be calculated from:

$$G = \frac{16Kl(180 - \alpha)}{ld^3\,\mu\alpha}$$

where: K = torsional constant of the wire (N m),
l = free length of test piece (m),
α = angle of twist of the test piece (degrees),
d = thickness of test piece (m),
μ = a factor based on the ratio b/d, and
b = width of test piece (m).

A table of values of μ for various ratios of b/d is given in the standard.

This is the same relationship, but different symbols have been used, as given in Section 8.4.2 for torsional modulus tests and in Section 9.2.2 for dynamic torsional tests. The factor $K(180 - \alpha)$ is the applied torque.

Although in theory this method gives an absolute measure of shear modulus, the actual result is dependent on details of the procedure and G is best regarded as an apparent torsional modulus. Among the factors which influence the result are the time between applying the torque and reading the angle of twist and the introduction of even small amounts of tensile strain. Interlaboratory variability with torsion tests is sometimes disappointing and tends to be worse for modulus results than for modulus ratios and for this reason specifications have tended to prefer the ratio method of presentation.

The interlaboratory differences can be particularly high, and the reasons not easy to identify, if different types of torsion apparatus are used. It is therefore not advisable to compare results from the standardised Gehman apparatus with, for example, Clash and Berg type instruments.

The Gehman test is also standardised in BS 903:Part A13[9] and ASTM D1053.[10] The British standard is identical to the international method but the ASTM has a rather different layout as it covers coated fabrics as well as rubbers and a single-point procedure is added for routine inspection. Be careful if calculating modulus to use the correct units for torsional constant.

A flexural procedure to obtain Young's modulus at low temperatures is given in ASTM D797.[11] This is a three-point loading method using dead weights to apply the load. This is one way of making use of a relatively simple apparatus to measure stiffness, and a somewhat similar approach has been taken by some workers using deformation in tension. Generally, the torsional methods have been much more popular and it is not immediately apparent what intrinsic advantages the other modes of deformation may have. When measures of stiffness in tension or compression are required then the standard methods for these properties can be used with a temperature controlled chamber on the tensile machine, the only disadvantage being the higher cost of the test equipment. Of all the modes of deformation, flexure is the least often used for normal ambient temperature testing.

The most obvious simple measure of stiffness, hardness, has in the past not often been used at low temperatures because of experimental difficulties due to icing up of the moving parts of the apparatus. There were no real fundamental reasons why this problem could not be overcome and suitable apparatus is now available although low temperature hardness tests seem to be mostly restricted to the detection of crystallisation (see Section 15.3.5).

One approach using deformation in tension is worthy of note. When the deformation at low temperatures is applied repeatedly the apparent modulus becomes lower until an equilibrium level is reached. Eagles and Fletcher[12] described a 'dynamic' low temperature test in which the test piece is continuously cycled in tension and the force monitored so that both initial and equilibrium moduli can be calculated, and, furthermore tests can be made at different applied strains. This method is undoubtedly a useful approach for development work providing both more comprehensive and probably more precise data but, despite claims of better reproducibility, it has not as yet been adopted as a standard method, principally because it involves rather more expensive apparatus than, for example, the Gehman test, and is not a multi-station apparatus.

Deformation in compression, or a mixture of bend and compression, is quite often used very successfully in *ad hoc* tests on complete products, for example rubber bellows. With low temperature tests, great efforts seem to have been made to look for relatively inexpensive methods. Even the self-contained torsional apparatus, which is cheaper than a tensile machine and cabinet, can be very time consuming in use. Care must be taken when formulating product tests using simple and inexpensive apparatus, that such 'details' as the rate of application of the force or the dwell time before noting deflection are carefully standardised because these can have a large effect on the result obtained.

15.3.3 Brittleness Temperature

Perhaps the most simple approach to measuring the low temperature behaviour of rubbers is to find the temperature at which it has become so stiff as to be glassy and brittle. The main disadvantage of this approach is that only one facet of low temperature behaviour is measured and that is at a point where for many purposes the rubber has long since become inadequate for its job. Nevertheless, brittleness temperature has been found to be a useful measure and innumerable *ad hoc* brittleness tests have been devised. These tests usually take the form of quickly bending a cooled strip of rubber and are almost inevitably very operator dependent and do not define the strain rate or the degree of strain precisely. Hence they show poor interlaboratory variability.

The best known of the bending tests using a simple jig was probably the method at one time standardised as ASTM D736. The test consisted of bending a strip between two platens and in its original form air was specified as the heat transfer medium but more often a liquid bath was used in practice. Because of the simplicity of apparatus it is still occasionally referred to under the D736 number even although this was discontinued in 1967. The test is still standardised as an appendix to ASTM C509 for cellular gasket material.

A more satisfactory method of measuring brittleness point, although still an arbitrary method, is that standardised in ISO R812.[13] A strip test piece, held at one end to form a simple cantilever, is impacted by a striker as shown in Fig. 15.4. The test piece can be either a strip or a T50 dumb-bell with one tab end removed. The critical dimensions are the test piece thickness, which is given as 2 ± 0.2 mm in each case, and the distance between the end of the grip and the point of impact of the striker. The striker radius is specified as 1.6 ± 0.1 mm and the clearance between the test piece clamp is 6.4 ± 0.3 mm. With these tolerances the maximum

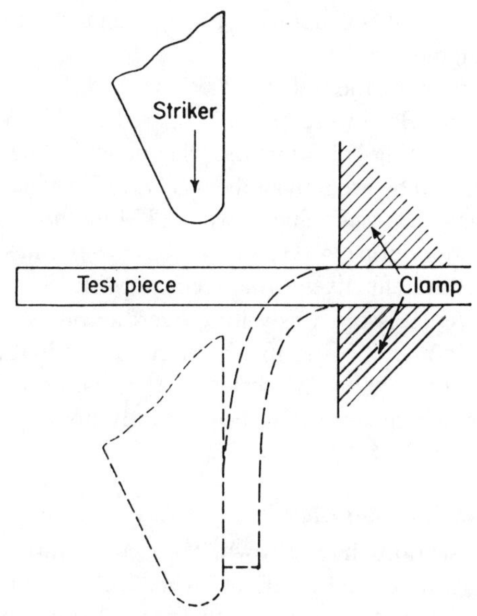

FIG. 15.4. Low temperature brittleness test; broken lines show position of striker and test piece (assumed unbroken) after impact.

surface strain in the test piece is held to almost $\pm 10\%$ and with the velocity of the striker controlled to between 1.83 and 2.13 m/s the rate of straining is constant to about $\pm 17\%$, which will give adequate reproducibility for most purposes.

Two examples of apparatus design are illustrated, although not very clearly, in the standard. If a pendulum or weight falling under gravity is used there is a possibility that the striker will be slowed up on first contacting the test pieces (it is usual to test four test pieces at the same time). The most satisfactory form of apparatus uses a powerful electric motor to continuously drive the striker and the test pieces are moved into its path by a solenoid. (This is not the solenoid apparatus illustrated in the standard.)

The heat exchange medium may be liquid or gaseous but for some curious reason the temperature control tolerance is given as $\pm 1\,°C$ for air and $\pm 0.5\,°C$ for liquid.

By impacting test pieces at a series of temperatures the brittleness temperature can be found as the lowest temperature at which none of the specimens tested failed. Failure is defined as the occurrence of any crack, fissure, hole or complete breakage visible to the naked eye. For specification purposes it is usual to test at a given temperature and record a pass or fail judgment.

The British version of the test, BS 903:Part A25,[14] is virtually identical to the ISO standard. Both were originally based on ASTM D746[15] which is designated for testing 'plastics and elastomers', whereas the ASTM method which should be used on rubbers is D2137[16] which covers 'flexible polymers and coated fabrics'. These two ASTM methods are not identical and their titles are confusing. However, D2137 specifies apparatus, test pieces and procedure effectively in agreement with ISO R812.

A revision of R812 is obviously long overdue but a new draft is at a late stage of development. It specifies both the determination of brittleness temperature and the use of the test at set temperatures on a pass/fail basis. There is much rewording but apparently no very large change in conditions.

15.3.4 Comparison of Methods

As the previous sections have shown, there are a large number of low temperature tests in existence. Even when *ad hoc* bending tests are disregarded together with the use of the normal range of physical tests, such as tensile modulus and resilience and the automation of a mechanical test by thermal analysis,[17] there remain several types of specially developed low temperature tests. The various tests do not all have equal relevance to a given product. A test, or tests, should wherever possible, be chosen to provide the information most relevant to the particular application, but for many quality control purposes a test is used simply as a general indication of low temperature behaviour. Whatever the relative merits of the different methods in any situation, the question of correlation between the methods is frequently asked.

A comparison between torsional modulus, 'dynamic' stiffness, retraction and brittleness which covers seven rubbers has been reported by Boult and Brown.[18] They found that the ranking of compounds was not always the same for all the tests but these results give a good general guide as to how the tests compare. Markova[19] also tested seven rubbers using brittleness, retractions, hardness and torsional tests and presented very detailed results showing the effects of time on hardness results and elongation on retraction. He found good linear correlations between brittleness point and TR10, and brittleness and T10 (Gehman) and reasonable linear correlation between modulus (hardness) and modulus (Gehman T2) and T10 and TR10. Kawamura *et al.*[20] also give correlations, including those between standard tests and a leakage test on seals and a flexing test. Kucherskii *et al.*[21] go to considerable lengths to analyse the reasons why the apparent modulus measured in a Gehman test is dependent on the test conditions and make a limited comparison with Russian standard methods in tension and compression. The brittleness and retraction methods have been compared with German standard dynamic methods by Englemann[22] who also investigated the effect of filler, softener and hardness on the low temperature properties of natural, nitrile and polychloroprene rubbers. Results for 15 compounds by the Gehman and temperature retraction tests are compared with results from a hybrid of the two, a 'twist recovery' test, by Wilson.[23]

Differences in results can occur between tests in a liquid and a gaseous medium. This is often because different times are required to reach equilibrium temperature, and if crystallisation is occurring, for example, the stiffness will be dependent on time of conditioning. It is also essential that if a liquid medium is used the liquid does not affect the rubber by swelling it or removing extractables, as either process can have a considerable effect on low temperature behaviour. Ethanol is most widely used but acetone, methanol, butanol, silicone fluid and *n*-hexane are all suggested in ISO 2921. Not all of these will be suitable for all rubbers and the suitability of any proposed liquid must be checked by preliminary swelling tests.

15.3.5 Crystallisation

In principle any of the low temperature tests can be used to study crystallisation effects by conditioning the test pieces at the low temperature for much longer times than is usual. In fact most of the standard methods include a clause to the effect that the method can be used in this way. In the temperature retraction test it is suggested that the greater degrees of applied elongation are used when the effects of crystallisation are to be considered, because crystallisation is more rapid in the strained state.

It would appear that the common standard low temperature tests are not thought totally suitable for measuring effects of crystallisation because a hardness test has been standardised for this purpose even although hardness tests are not so commonly used for measuring the immediate effect of low temperature. The international method ISO 3387[24] and the British method BS 5294[25] are essentially the same and are applicable to unvulcanised as well as vulcanised rubber. This is probably one reason why the hardness test has been introduced because the other methods would not be satisfactory with an unvulcanised compound. The unvulcanised materials often have very low hardnesses at the beginning of the test and the standards take account of the fact that the values after exposure may be in a different hardness range than at the beginning (see Section 8.1) by stipulating that the same range (instrument) should be used throughout. Results may be presented as the hardness increase in a given time or the time for the hardness to increase half-way to its final, equilibrium value.

In the hardness increase method for measuring effect of crystallisation the rubber is in the unstrained condition. For rubber in the strained state it would seem reasonable to suppose that the retraction or compression set methods could be adopted with longer exposure times. A method has been standardised as ISO 6471[26] which operates in compression but this uses much greater degrees of compression than are usual and requires a special apparatus. This method is derived from a Russian standard and essentially involves measuring recovery from low strain and from high strain, the degree of crystallisation being deduced from the difference between the two. It has the dubious distinction of being the only international physical test method for rubber on which the UK and USA abstained.

15.4 HEAT AGEING

Heat ageing on rubber is taken to mean the effect of elevated temperatures on rubber for prolonged periods but heat ageing tests are carried out for two distinct purposes. First, they can be intended to measure changes in the rubber at the (elevated) service temperature or secondly they can be used as an accelerated test to estimate the degree of change which would take place over much longer times at normal ambient temperature. The degree to which accelerated tests are successful in predicting long-term life at ambient temperature is highly debatable but nevertheless such tests are very widely used in specifications as a quality control test.

As with tests to measure the short-term effect of temperature, ideally, the properties most relevant to service should be used to monitor the effect of ageing but in practice the simpler mechanical tests are most often used as a matter of convenience. Heat ageing is not normally carried out in a vacuum or an inert atmosphere but in air or oxygen so that the ageing effect is caused by a combination of heat and oxygen. Most standard tests expose prepared test pieces rather than a whole product, which generally results in a relatively large surface area to volume ratio which assists in rapid diffusion of oxygen throughout the rubber to give more uniform ageing. At high temperatures the rate of oxidation of the rubber may be faster than the rate of oxygen diffusion so that uneven degradation occurs and hence it is undesirable to compare results from test pieces of markedly different size and shape.

15.4.1 Standard Tests

The international standard for heat ageing is ISO 188[27] which specifies an air oven and an oxygen bomb method. The principle is simply that prepared test pieces are aged for a given period at a given temperature and then tested for whatever physical properties have been selected and the results compared to those obtained on unaged test pieces. The standard does separate the two purposes for which ageing tests are carried out but warns of the difficulties and dangers of predicting room temperature performance from accelerated tests. It recommends that physical properties concerned in the service application are used to monitor ageing but suggests tensile properties and hardness in the absence of any specific information.

In the oven method the test pieces are exposed to air at atmospheric pressure in either the usual single chamber type of oven or a multi-cell oven. The advantage of a multi-cell oven is that by placing one compound only in each cell there is no danger of migration of plasticisers , etc., from one material to another. If a single chamber oven is used only very similar compounds should be heated together. With either type of oven there must be a steady flow of air through the oven giving between 3 and 10 complete changes of air per hour and no copper or copper alloys which may accelerate ageing should be used in the oven construction. Because oxygen is being used up in ageing processes it is important that the air flow is maintained and also that the rubber is exposed to air on all sides. Royo[28] has shown that even within the range of 3–10 air changes per hour differences in degree of degradation are found.

The temperature of test is left for the product or material standard to specify but 70°C or 100°C are those most commonly used for general purposes. The length of test is recommended as 1, 3, 7, 10, or a multiple of 7 days but the advantage of ageing for a series of times and constructing a graph of property level against time is not mentioned.

In the oxygen bomb method, test pieces are exposed to oxygen at above atmospheric pressure and at elevated temperature but otherwise the procedures are the same. The specific conditions of 70°C and 2.1 MPa pressure are given. Although the use of a high pressure speeds oxygen diffusion and hence helps to ensure uniform oxidation the increased acceleration of oxidation reduces the probability of the artificial ageing correlating with natural ageing.

The equivalent British standard, BS 903:Part A19[29] specifies only the air oven method as the oxygen bomb is now little used in Britain. The oven method is technically very similar to that given in ISO 188 but rather more detailed notes on the effect of test piece thickness are included. It is planned to make Part A19 identical with ISO 188 hence re-introducing the oxygen method.

ASTM finds need for four separate heat ageing tests. ASTM D573[30] is an air ageing method using a single chamber oven, and is hence similar to the parts of the British and ISO standards which deal with this type of oven, but differs appreciably in the description of the apparatus. Such differences in detail of oven construction which affect air flow can cause significant differences in ageing performance. Only standard tensile stress–strain tests are specifically covered as measures of degree of deterioration and perhaps because of this there is no discussion of the effect of oxygen diffusion rate and test piece thickness. The nearest ASTM method to the ISO and British multi-cell oven is D865[31] which exposes the specimens in test tubes which are in turn placed in a heat exchange medium. This results effectively in a very restrictively described, simple multi-cell apparatus but without the air flow past the test pieces being adequately specified.

An oxygen bomb procedure is given in ASTM D572[32] which uses the same standard conditions of 70°C and 2.1 MPa as in ISO 188. An air bomb method, which is perhaps a logical compromise between oxygen pressure and atmospheric air, is given in ASTM D454.[33] The standard conditions of 125°C and 0.55 MPa must give a high degree of acceleration and a lower temperature might be more appropriate for many purposes.

15.4.2 Correlation with Natural Ageing

The main use of such standards as ISO 188 is for quality control when the details of apparatus requirements and procedure must be observed to obtain good reproducibility. It is a fact, acknowledged in most standards, that no universal correlation between accelerated tests and natural ageing has been found. Even when heat ageing tests are used to simulate high temperature applications correlation may not be good unless care is taken to align test piece geometry, air flow and pressure and relevant tests are chosen to monitor changes. When the standard procedures are used to estimate performance at elevated temperatures or even, rather hopefully, to predict long-term room temperature performance, multi-point data should be obtained as a function of time of ageing and at a series of temperatures. This involves far more work than a test at a single temperature and time and consequently is not often carried out adequately. The usual procedure is to select a measure of the rate of change of the chosen property which can be termed the reaction rate and construct an Arrhenius type plot of log(reaction rate) against reciprocal of time. If this yields a straight line it is reasonable to estimate the rate of change at a lower temperature by extrapolation. An example of this approach is given for stress relaxation data in Section 15.4.3.

One difficulty when investigating the correlation between natural and accelerated ageing is the necessity to obtain the natural ageing data over a very long period of time. A major study of long-term behaviour covering 19 rubbers stored for up to 20 years has been reported.[34] The common accelerated ageing tests made for comparison were shown to be inadequate but no data have yet been published for these rubbers which used the Arrhenius plot approach. Mandel *et al.*[35] have given results for five rubbers in which elongation at break was used as the measure of degradation. An Arrhenius plot of accelerated test data appears to be in good agreement with natural ageing results up to 8 years. In such comparisons the natural ageing results are from 'shelf ageing', i.e. only the effect of oxygen and temperature are considered and the rubbers protected from light and other weathering effects during storage. Bergstrom[36,37] gives comparison of single-point accelerated tests with outdoor exposure and also considers the merits of using the product of tensile strength and elongation at break as the measure of change. Dlab and Kontry[38] suggest that elongation is the best measure to take. They describe an alternative procedure to an Arrhenius plot for predicting changes at lower temperatures but do not make comparisons with natural ageing.

15.4.3 Stress Relaxation

Obtaining multi-point ageing data is very time consuming. In addition, there is uncertainty as to the value of the simpler mechanical properties in relation to service performance and variability is increased by the use of separate test pieces for each point in the time and temperature sequence. Stress relaxation measurements in tension show some promise as a general guide to ageing performance by reducing or eliminating some of these difficulties. These measurements should not be confused with the stress relaxation measurements in compression used to study sealing force (see Section 11.2).

The concept of using tension stress relaxation measurements to investigate rubber networks which are undergoing chemical changes was originated by Tobolsky *et al.*[39,40] The measurement consists basically of monitoring the stress in a sample whilst subjecting it to an ageing procedure, usually accelerated. There are two variants of the technique, continuous relaxation in which the sample is held stretched throughout the test and intermittent relaxation in which the sample is stretched only periodically for short times to enable measurements to be made.

Under suitable conditions when viscous flow is not dominant it has been proposed that the reactions within the rubber network may be related to stress changes as follows. The decay of stress in continuous relaxation measurements provides a measure of the degradative reactions in the network whilst intermittent relaxation measures the net effect of both degradative and crosslinking reactions. In the continuous measurement any new networks formed are considered to be in equilibrium with the main network and do not impose any new stress.

The intermittent measurements are hence in effect a measure of the change in stiffness with time, with the advantage over the standard tensile measurements, that low strains more compatible with service conditions are used, very thin test pieces eliminate the effect of rate of oxygen diffusion and the same test piece is used for all measurements at a given temperature. The technique is attractive but experimentally fairly difficult and interpretation of the results is generally far from easy as was demonstrated by Brown *et al.*[41] Normally relaxation measurements are made at the ageing temperature but Brown *et al.* suggested a more simple test in which intermittent modulus measurements can be made at ambient temperature whilst ageing takes place at elevated temperature. This procedure enables tests to be made without specialised and expensive relaxation apparatus. A commercially available stress relaxometer for use in a cell oven is shown in Fig. 15.5. A miniature stress relaxometer using a strain gauge stress measuring device has been developed[42] which can be used in a gas absorption apparatus so that the relaxation can be related to the oxygen absorbed during degradation.

There is now an international standard, ISO 6914[43] which covers both the continuous and intermittent procedures plus the simplified intermittent method. The term stress relaxation, although commonly used, is not adopted in this standard on the basis that increases in stress as occur

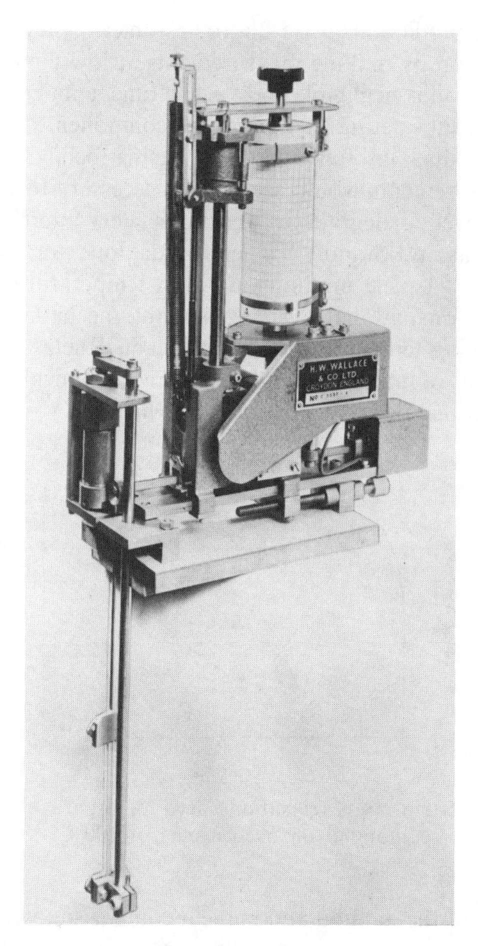

FIG. 15.5. Wallace–Shawbury age tester.

with intermittent tests cannot be called relaxation. Strip test pieces are used, 1 mm thick to minimise oxygen diffusion effects. Measurement at a series of temperatures is recommended for development purposes and it is preferred that results are presented in graphical form but no consideration is given to interpretation.

Many studies of tensile stress relaxation measurements have been reported but the majority have been concerned with continuous relaxation and particularly the practical difficulty of distinguishing between physical and chemical relaxation, for example the work of Aben.[44] Salazar *et al.*[45]

give results and predictions for a fluoroelastomer. They compensate for physical relaxation by making measurements at low temperatures where chemical relaxation is negligible and via the time–temperature superposition technique subtracting the physical component from their high temperature results. To fully describe ageing behaviour intermittent relaxation measurements would seem to be necessary to take account of crosslinking as well as degradative reactions. Very interesting results are given by Thomas and Sinnott[46] where predictions are obtained for the rate of change of tensile modulus at room temperature by both stress relaxation and conventional oven ageing, the mechanical testing being done at the ageing temperature. Clamroth and Ruetz[47] made a careful study of the value of intermittent relaxation measurements with particular reference to antioxidant evaluation in which they demonstrate the

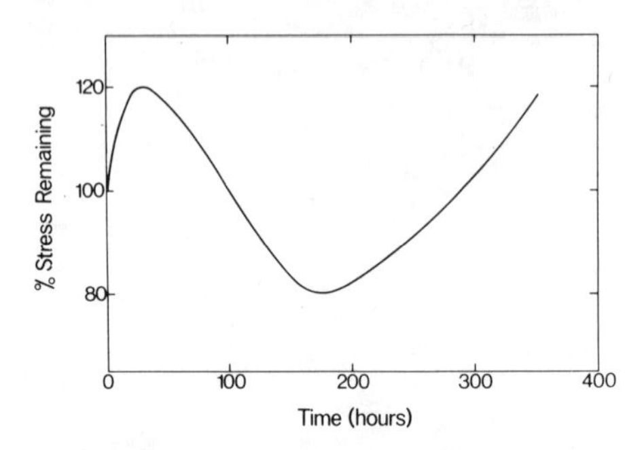

FIG. 15.6. Intermittent stress relaxation curve for a particular natural rubber compound on oven ageing at 100°C.

reproducibility of the method and make a comparison with conventional ageing techniques and practical experience. It is possible to make simultaneous measurements of continuous and intermittent relaxation and comment on the procedures is given by Ore.[48]

As an illustration of the possible treatment of stress relaxation results (or other accelerated tests) to predict long-term behaviour at lower temperatures, Figs 15.6 and 15.7 show data taken from Brown *et al.*[41] In Fig. 15.6 the change of modulus with time measured by their simplified intermittent procedure is given for a TMT/CBS cured natural rubber. The complicated shape of the curve emphasises the need to measure multi-point data in ageing tests. They took the rate of fall of stress in the fairly linear part of the curve as the reaction rate and from results obtained

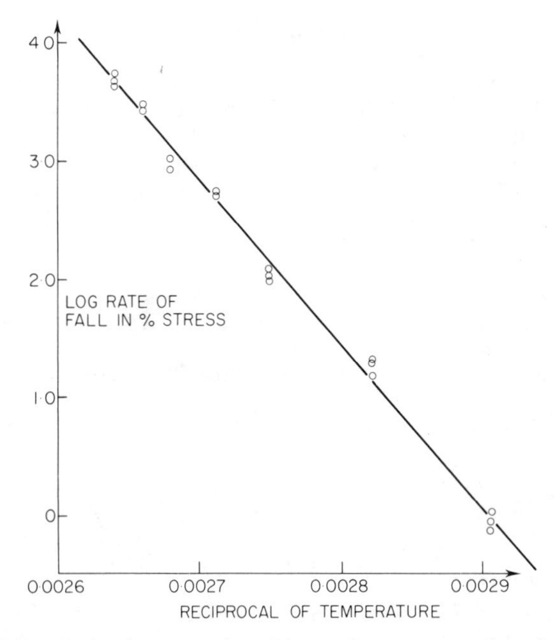

FIG. 15.7. Arrhenius type plot of intermittent stress relaxation data.

at several temperatures constructed the Arrhenius plot shown in Fig. 15.7. This gave a straight line fitting the equation:

$$\log(k) = \frac{-U}{RT} + C$$

where: k = reaction rate (fall in modulus as a percentage of original modulus over 50 h),
U = reaction energy,
R = gas constant,
T = absolute temperature, and
C = constant.

Extrapolation of this line to room temperature gives the reaction rate at that temperature. By assuming that this extrapolation was valid and also that the fall of stress was linear with time they predicted a time of some 200 years for this compound to drop 10% in stress at 50% extension and presumably will not be around to be proven wrong. Extrapolation is of course only valid if the same reactions are involved in changes at the lower temperature as at the test temperatures.

REFERENCES

1. ASTM D864-52. Coefficient of Cubical Thermal Expansion of Plastics.
2. Bekkedahl, I., Research, NBS, 42 RP2016, August 1949, p.145.
3. Griffiths, Marion D. (July 1974). *RAPRA Members J.*, 187.
4. Burfield, D.R. and Lim, K.L. (1983). *Macromolecules*, **16**(7).
5. ISO 2921, 1982. Determination of Low Temperature Characteristics–Temperature–Retraction Procedure (TR Test).
6. BS903, Part A29, 1984. Determination of Low Temperature Retraction (TR Test).
7. ASTM D1329–79. Retraction at Lower Temperatures (TR Test).
8. ISO 1432, 1982. Determination of Stiffness at Low Temperature (Gehman Test).
9. BS903:Part A13, 1983. Determination of the Stiffness of Vulcanised Rubbers at Low Temperature (Gehman Test).
10. ASTM D1053–85. Stiffening at Low Temperature Using a Torsional Wire Apparatus.
11. ASTM D797–82. Young's Modulus at Normal and Subnormal Temperatures.
12. Eagles, A.E. and Fletcher, W., ASTM Special Publication 553, Symposium Philadelphia, PA, June 1973, p.104.
13. ISO R812, 1968. Method of Test for Temperature Limit of Brittleness for Vulcanised Rubbers.
14. BS903:Part A25, 1968. Determination of Impact Brittleness Temperature.
15. ASTM D746–79. Brittleness Temperature of Plastics and Elastomers by Impact.
16. ASTM D2137–83. Brittleness Point of Flexible Polymers and Coated Fabrics.
17. Brazier, D.W. and Nickel, G.H. (1978). *Thermochimica Acta*, **26**(1–3).
18. Boult, B.G. and Brown, R.P. (Oct. 1974). *RAPRA Members J.*
19. Markova, L. (1985). *Int. Polym. Sci. Tech.*, **12**(2).
20. Kawamura, T., Kanayama, N. and Harashima, N. (1979). *Int. Polym. Sci. Tech.*, **7**(1).
21. Kucherskii, A.M. Fedyukina, L.P. and Gleizer, L.G. (1975). *Kauchi i Rezina*, No. 10, 46.
22. Englemann, E. (Nov. 1972). *Kaut. u. Gummi Kunst.*, **25**(11), 538.
23. Wilson, A. (Apr. 1967). *Rubber World*, 81.
24. ISO 3387, 1978. Determination of Crystallisation Effects by Hardness Measurements.
25. BS 5294, 1976. Determination of Crystallisation Effects in Rubbers by Hardness Measurements.
26. ISO 6471, 1983. Determination of Crystallisation Effects under Compression.
27. ISO 188, 1982. Accelerated Ageing or Heat Resistance Tests.
28. Royo, J. (1982). *Polym. Test.*, **3**(2).
29. BS903:Part A19, 1975. Heat Resistance and Accelerated Air Ageing Tests.
30. ASTM D573–81. Deterioration in Air Oven.
31. ASTM D865–81. Deterioration by Heating Air (Test Tube Enclosure).
32. ASTM D572–81. Deterioration by Heat and Oxygen.
33. ASTM D454–82. Deterioration by Heat and Air Pressure.
34. Brown, R.P. and Price, C, RAPRA Members Report No. 55, 1980.

35. Mandel, J., Roth, F.L., Steel, M.N. and Stiehler, R.D., Proc. Int. Rubb. Conf. Washington, Nov. 1959, p.221.
36. Bergstrom, E.W. (1977). *Elastomerics*, **109**(2).
37. Bergstrom, E.W. (1977). *Elastomerics*, **109**(3).
38. Dlab, J. and Kontry, F. (1985). *Int. Polym. Sci. Tech.*, **12**(4).
39. Tobolsky, A.V., Prettyman, I.B. and Sillon, J.H. (1944). *J. Appl. Phys.* **15**, 380.
40. Tobolsky, A.V., Andrews, R.D. and Hanson, E.E. (1944). *J. Appl. Phys.* **17**, 352.
41. Brown, R.P., Morrell, S.H. and Norman, R.H. (1973). *J. IRI*, **7**(4).
42. MRPRA (1974). *Rubb. Dev.*, **27**(1).
43. ISO 6914, 1985. Determination of Ageing Characteristics by Measurement of Stress at a Given Elongation.
44. Aben, W.J.G.M. (1974). *Med. Rubberind*, **35**(2).
45. Salazar, E.A., Curro, J.G. and Gillen, K.T. (1977). *J. Appl. Phys. Sci.*, **21**(6).
46. Thomas, D.K. and Sinnott, R. (Aug. 1969). *J. IRI*, 163.
47. Clamroth, R. and Ruetz, L., ACS Meeting, Philadelphia, 1982, Paper 57.
48. Ore, S. (1980). *J. Appl. Polym. Sci.*, **25**(10).

Chapter 16

Environmental Resistance

Temperature plays a part in all environmental tests but it is convenient to separate into Chapter 15 those tests which are particularly concerned with thermal resistance and to consider here the other types of 'environment' to which rubbers may be exposed. This includes resistance to liquids and gases other than permeability tests, which are discussed in Chapter 17.

16.1 MOIST HEAT AND STEAM TESTS

Although it is very important to condition at a known humidity for such tests as electrical properties there is not generally great concern over the long-term effects of moist heat judging by the lack of standard test methods. In some circumstances the effect of high humidity may be important; Soden and Wake[1] found a near doubling of the rate of deterioration of natural rubber by increasing the humidity to 100% in a 70°C air ageing test and polymers containing hydrolysable bonds (e.g. polyurethanes) can be especially liable to breakdown under humid conditions.

If the effects of humidity are of interest then tests along the same lines as the heat ageing tests would be used but both temperature and humidity would be controlled. A standard method is given in ASTM D3137[2] for determining the effect of moisture on tensile strength, it being recommended that a similar dry heat ageing test be conducted so that the effect of humidity can be isolated. The test pieces are suspended above water in a loosely capped container in an oven at 85°C for 96 h. More generally, a range of physical properties could be monitored after exposure and an injection type humidity cabinet (see Section 5.4) would give a range of humidities up to 100%.

The use of steam at 100°C or above would provide an accelerating effect although this would probably be considered too severe for most applications. Such a test would, however, be relevant for a product such as hose intended for use with steam and the particular test procedure would be found in the product specification. In designing any tests for exposure to steam it is necessary to control the amount of air (if any) present since oxygen at the temperatures used would have a strong deteriorating action.

16.2 EFFECT OF LIQUIDS

Tests in which rubbers are exposed to liquids are often called 'swelling tests' simply because the resulting change in volume of the test piece is by far the most commonly used measure of the effect of the liquid. Similarly, the tests are also referred to as 'oil ageing' because standard grades of mineral oil are the liquids most often specified.

Volume change is a very good measure of the general resistance of a rubber to a given liquid, a high degree of swelling clearly indicating that the rubber is not suitable for use in that environment. In addition, the degree of swelling can be related to the state of cure of the rubber, the crosslink density being estimated by use of the Flory–Rehner equation:[3–5]

$$\frac{1}{M_c} = \frac{\log_e(1 - V_r) + V_r + \mu V_r^2}{\rho V_1(V_r^{1/3} - \frac{1}{2}V_r)}$$

where: M_c = number average molecular weight of network chains,
V_r = the volume fraction of rubber in the swollen material,
μ = a solvent–rubber interaction constant,
ρ = density of the network, and
V_1 = molar volume of the swelling liquid.

The concentration of effective crosslinks is $1/2M_c$.

The standardised test procedures are concerned with the resistance of the rubber to the liquid, not the estimation of degree of cure, and generally recommend the measurement of change in dimensions, tensile properties and hardness as well as volume change. The action of a liquid on rubber may result in absorption of liquid by the rubber, extraction of soluble constituents from the rubber and chemical reaction with the rubber. Usually absorption is greater than extraction and an increase in volume results, but this is not always the case. For some products a decrease in volume or dimensions could be more serious than swelling and if there is significant chemical reaction a low swelling may be hiding a large deterioration in physical properties. Consequently, although degree of swelling is a good general indication of resistance it is important to also measure the change in other properties. Swelling, being relatively simple to measure fairly precisely, is particularly useful as a quality control test. Indeed its use has been suggested for general routine checks on composition.[6]

In planning an exposure to liquid test there are a number of general points which need to be considered. The degree of volume change with time will, in general, follow the form shown in Fig. 16.1 and it is preferable to take several readings to ensure that the full curve is recorded. If only a single time of exposure is used this should not fall on the early part of the curve where the degree of swelling is changing relatively rapidly. The time to reach equilibrium or 'maximum' swelling will increase with increased test piece thickness, in a manner roughly proportional to the square of the thickness. The time to 'maximum' swelling will also be roughly proportional to the viscosity of the liquid.

Curve III of Fig. 16.1(a) does not show a maximum swelling but a continued slow rise in volume which can be attributed (in natural rubber at least) to oxidation from air not being totally excluded. Particularly at high temperatures, consideration should be given as to whether air is present in the intended application because oxidation is likely to affect mechanical properties rather more than it does swelling. If there is some extraction the level of swelling may fall slightly before reaching equilibrium, as shown in curve II. If extraction is greater than swelling then the curve would show a reduction in volume, reaching an equilibrium (negative swelling) level.

16.2.1 Standard Methods

The international method for resistance to liquids is ISO 1817.[7] It has general clauses covering choice of liquid, conditioning, temperatures and duration of test and then details separate routines for determining the changes in various physical properties. The result is a somewhat repetitious document which is not particularly easy to follow when changes in more than one property are to be determined at the same time. It would be

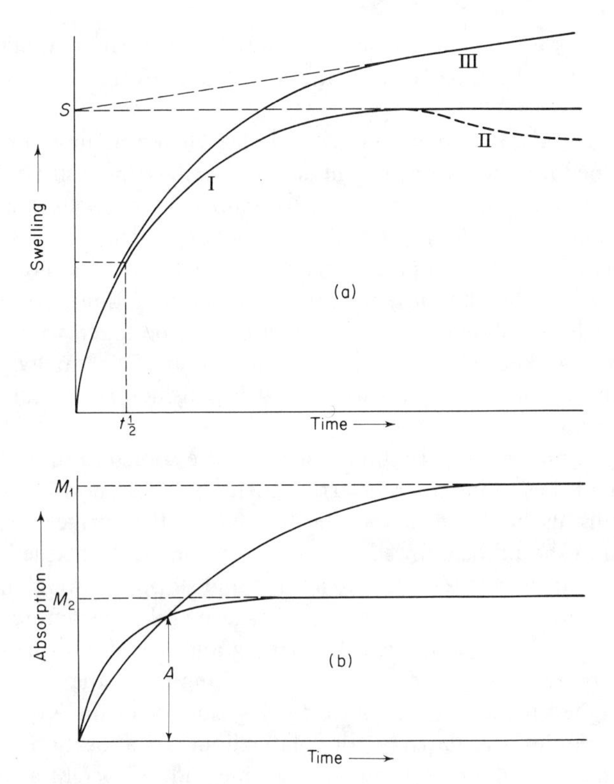

FIG. 16.1. (a) Time–swelling curves (diagrammatic). I, in absence of oxidation; II, in absence of oxidation with extraction of soluble matter; III, with oxidation. S = swelling 'maximum', $t_{1/2}$ = time to reach $\frac{1}{2}$ swelling maximum; (b) water-absorption curves differing in maximum absorption (M_1, M_2) and in absorption rate. Absorption A can arise from either a high maximum absorption M_1 and a slow approach or a low maximum absorption M_2 and a more rapid approach.

difficult to simplify the standard because although essentially the tests consist only of measuring properties before and after immersion there are a number of variants which must be included and fine detail must be observed if good reproducibility is to be achieved.

The preferred method for volume change is the gravimetric method which consists of weighing the test piece in air and in a liquid (usually water) before and after immersion and calculating volume change on the basis that volume is proportional to weight in air – weight in water. The test piece should be between 1–3 cm^3 in volume and 2 ± 0.2 mm thick.

The thickness is very important, especially if equilibrium swelling is not reached within the time scale of the test, but the other dimensions are not critical.

Care must be taken to exclude air bubbles when weighing in water and this is helped if a trace of detergent is added and/or the test piece quickly dipped in ethanol before weighing. If the rubber is less dense than water, then a sinker must be used in the same manner as for density measurements (see Section 7.1). The test piece is then immersed in the test liquid for the chosen time at the chosen temperature. At least 15 times the test piece volume of liquid should be used and care must be taken to ensure that the rubber is exposed on all sides to the liquid. This can be done by suspending the test pieces on wires or it is satisfactory to rest them on glass marbles.

After the immersion period the test pieces are cooled to room temperature which is best done by transferring them to a fresh portion of the test liquid. Surplus liquid must then be blotted off the surface and yet no evaporation should be allowed. When the test liquid is volatile it is usual to rapidly transfer the test piece to a tared and stoppered weighing bottle, but if the liquid was, for example, a lubricating oil rather more time can be allowed, and will be needed, for wiping and it is doubtful whether a weighing bottle is necessary. After the weighing in air the test pieces are again weighed in water, the transfer being done very quickly if the test liquid is volatile. Alternatively, for volatile liquids evaporation curves can be plotted by weighing as a function of time and the weight at zero time estimated by extrapolation.

The change of mass is sometimes used as a quality control measure and is simply obtained by weighings in air only.

Volume change can also be calculated from measurement of dimensions but the emphasis in ISO 1817 is on obtaining change in length, etc. The method for change in length, width and thickness uses a test piece 50 mm × 25 mm × 2 mm and the dimensions are measured before and after immersion using a dial gauge for thickness and, preferably, an optical system for length and width. A second dimensional change method monitors the area of a rhomboid shaped test piece by measurement of its diagonals. This method is semi micro, the rhombus usually having sides of about 8 mm length and can conveniently be used with very thin test pieces which will reach equilibrium swelling quickly. As originally described,[8] a magnified image of the test piece is projected to enable greater accuracy and convenience of measurement to be achieved. By assuming that swelling is isotropic, i.e. swelling in the thickness direction is equal to that in the other directions volume change can be calculated from:

$$V = \left[\left(\frac{AB}{ab} \right)^{3/2} - 1 \right] 100\%$$

where: V = volume change,
A and B = lengths of diagonals after swelling, and
a and b = lengths of diagonals before swelling.

It may be of interest to know the amount of matter which a liquid extracts from the rubber. Despite the fact that neither is very accurate, two procedures are given in the standard: drying the treated rubber to find loss in mass from the original or drying off the test liquid and weighing any residue. It is difficult to see the value of attempting to standardise this sort of procedure, as it is unlikely to be used in specifications.

ISO 1817 specifies two procedures for determination of tensile properties and hardness, one immediately after immersion and one after immersion and subsequent drying. The exposure procedure is similar to that for volume change, dumb-bells or rings being immersed for tensile measurements and a piece of sheet for micro hardness tests. (Normal hardness test pieces would take too long to reach equilibrium.)

The cross section of tensile test pieces is measured before immersion but the gauge length for elongation measurement marked after immersion. This is simply the most convenient practical arrangement, although the calculated results, especially stress at given elongation, are to say the least arbitrary. After removal from the test liquid it is necessary to wipe off surplus liquid, mark the gauge length and make the test within three minutes. A little practice is needed to achieve this, and a little more time would probably have little effect if the test liquid was not volatile at room temperature. The alternative procedure, which cannot be expected to give the same results, is to dry the test pieces to constant mass at 40°C and at reduced pressure, recondition at standard laboratory temperature and then test. The question to dry or not to dry should be answered on the basis of the relevance to service and quite possibly both figures would be of interest.

The final procedure in ISO 1817 is for exposure of a test piece to a liquid on one side only. This is applicable to relatively thin sheet materials whose service conditions involve one sided exposure. A suitable jig is used to contain the liquid and the change of weight measured. The result is expressed as change in mass per unit surface area.

The British standard BS 903:Part A16[9] is essentially the same as ISO 1817 but includes additional procedures. The method for measuring surface absorption is the volumetric procedure with the results expressed as a fraction of the total surface area exposed and is intended to be used only when the viscosity of the test liquid is such that with the thickness of test piece and length of immersion used less than 1/10th of the equilibrium swelling level is reached, i.e. the absorption of liquid is largely confined to the surface layers. There is no advantage in this method of expressing results if all the test piece dimensions are defined but if the dimensions vary from test to test it is more sensible to relate the volume change to surface area at times less that than to reach equilibrium. This may be the case in quality control tests when a long period of immersion could not be afforded.

The second addition in BS 903 is the inclusion of a procedure for measuring volume or dimensional change after drying, alongside the methods for change in tensile properties and hardness. It is to be expected that the long overdue revision of BS 903:Part A16 will be identical to ISO 1817.

The general ASTM method for effect of liquids on rubbers is ASTM D471.[10] It contains procedures for change in volume and change in mass by volumetric or gravimetric methods similar to the ISO and British methods but the specification of the exposure containers is more restrictive and more emphasis is put on the use of reflux condensors. The procedure for change in dimensions is similar to the first ISO method and an area change procedure is not mentioned. The procedures for extractable matter are included but rather curiously no mention is made in the calculation section and no warning as to the inaccuracy of the procedures is given. The methods for one sided exposure and for change in tensile properties and hardness before and after drying are similar to those in ISO 1817. In addition methods are given for determination of change in strength, burst strength, tear and adhesion of coated fabrics.

A second ASTM method D1460[11] gives a procedure for change in length after immersion using a long, relatively thin test piece. This used to be a much more widely standardised method but the volumetric method has generally proved far more convenient. One advantage of the length change method is that measurement can be made through the transparent wall of the container and hence can be used with liquids (or gases) under pressure.

16.2.2 Standard Liquids

Although for any particular application the liquid to be found in service should be used for testing, it has long been common practice to use standard liquids representative of the various types of liquid to which the product should be resistant. There is obvious advantage in this approach when considering interlaboratory reproducibility and quality control generally, particularly as commercial liquids are often not well-defined.

The principal standard liquids defined in the international and other standards are the oils and fuels originating from ASTM. It has generally been accepted that the ASTM oils, although defined in the standards, are only really satisfactory if obtained from a single source. In recent standards efforts have been made to improve the definitions, or specifications, for these oils but it is likely that the single source of supply situation will remain. In Britain, reference oils 1, 2 and 3 as well as the various service liquids specified in ISO 1817 and other standards are stocked by Rapra Technology Limited. Oils 1, 2 and 3 as specified in ISO 1817 are the same as those specified in ASTM D471 but ASTM also includes a No. 5 oil which has an aniline point intermediate between oils 1 and 2.

The standard reference fuels in ISO 1817 are intended to simulate the range of swelling induced by commercial petroleum derived fuels. In recent times motor fuels containing alcohols have been used in some areas but because the proportions have varied considerably suggested compositions are only given in a note.

Three simulated service liquids are given in ISO 1817, to simulate a diester type lubricating oil, and two hydraulic oils.

16.2.3 Non-Standard Methods

To determine the resistance of rubbers to liquids, as opposed to using the increase in volume as a measure of state of cure, the standard methods discussed above are almost universally used, at least in principle. When volume change and not change in other physical properties is wanted, either for quality control purposes or to estimate degree of cure, a variety of non-standard experimental procedures have been reported. Generally the intention is to simplify the test, speed up quality control or to use very small non-standard test pieces. An example which has gained acceptance as a standard method is the area change procedure described in Section 16.2.2 and indeed many of the non-standard procedures are based on dimensional change. It is not necessary to consider all the procedures here but they have been extensively reviewed by Brown and Jones[12] and for thickness increase measurement by Brown and Hughes.[13]

16.2.4 Water Absorption

The penetration of water into rubber is very slow compared with most organic liquids and hence with the usual test piece for the standard volumetric method a very long time is required to reach equilibrium. For this reason the surface absorption procedure is included in BS 903:Part A16 for measurements where equilibrium swelling is not reached. It would be preferable to measure equilibrium absorption and to achieve this in a reasonable time it is necessary to use a test piece with a very large surface area to volume ratio.

A standard procedure is given in BS 903:Part A18[14] which uses a test piece composed of small particles which will pass through an 850 μm sieve. The sample can be prepared by cutting, rasping or grinding. Unlike the more conventional swelling tests, the rubber is not exposed to 'liquid water' but to water vapour in a controlled humidity cabinet. There is then no problem of drying the surface of the rubber, which with small particles would be impossible, but the absorption measured is that relating to the level of humidity during test. Humidities of less than 100% are used because of the virtual impossibility in practice of maintaining exactly 100% RH. Because of its effect on humidity, it is necessary to control temperature during exposure very closely (especially at higher humidities) and 25 ± 0.2°C is specified in the standard. Reference can be made to Chapter 5 for methods of attaining and measuring given humidity levels.

The standard states that the measured equilibrium water vapour absorption is substantially the same as the equilibrium absorption which would be obtained by immersion in an aqueous solution and which would be in equilibrium with the vapour (apart from effects due to extraction of water soluble constituents), i.e. in a solution which would maintain the test humidity. As this implies, the equilibrium water absorption of rubber is reduced if the water is not pure and this test method, because 100% RH is not readily maintained, does not measure the absorption of pure water. When approaching 100% RH the effect is rapid and even very small amounts of a salt in solution will significantly lower the equilibrium absorption. Hence, tests intended to simulate the use of rubber in contact with an aqueous solution, rather than pure water, should be made with that solution or with one having the same equivalent relative humidity.

16.3 EFFECT OF GASES

In comparison with the effect of liquids on rubber very little testing is carried out on the effect of gases, with the most notable exceptions of exposure to air or ozone. (Permeability is considered in Chapter 17.) This presumably reflects the relatively small number of applications where the effect of gases other than atmospheric oxygen and ozone is important. There do not appear to be any general standard methods and for any particular gas and product a specialised test would need to be devised. For vapours or gases fairly readily obtainable in the liquid state the liquid is often used, but for other gases *ad hoc* tests, using an exposure chamber through which the gas is circulated, are necessary. The main practical difficulty is safely disposing of the used gas if this is toxic or an explosion risk.

Exposure to air or oxygen has been covered as regards laboratory tests by Section 15.4 on heat ageing. Natural weathering (Section 16.4) includes the effect of oxygen as well as that of sunlight, rain and ozone. Also in Section 16.4, artificial weathering tests are considered which again include the effects of air, light and water. It is apparent that all these environmental effects from heat ageing to artificial weathering are very much interconnected and it is largely for convenience that they have been separated in the present manner.

16.3.1 Effect of Ozone

Ozone exists in small quantities in the atmosphere but even levels of less than 1 part per hundred million (pphm) can severely attack non-resistant rubbers if they are in the strained condition. Hence, ozone attack is often the most important effect of exposure to the atmosphere and, not surprisingly, specialised laboratory tests have been developed which are more commonly used than general weathering tests. The effect of ozone is to produce clearly visible and mechanically very damaging cracking of the rubber surface, and although the importance of ozone may seem strange to anyone more familiar with other materials, the resistance of the polymer to ozone is the parameter considered of paramount importance in atmospheric applications involving tensile strains.

The laboratory tests are very much accelerated in that the levels of ozone used are much higher than those existing naturally in most parts of the world. In essence, they consist of exposing strained test pieces to air containing ozone and observing any cracking. In the simplest case, rubbers can be divided into those that will crack and those that do not, but because the common general purpose rubbers fall into the first category vast effort has been expended on finding anti-ozonants, etc., which will improve their resistance. The result as regards testing is that much time has been spent on trying to develop precise, reproducible and meaningful test methods. There is the inevitable problem of correlating an accelerated test with natural exposure, such minute quantities of an

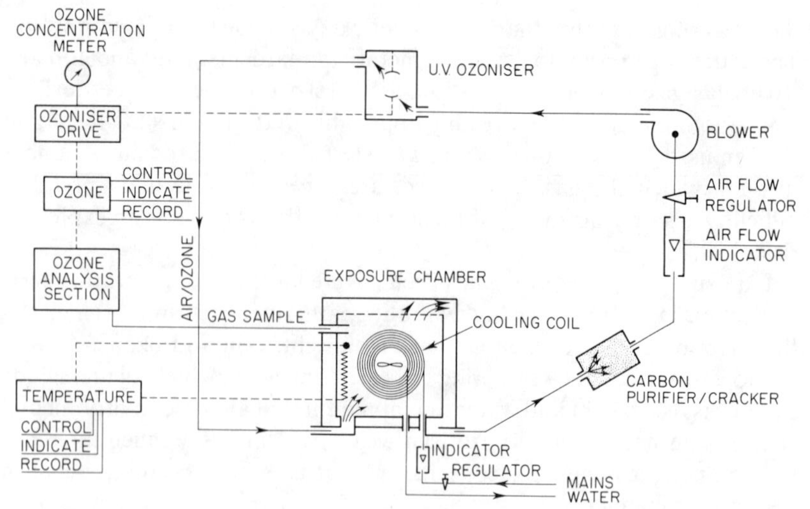

FIG. 16.2. Circuit of modern automatic ozone exposure apparatus.

unstable gas are extremely difficult to measure and control, the observation of cracks is by nature subjective and the pattern of cracking as a function of strain, time and ozone concentration is complex.

Very good standard test methods were established by British standards in 1963 and by ISO in 1972 but the revision of these methods has been a long, hard struggle which even now is not complete. The international standard, ISO 1431,[15,16] is in three parts, only two of which are published. Part 1 covers static tests, Part 2 dynamic tests and the third part will eventually cover the measurement of ozone concentration. BS 903:Parts A43[17] and A44[18] are identical to 1431:Parts 1 and 2.

OZONE CABINET

This is essentially a closed non-illuminated chamber containing the test pieces at constant temperature and through which ozonised air at a known concentration is passed. The principle is shown in Fig. 16.2. It must be constructed of a material such as aluminium which does not decompose ozone. The dimensions of the cabinet are not intrinsically important but the flow rate and velocity of the ozonised air do affect the severity of attack and must be controlled.

The current British and ISO standards settle on a gas velocity between 12 and 16 mm/s but ASTM D1149[19] requires any velocity above 0.6 m/s and suggests the use of a fan to achieve this. Cheetham and Gurney[20] demonstrated the dependence of ozone attack on gas velocity but more investigation is necessary. The ISO level is very reasonable for a standard procedure but the ASTM conditions can lead to very high velocities and increased severity of attack. Too low a flow rate must be avoided or destruction of the ozone by the test pieces will, at least locally, reduce the concentration. The ISO and British methods now define the flow rate in terms of the test piece area and say that the ratio of area to rate should not exceed $12\,s\,m^{-1}$. ASTM suggests a gas replacement rate of 3/4 cabinet volume per minute.

The ozonised air must be evenly distributed throughout the chamber and the fan in the ASTM cabinet achieves this but at the expense of an uncontrolled air velocity. Probably a better procedure is to use a diffuser at the gas inlet and also to attach the test pieces to a mobile test piece carrier such that each test piece 'visits' every part of the cabinet at intervals. Such a carrier is recommended in the ISO standard.

The ozone can be produced by an ultra-violet lamp or a silent discharge tube. The latter is usually necessary if very high concentrations are required but is not as convenient for normal use because oxides of nitrogen are also produced unless the tube is fed with pure oxygen. The nitrogen oxides may affect the degree of cracking and would interfere with chemical methods of measuring ozone concentration.

MEASUREMENT OF OZONE CONCENTRATION

Most ozone tests are carried out at concentrations in the range 25 pphm–200 pphm and a very sensitive method is necessary to precisely measure these low levels. The traditional chemical methods rely on the reaction of ozone with potassium iodide to produce iodine, the iodine being estimated volumetrically by reaction with sodium thiosulphate. There are many variants on this basic method, including different arrangements for passing the gas through the solution, the type of buffer used and different methods for estimating the end point of the titration. These chemical methods are not suitable for continuous monitoring or automatic control and instrumental methods are widely used in practice, notably the electrochemical and UV absorption methods. The former utilises the same basic reaction as the chemical methods but estimates the iodine by change in electric current passing through the iodide solution. All these methods are in theory absolute and do not require calibration.

Unfortunately the numerous variations on the chemical methods, the electrochemical and the UV methods do not all agree and, despite considerable investigation, the problem has not yet been fully elucidated. Here lies the difficulty in reaching agreement on the third part of ISO 1431.

It is perhaps interesting to remember that an early 'ozometer' used the decay of stress in a strained piece of rubber to indicate concentration!

There is not space to detail all the theories, experiments and arguments which have been put forward. In earlier standards and draft revisions variations on the original chemical method due to Crabtree and Kemp[21] were used. In Britain and elsewhere variations on the electrochemical method of Brewer and Milford[22] are much more commonly used because they are continuous and may be automatic. More recently the UV instrumental method which has the same advantages has become increasingly popular.

Evaluation and comparison of chemical and electrochemical methods by Brown *et al.*[23–25] firstly concluded that the leading chemical methods of the time read something of the order of 40% higher than the electrochemical procedure. Thelamon[26] has shown that the UV method also reads lower than the chemical method with the traditional buffer but also that results from chemical methods are lowered by changing to a basic acid buffer. These conclusions, generally found also by other workers, led to a swing towards a standard using a basic acid buffer and then to the UV method. At the time of writing attention has centred on the careful investigation made by Wundrich and Hentrich[27] who conclude that the chemical methods with the original buffer yield the correct figures and also give good reasons for the lower results with other chemical methods and the electrochemical method, but do not explain the UV results!

You may question whether the absolute value of ozone concentration is of importance, it being reproducibility between laboratories and the establishment of a standard that matters. Probably the main reason why this attitude has not been carried through before now is that it would be logical to settle on an instrumental method as these are used in practice but they are also rather difficult to standardise precisely, being commercial instruments with particular, perhaps arbitrary, characteristics. So, we all take enormous care over every aspect of an ozone test, agonise over the effect of air velocity, the time of conditioning and whether or not we can see a crack whilst probably using a very different concentration to the next fellow.

TEST PIECE

Ozone only attacks rubber in the strained condition although with the less resistant rubbers the 'threshold strain' for attack may be very low. The most obvious test piece is a thin strip held in tension between clamps

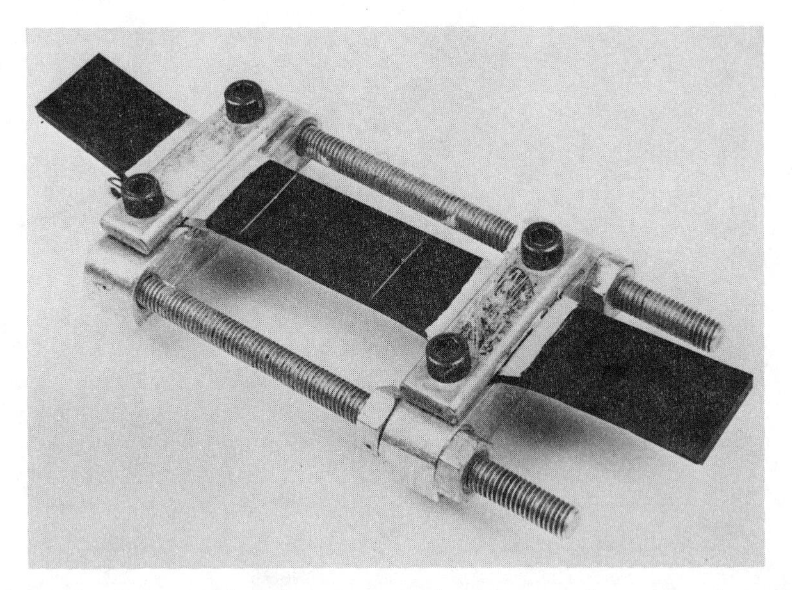

FIG. 16.3. Strip test piece for ozone test showing protective paint and gauge marks.

made of a material which does not decompose ozone (Fig. 16.3). This type of test piece is specified in ISO 1431, BS 903 and ASTM D1149. A variation of this is to add tab ends to the strip to facilitate gripping and it is expected that ISO 1431 will be modified to include this form of test piece. A particular form, the T50 dumb-bell is already included as a note and has the advantage of small size when cutting from products. Stretching tab-ended test pieces is made particularly easy by hooking the ends over suitable frames (Fig. 16.4).

It is often more convenient with extrusions to wrap them around a mandrel, although the resulting strain is generally less well-defined than it is with strips. ASTM D1171[28] is complementary to D1149 and specifically covers a triangular cross section test piece which may be moulded or extruded and is bent around a mandrel.

It is usually desirable to expose test pieces at a number of different strains. The small T50 test pieces are economical when this is the case but in theory at least it would be advantageous to have a form of test piece which covered several strains simultaneously. An annulus test piece (Fig. 16.5) was developed by Amsden[29] specifically to give a graduated range of strains and this is noted in ISO 1431. The only objections to this

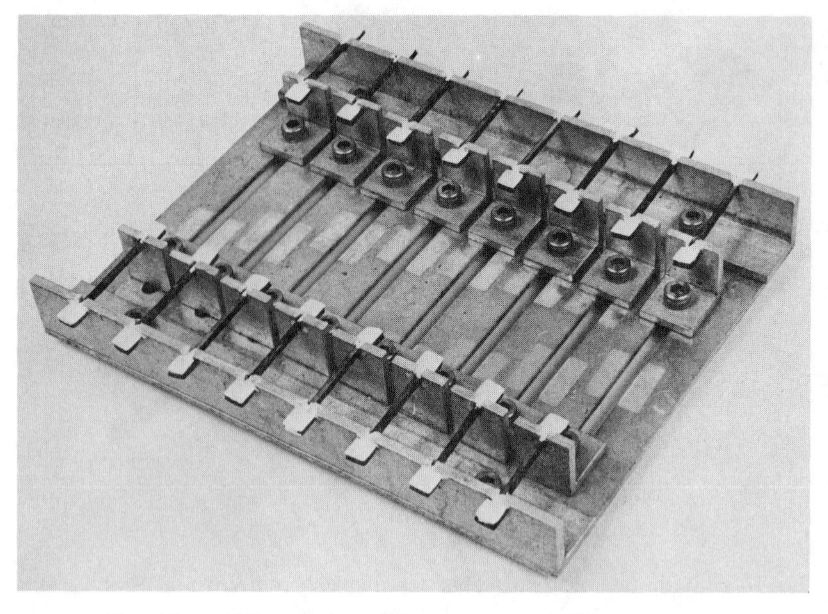

FIG. 16.4. T50 test pieces for ozone tests at different strains.

test piece are that the strain is slightly different on the two sides and a fear has been expressed that cracks in the high strain region may affect the strains remaining in the lower strain regions. Because of this, and its relatively large size, a number of T50 test pieces or even a number of strips are preferred by many workers for multi-strain exposure. ASTM D1149 includes a bent loop (without a mandrel) and a tapered strip which give varying strains along their lengths, although one would not expect these strains to be very precisely defined. A further multi-strain test piece is the trapezoidal specimen proposed by Dlab[30] but this does not appear to have been evaluated widely.

The areas where a test piece is attached to clamps and cut edges are preferential sites for cracking. It is generally good practice to coat clamped areas with an ozone resistant paint (which does not affect the rubber in any way) but cut edges are best left. For most purposes a Hypalon-based paint is satisfactory. Clamps, even when made of material such as aluminium, should be 'soaked' in ozone prior to use. Any pattern or flaws on the test piece surface will also tend to act as stress raisers and show preferential cracking.

FIG. 16.5. Annulus test piece for ozone tests, both strained and unstrained.

CONDITIONING

Because anti-ozonants and waxes, which to be effective must form a surface bloom, are used to enhance ozone resistance it is usual to condition test pieces in the strained state before exposure. The usual conditioning period is between 48 and 96 h and the test pieces should be kept in the dark and in an ozone-free atmosphere. For this treatment to be effective the test piece surface must not of course be touched in the course of subsequent handling. Where specifications wish to specifically exclude compounds which rely on an adequate wax film for protection the conditioning period is dispensed with. Hill and Jowett[31] in a criticism of ozone test methods strongly make the point that the conditioning process should be relevant to service conditions if a discriminating evaluation of waxes is to be made.

TEST CONDITIONS

Preferably, a series of strains should be used but if for specification or quality control purposes a single strain level is used this is usually 20%.

The most widely used standard ozone concentration is 50 ± 5 pphm. This is much higher than levels found in most parts of the world and a lower test level such as 25 ± 5 pphm is perhaps better but still very high. Any lower levels would be virtually impossible to control with present equipment. Despite 50 pphm being relatively high compared to average ambient levels it is, with exposure periods of a few days, quite convenient for discriminating between poor and good ozone resistance, although no correlation with service can be implied. To eliminate all but the most ozone resistant rubbers a level of 200 ± 20 pphm is often used. Extremely high levels such as the 15 000 pphm which used to be specified in BS 903 are nowadays very rarely met with in specifications. If a rubber is not completely ozone resistant it will fail at much lower levels than this.

Although the ISO and BS standards express the ozone concentration in pphm ASTM D1149 now uses partial pressure in mPa. The significance of this has been demonstrated by Veith and Evans[32] and it is expected that when the third part of ISO 1431 is eventually agreed it will introduce this measure. Basically, the rate of cracking is a function of the collision rate of ozone molecules with the rubber. At different atmospheric pressures in the cabinet the collision rate and hence the cracking will be different at the same concentration expressed in pphm. Clearly, the effect is important in locations with, relatively speaking, extremes of pressure from standard.

Temperature does affect the rate of ozone cracking but it cannot be said simply that higher temperatures accelerate the effect. The blooming characteristics of different waxes can make an increase in temperature increase or decrease ozone resistance. Above about 70°C all ozone is destroyed. In the current major standards 40°C is specified, just a small degree of acceleration above ambient and practically the lowest level which can be controlled without cooling.

Hill and Jowett[31] have demonstrated that a test at about 0°C is much better for discriminating between protective wax systems and coincides with a temperature at which protection is most difficult. It can only be for reasons of inconvenience that standards bodies have not adopted this suggestion and their procedure for conditioning mentioned earlier.

It has been shown[33] that the humidity of the ozonised air can affect the rate of ozone attack. Generally any significant change is restricted to very high humidities and ISO 1431 states that normally the humidity shall be less than 65% at the test temperature.

TEST PROCEDURE

Briefly, the test pieces are placed in the chamber at the required concentration and inspected at intervals. If opening the cabinet reduces the concentration for appreciable periods this will affect the results. An automatically controlled cabinet will probably show a much faster response than a manual one. Some workers observe the test pieces through a window in the cabinet. This avoids disturbing the concentration and any malhandling of the test pieces but it is doubtful whether the inspection can be as thorough as when the test pieces are removed.

Most specifications give a set exposure period but it is preferable to examine test pieces at a series of times such that data can be obtained on the relationship between strain and time to appearance of cracks. ISO 1431 requires examination to be carried out with a lens of ×7 magnification but, unfortunately, any examination of cracks is to some extent dependent on the eyesight of the operator. In practice many workers say a crack is only a crack if they can see it with the naked eye. The alternative procedure of measuring relaxation in stress will be discussed later. An optical method of automatically detecting cracks has been described by Zeplichal[34] but this is relatively complicated and has not been considered for standardisation.

EXPRESSION OF RESULTS

When only a single strain and exposure period has been used the result is simply expressed as either cracking or no cracking. The degree of cracking can be described and a number of arbitrary scales have been used, but they are all terribly subjective. The most widely used is the 0–3 scale where 0 is no cracking, 1 is cracks only seen under magnification, 2 is very small cracks and 3 anything worse. Even this simple rating scheme falls down when there are one or two large cracks only.

Alternative approaches are based on recording the time until the first appearance of cracks. Regular inspections are necessary but more information is gained than in a 'go/no go' test. The real advantage of recording time to failure is realised when a number of strains are used. It is then possible to observe the relationship between time to crack and applied strain. In some cases a linear plot will show the existence of a limiting threshold strain as shown in Fig. 16.6. For other rubbers a log–log plot will yield a straight line but it is dangerous to extrapolate this to much longer times.

The first criterion for describing a material as ozone resistant is total freedom from cracking. Therefore, the higher the threshold strain after a given exposure period, or the higher the limiting threshold strain if this exists, or the longer the time before cracks appear at a given strain, the better is the ozone resistance. However, when materials with relatively low ozone resistance are being compared such that cracking is inevitable during service life, then the severity of cracking is important. Very small

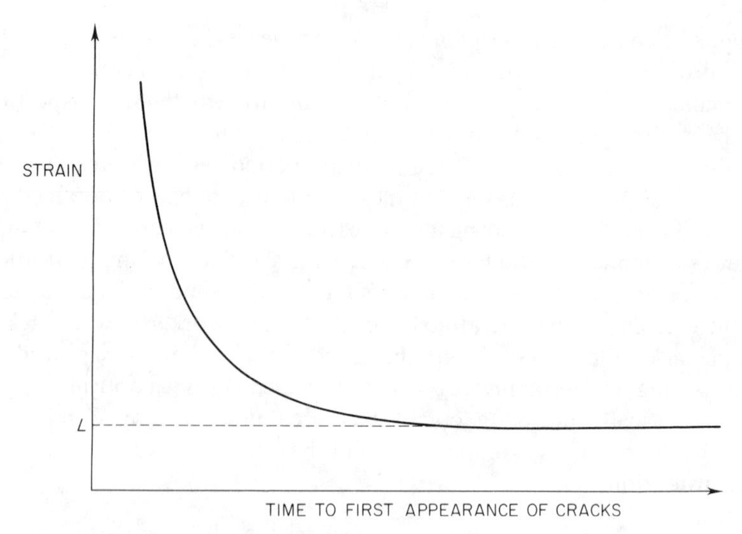

FIG. 16.6. Threshold strain. L = limiting threshold strain.

cracks may be of little consequence apart from a cosmetic point of view. This is usually the case when thick sections of rubber are involved and cracking is confined to the surface.

The way in which the severity of cracking is related to strain is not simple. The usual trend is shown in Fig. 16.7; by definition there are no cracks below the threshold strain for any given exposure period. A few cracks, often large, are found at strains slightly above the threshold and the cracks will become more numerous and smaller at progessively higher strains. It is quite possible for the cracks at very high strains to be so small as to be invisible to the naked eye. As exposure time increases numerous very small cracks may coalesce to form larger but relatively shallow cracks. Hence, a non-resistant rubber at high strains could be more suitable than a 'better' resistance rubber just above its threshold strain. This illustrates the futility of protecting a rubber such that it will just pass a single strain and period standard test when it will exhibit large cracks in service.

DYNAMIC OZONE TESTS

All the previous discussion was referring to test pieces exposed to ozone whilst held at a static strain. Because many products are subjected to

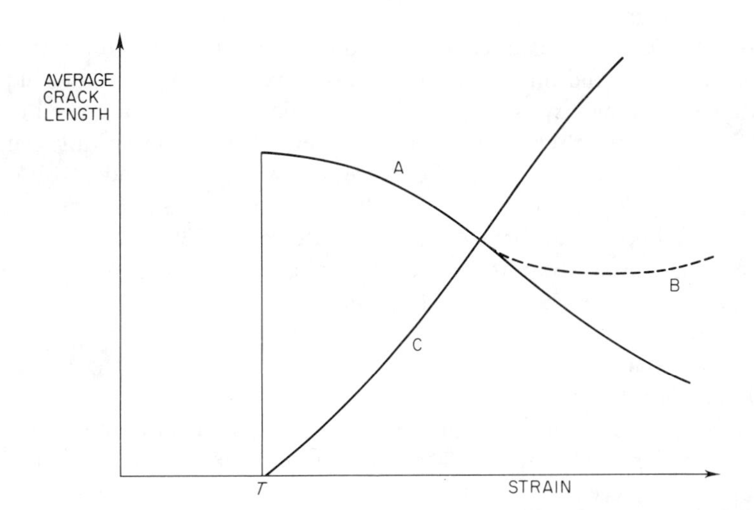

FIG. 16.7. Relation between crack size and strain (diagrammatic). *T* is the threshold strain; curve A, average crack length; curve B, average crack length with coalescence of cracks; curve C; crack density.

cyclic strain in service and because protective wax coatings, which are easily removed by mechanical contact, cannot withstand cycling there is much logic in using a dynamic exposure test. A standard method is given in ISO 1431:Part 2[16] and BS 903:Part A44[18] in which strip (or T50) test pieces are cycled in tension at 0.5 Hz. The low frequency is used so that there is little contribution from fatigue mechanisms. The exposure and expression of results is generally the same as in the static standard method but either continuous cycling of a sequence of dynamic cycles and periods of static strain is specified. Various views have been expressed as to which sequence correlates best with particular service applications[35,36] but no sequence is given in the standard reflecting the fact that no one sequence has attained widespread acceptance. For general purposes continuous cycling would be used.

It has also been shown[36] that T50 test pieces can be successfully used with complete fracture as the criterion of failure. This obviously results in a longer test but the means of assessment is much easier and not subjective. Although rupture is not widely used as a measure of ozone attack it is used in fatigue tests (Chapter 12) and could possibly be used for static ozone tests.

ASTM D3395[37] has a similar method to ISO 1431 but with a higher standard strain and only covers continuous cycling. D3395 also includes a second dynamic exposure procedure in which test pieces are fixed to a fabric belt which runs over a pair of pulleys. The test pieces are hence strained by bending so that the degree of strain is dependent on the thickness of the test piece as well as the pulley diameters. The advantage of this method is that there are no clamps to cause preferential cracking, but the maximum strain is less well-defined and it is less easy to vary the strain than it is with a test in the tensile mode.

STRESS RELAXATION

It has been commented earlier that the observation of cracking is a subjective measure and using the time to the onset of cracking still involves detecting the cracks by eye. Measuring the relaxation of stress in the strained test piece as ozone attack proceeds offers an attractive alternative which eliminates any assessment by the operator. Although this approach was suggested at least as far back as 1956 it has not as yet been widely accepted, perhaps because of the extra apparatus, expense and complexity and because a sensitive force detection system is needed to detect early symptoms of attack. Several workers, notably in France, have described apparatus.[38-42] Proposals have been made to ISO that a standard method should be developed but this has not advanced.

16.4 WEATHERING

The most catastrophic cause of deterioration of susceptible rubbers when exposed to the atmosphere out of doors is ozone. Because it is hopeless to try to find a consistent level of ozone in the atmosphere the usual approach is to use the laboratory ozone tests discussed previously. Oxygen, temperature, moisture and sunlight also affect rubber. At normal temperatures the rate of degradation by oxygen and heat is rather slow and accelerated ageing tests are usually used, despite difficulties of correlation (see Section 15.4). The effect of light on rubbers is much less important than it is on, for example, plastics and, degradation being generally restricted to the surface layer, is of most consequence in the case of coated fabrics and very thin-walled articles. The exception is change of colour in non-black rubbers. The nett result is that weathering tests on rubbers are carried out relatively infrequently and international standard methods have only recently been published.

The standards now adopted by ISO have been developed from, and are very similar to, the plastics equivalents which have been in existence for many years. It is intended that they should be combined at the next revision. ISO 4665[43–45] is in three parts; the first covers assessment of changes in properties after either natural or artificial weathering, the second deals with natural weathering and the third with artificial weathering. At the time of writing this third part is awaiting publication.

Part 1 is not a very specific document; in essence it says select the relevant properties and test before and after exposure in accordance with the relevant international test method. It does, however, set procedures for conditioning, expression of results, etc., which will allow sensible comparison of results with other workers and describes the use of the grey scale for colour change.

Part 2 gives fairly comprehensive information on the construction of exposure racks, exposure sites and measurement of radiation dosage. It includes provision for the exposure of strained test pieces to measure resistance to ozone cracking in the same manner as used in laboratory accelerated tests. ASTM D1171[28] and D518[46] also cover outdoor exposure but only in respect of measuring resistance to ozone. However well you standardise the procedures for natural weathering exposure the great advantage of obtaining deterioration data under 'real' conditions is to some extent nullified by the enormous variation of those real conditions from site to site and time to time. Also the tests will usually need to be continued for very long times. In this respect it is often recommended that exposure trials of any new product should begin as early as possible so that the experience or data is always ahead in time of actual use and may be used to give advance notice of any possible trouble.

When we come to artificial weathering it should be remembered that the aspects of weathering generally of most concern with rubbers, ozone attack and oxidative ageing, are catered for by the specific accelerated tests already discussed. ISO 4665:Part 3 covers general artificial weathering in which the main consideration is the effect of light and, if required, rain cycles but where the temperatures are usually fairly modest by rubber ageing standards and ozone is deliberately excluded. For special purposes more complex equipment is occasionally used (not covered by 4665) which additionally attempts to simulate corrosive atmospheres, for example salt spray.

The first objective in an accelerated test is to use a light source which simulates natural sunlight, particularly in the UV region, and ISO 4665 specifies only a xenon arc lamp with filters which, although far from perfect, is considered the best available at present. The standard also has an appendix which advises on the correlation with exposure to natural daylight and an appendix which describes fluorescent tube lamps. These have potentially many advantages over the xenon arc and as experience is gained with this type of light source it is likely that their use will become more widespread.

Although ISO 4665 specifies the irradiance of the light source it does not say how it should be measured. For measuring radiation dose it specifies the well known blue wool standards, other physical standards or instrumental methods. This is not complete or precise standardisation but reflects the somewhat uncertain present state of the art.

More detailed information on weathering tests, natural as well as artificial can be found in the large numbers of published papers which refer to plastics. For earlier work, an exhaustive bibliography was compiled by the Building Research Establishment[47] which follows on from the RAPRA review[48] of the subject up to 1968. A comparison of natural weathering with fluorescent tube accelerated tests, which included some rubbers has been reported by Wright.[49] Effect of various climatic conditions on rubbers are given by Bergstrom[50] and Sourisseau and Ehrhardt.[51] General discussion of accelerated weathering methods is given by Capron *et al.*[52] and by Nishizawa.[53] Morse[54] has evaluated specialised exposure racks and references 55–58 give comparisons of light sources and correlation with natural weathering.

The blue wool actinometric standards for estimating integrated radiation dose are well known but a number of polymeric standards have also been developed. Standardisation of these has unfortunately not proceeded but they offer a cheap, if less precise, alternative to instrumental methods but with advantages over the blue wool. Polysulphone[59] seems to have emerged as the preferred material. Instrumental methods are considered in references 60–65.

16.5 BIOLOGICAL ATTACK

Rubber (and/or the additives in it) can under some circumstances prove a tasty morsel for living organisms, particularly micro-organisms. Fortunately their interest is not so great that no piece of rubber is safe, and significant attack is comparatively rare. There are however circumstances in tropical countries where biological attack on rubber is a serious problem and recently concern has been expressed that rubber seals for water pipes are susceptible in temperate climates.

Exposure to living organisms is hardly a physical test (although measurement of the damage would be) and it is a very specialised subject. It is generally agreed that 'biological ageing' is a subject best entrusted to the experts and if a rubber is to be tested the assistance of industrial or academic establishments specialising in that field should be sought.

For particular products or circumstances where biological attack is very important, the problem has received careful consideration. Pipe joints, mentioned above, is a good example. After quite extensive investigations the British standard for pipe joint rings[66] now includes a requirement for resistance to microbiological degradation. The standard also specifies requirements for effect of materials on water quality which includes microbial growth. The introduction of these requirements has however been the subject of much controversy. Two fairly extensive reviews[67,68] cover microbiological deterioration and attack by insects and rodents respectively and references 69–71 are further examples of discussion of microbiological degradation. Cundell and Mulcock[72,73] describe methods they used to assess microbiological attack of natural rubber and there is an international standard, ISO 846,[74] for the resistance of plastics to fungi and bacteria.

16.6 FIRE

Most rubbers burn, although this fact has caused nothing like the alarm which the flammability of plastics, and particularly foams, has given rise to. The fire hazards of plastics and the methods of test have received enormous attention in recent years, including careful attention to the 'philosophy' of fire testing. Rubber has received the same attention on a much smaller scale and in general the principles of the fire testing of plastics apply to rubbers. Hence, fire testing will be dealt with briefly here, bearing in mind that a more detailed study of the subject as applied to polymers can be made by reference to the quantities of published information on plastics.

Some form of agreed and understood philosophy of fire testing is necessary because it is fairly easy to invent a host of more or less *ad hoc* fire tests which are confusing as to which aspect of fire they are meant to cover, may give positively dangerous impressions because of ill-conceived presentation of results and in no way predict the performance of the material in a real fire situation.

The most important distinctions to make are between large-scale and small-scale tests and to clearly define which aspect of fire is being evaluated, for example ease of ignition, rate of burning, smoke production, etc. Large scale tests are necessary to evaluate the performance of a material or product in most real fire situations. The point has been illustrated by the example of wood. It is easy to set fire to matchsticks with a small flame which would never get a large log burning, but if that log is fired by a large flame it may burn very well and would burn better if there were several logs together. In addition, one log might produce far more smoke than another which may be far more dangerous than the flame or heat. Small scale tests are in general restricted to investigating the ease of ignition of small amounts of materials by small flames and for the quality control of materials.

The international and most national standards committees whose terms of reference are concerned specifically with the polymer industry deal only with small-scale tests. The large-scale tests, which are not usually specific to any one type of material, are covered by committees whose concern and expertise is fire.

At the time of writing there are no ISO TC 45 fire test method standards for solid rubbers although fire tests are specified for particular products and laboratory methods specifically for rubber are being considered. There are published international test methods for cellular materials and plastics, the majority of which have been applied to rubbers. Comprehensive coverage of flammability is given by the *International Plastics Flammability Handbook*[75] whilst *An Introduction to Fire Dynamics*[76] is a good text on the science of the subject.

16.7 RADIATION

Radiation is taken here to mean atomic and nuclear particles, i.e. gamma rays, electrons, neutrons, etc. The intensity of such radiation at the earth's surface is not high enough to significantly affect rubbers and tests are only required in connection with applications in atomic or nuclear plant or perhaps where radiation is used to induce crosslinking. Not surprisingly, such a specialised subject has not given rise to a wide scale standardisation of test methods. ASTM D1672[77] specifies a recommended practice for exposure of polymeric materials to various types of radiation and a detailed account of testing polymers for radiation resistance has been given by Metz.[78]

REFERENCES

1. Soden, A.L. and Wake, W.C. (1951). *Trans. IRI*, **27**, 223.
2. ASTM D3137–81. Hydrolytic Stability.
3. Flory, P.J. (1950). *J. Chem. Phys.*, **18**, 108.
4. Flory, P.J. and Rehner, J. (1943). *J. Chem. Phys.*, **11**, 521.
5. Kraus, G. (Oct. 1956). *Rubb. World*, 67.
6. Martin, J.W., Peterson, E.R. and McCormick, H.E., ACS Meeting, Cleveland, Oct. 13–16, 1981, Paper 3.
7. ISO 1817, 1985. Determination of the Effect of Liquids.
8. Brown, R.P., RAPRA Research Report No. 191, 1970.
9. BS 903:Part A16, 1971. The Resistance of Vulcanised Rubber to Liquids.
10. ASTM D471–79. Effect of Liquids.
11. ASTM D1460–81. Change in Length during Liquid Immersion.
12. Brown, R.P. and Jones, W.L. (Feb. 1972). *RAPRA Bulletin*, 42.
13. Brown, R.P. and Hughes, R.C. (Aug. 1973). *RAPRA Members J.*, 194.
14. BS 903:Part A18, 1973. Determination of Equilibrium Water Vapour Absorption.
15. ISO 1431:Part 1, 1980. Resistance to Ozone Cracking—Static Strain Test.
16. ISO 1431:Part 2, 1981. Resistance to Ozone Cracking—Dynamic Strain.
17. BS 903:Part A43, 1982. Resistance to Ozone Cracking—Static Strain Test.
18. BS 903:Part A44, 1983. Resistance to Ozone Cracking—Dynamic Strain Test.
19. ASTM D1149–81. Surface Ozone Cracking in a Chamber (Flat Specimen).
20. Cheetham, K. and Gurney, W.A. (1961). *Trans. IRI*, **37**, 35.
21. Crabtree, J. and Kemp, A.R. (1946). *Ind. Eng. Chem. Anal. Ed.*, **18**, 769.
22. Brewer, A.W. and Milford, J.R. (1960). *Proc. Roy. Soc.*, **A256**(1287), 470.
23. Bonell, P. and Brown, R.P. (Sept. 1971). *Rubb. J.*, 64.
24. Brown, R.P. (April 1974). *Rev. Gen. Caout. Plast.*, **51**, 215.
25. Brown, R.P. and Peppiatt, A. RAPRA Members Report No. 13, 1978.
26. Thelamon, C. (1982). *Polym. Test.*, **3**(2).
27. Wundrich, K. and Hentrich, H. (1986). *Polym. Test.*, **6**(3).
28. ASTM D1171–83. Surface Ozone Cracking Outdoors or Chamber (Triangular Specimens).
29. Amsden, C.S., *The Annulus Ozone Test*, ICI Ltd, Dyestuffs Div., 1967.
30. Dlab, J. (Aug. 1975). *Plasty a Kaucuk*, **12**(8), 253.
31. Hill, M.L. and Jowett, F. (1980). *Polym. Test.*, **1**(4).
32. Veith, A.G. and Evans, P.L. (1980). *Polym. Test.*, **1**(1).
33. Kempermann, T. (June 1974). *Rubb. News*, **13**(9), 26.
34. Zeplichal, F. (1972). *Rev. Gen. Caout. Plast.*, **49**(11), 1047.
35. Lamm, G. and Thelamon Meynard, C., LRCC Tech. Bull., No. 85, 1970.
36. Fletcher, W., RAPRA (Unpublished work).
37. ASTM D3395–82. Dynamic Ozone Cracking in a Chamber.
38. Emin, G. (1971). *Rev. Gen. Caout. Plast.*, **48**(2), 145.
39. Bertrand, G. and Sebonc, J. (Sept. 1973). *Rev. Gen. Caout. Plast.*, **50**(9), 719.
40. Zuev, Yu. S. and Postorskaya, A.F. (1974). *Kauch i Rezina*, No. 5, 26.
41. LRCC Technical Bulletin No. 92, 1973.
42. Ganslandt, E. and Svensson, S. (1980). *Polym. Test.*, **1**(2).

43. ISO 4665, Part 1, 1985. Assessment of Change in Properties after Exposure to Natural Weathering or Artificial Light.
44. ISO 4665, Part 2, 1985. Methods of Exposure to Natural Weathering.
45. ISO 4665, Part 3. Methods of Exposure to Natural Weathering.
46. ASTM D518-83. Rubber Deterioration—Surface Cracking.
47. BRE Library Bibliography No. 247, Feb. 1975.
48. Matthan, J., Scott, K.A. and Wiechers, M. (1970). *Ageing and Weathering of Plastics*, RAPRA.
49. Wright, I., ERDE Technical Report No. 197, March 1972.
50. Bergstrom, E.W., US, NTIS, Rept. AD B000996., 1974.
51. Sourisseau, R. and Ehrhardt, D., Int. Rubber Conference, Kiev, 1978, Preprint C12.
52. Capron, E., Crowder, J.R. and Smith, R.G. (Mar. 1973). *PRT Polym. Age.*, **4** (3), 97.
53. Nishizawa, H. (1976). *Int. Pol. Sci. Tech.*, **3**(7), T/18.
54. Morse, M.P., ASTM Special Publication 781, 1981.
55. Ellinger, M.L. (1979). *J. OCCA*, **62**(4).
56. Voigt, J. (1981). *Off. Plast. Caout.*, **28**(292).
57. Kinmouth, R.A. (1981). *Polym. Paint Co. J.*, **171**(4060).
58. Aumasson, M. (1983). *Rev. Gen. Caout. Plast.*, **60**, 632.
59. Davis, A. and Gardiner, D. (1982). *Polym. Degrad. Stabil.*, **4**(2).
60. Minematu, Y. (1985). *Nippon Gomme Kyokaishi*, **58**(1).
61. Zerlaut, G.A., ASTM Special Tech. Publication 781, 1981.
62. Batty, T. (1979). *Polym. Paint Col. J.*, **160**(4013).
63. Suga, S. and Katayanagi, S. (1981). *Polym. Test.*, **2**(3).
64. Harris, P.B., BRE Current Paper No. 17/73, June 1972.
65. Capron, E. and Crowder, J.R., BRE Current Paper No. 17/75, Feb. 1975.
66. BS 2494 1986. Specification for Elastomeric Joint Rings for Pipework and Pipelines.
67. Heap, W.M., RAPRA Information Circular No. 476, 1965.
68. Pacitti, J., RAPRA Information Circular No. 475, 1965.
69. Rook, J.J. (1955). *Appl. Microbiol.*, **3**, 302.
70. Heap, W.M. and Morrell, S.H. (July 1968). *J. App. Chem.*, **18**, 189.
71. Kilkarni, R.K. (Oct. 1965). *Polym. Eng. and Sci.*, **5**(4), 227.
72. Cundell, A.M. and Mulcock, A.P. (1973). *Int. Biodetn Bull.*, No. 9, 91.
73. Cundell, A.M. and Mulcock, A.P. (1973). *Int. Biodetn. Bull.*, No. 9 (1–2), 17.
74. ISO 846 1978. Determination of Behaviour under the Action of Fungi and Bacteria—Evaluation by Visual Examination or Measurement of Change in Mass or Physical Properties.
75. Troitzech, J., Haim, J. (transl.) (1983). *International Plastics Flammability Handbook—Principles—Regulations—Testing and Approval*, Carl Hanser.
76. Drysdale, D. (1985). *An Introduction to Fire Dynamics*, Wiley Interscience.
77. ASTM D1672–66. Exposure of Polymeric Materials to High Energy Radiation.
78. Metz, D.J. (1966). *Testing of Polymers*, Vol. 2, Chapter 5 (Ed. J.V. Schmitz), Interscience Publishers.

Chapter 17

Permeability

Rubbers are by no means impermeable to vapours and gases although in many cases the rate of transmission is low. In a number of applications even a small loss (or gain) of liquid or gas may be important, for example balloons, fuel tanks or water vapour barriers, and in consequence the rate of transmission needs to be measured.

The theoretical aspects of permeation through polymers have been considered in some detail by Hennessy *et al.*[1] and in a very comprehensive review by Lomax.[2,3] Only the basic concepts are necessary here and Lomax's review, which describes and comments on virtually all known test methods as well as considering the theory and providing a bibliography of almost 100 references is recommended for a more detailed consideration of the subject. Gas or liquid can flow through the holes in a porous material but even if there is no porosity or flaws permeation through the material will take place by a process of absorption and diffusion. In the ideal case the quantity of gas or vapour being transmitted builds up to a constant steady state level after a period of time and in the steady state:

$$q = \frac{QtPA}{d}$$

where: q = volume of gas transmitted,
Q = permeability coefficient,
t = time,
P = partial pressure difference across the test piece,
A = test piece area, and
d = test piece thickness.

In many cases Q is a constant for a given gas and polymer combination but for other combinations, particularly with vapours, Q varies with, for example, test piece thickness or pressure difference and hence it is necessary to know the dependence of Q on all possible variables in order to characterise the permeability of the material completely.

The preferred units for permeability coefficient are $m^4 s^{-1} N^{-1}$ ($m^3 m s^{-1} Pa^{-1} m^{-2}$) but the terms permeability coefficient or permeability constant are often applied to various transmission rates using a variety of units and care must be taken to avoid confusion. Useful conversion factors are given by Yasuda and Stannett.[4] When the permeability coefficient is dependent on test piece thickness it is convenient to use a transmission rate—the amount of permeant transmitted per unit time and area for a given test piece thickness—which may be in units of $m^3 s^{-1} N^{-1}$ ($m^3 s^{-1} Pa^{-1} m^{-2}$). Transmission rate is almost always used in the case of vapours and often in the units $g\,24h^{-1} m^{-2}$.

The permeation of a gas through a polymer (disregarding flaws) takes place in two steps, the gas dissolving in the polymer and then the dissolved gas diffusing through the polymer. The solubility constant is the amount of a substance which will dissolve in unit amount of the polymer under specified conditions and the diffusion constant is the amount of substance passing through unit area of a given plane in the polymer in unit time for a unit concentration gradient of the substance across the plane. It can be shown that:

$$Q = SD$$

where: S = the solubility constant, and
D = the diffusion constant.

Although this simple relationship holds for some gases, for other gases and most vapours it does not and, as noted above, the permeability 'constant' is then not a constant. It depends on the solubility and diffusion characteristics but these may vary with different conditions. The permeability constant varies with temperature and although simple theory predicts that the change will follow an Arrhenius type relationship this also is not true for many vapours.

The permeability constant has been defined for steady state conditions, and at time before this is reached a smaller apparent permeability constant will be measured. Hence, when measuring permeability constant or transmission rate it is necessary to wait until the steady state has been reached. For some vapours, particularly with thick films, equilibrium can take several days or even longer.

17.1 GAS PERMEABILITY

The traditional procedures for measuring gas permeability involve setting up a pressure differential across the test piece and measuring by change of pressure or volume the amount of gas passing to the low pressure side of the system. A number of variations on this theme have been used but two procedures are standardised for rubbers as ISO 1399[5] Constant Volume Method, and ISO 2782,[6] Constant Pressure Method. The equivalent British standard methods, BS 903:Parts A17 and A30,[7,8] are very similar. These standard methods are outlined below but for a critical understanding of the problems of operation and likely errors reference should be made to, for example, the review by Lomax.[2,3] At the time of writing a draft revision is being considered by ISO TC 45 in which both the constant volume and constant pressure methods have been combined into one standard.

17.1.1 Constant Volume Method

The apparatus for ISO 1399 and BS 903:Part A17 (Fig. 17.1) consists of a metal cell having two cavities separated by the test piece. The high pressure cavity is filled with the test gas at the required pressure and this pressure must be measured to an accuracy of 1%.

It is suggested that the usual test pressure is between 200 and 400 kPa and the cavity should be at least 25 ml volume to minimise pressure loss during the course of the test. The low pressure cavity should be of as small a volume as possible and this requirement is helped by the use of rigid porous packing to support the test piece against the pressure of the test gas. The low pressure side is connected to a pressure measuring device, usually a capillary U-tube manometer which has an adjustable height reservoir as shown in the diagram. The test cell is enclosed in a constant temperature bath or other device to maintain the temperature within ± 0.5°C. The close tolerance is necessary because the permeability of many gases is extremely sensitive to temperature.

The test piece is a disc, suitable dimensions being between 50 mm and 65 mm diameter and thickness between 0.25 mm and 3 mm. The lower the permeability of the rubber the more advantageous it is to use a thin test piece. It is essential that the means of clamping the test piece in the cell is such that there is no leakage of gas.

After the cell and test piece have been assembled and the high pressure side filled with gas at the test pressure the increase in pressure on the low pressure side is measured as a function of time. The standard suggests a

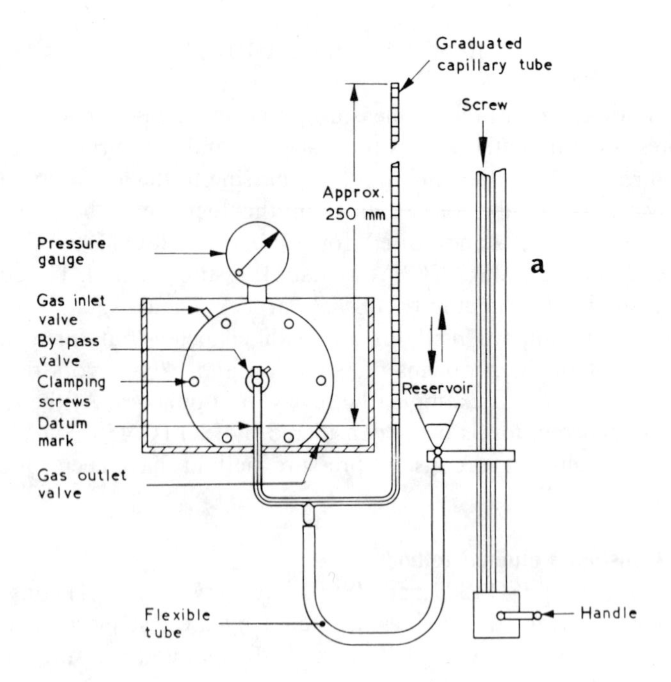

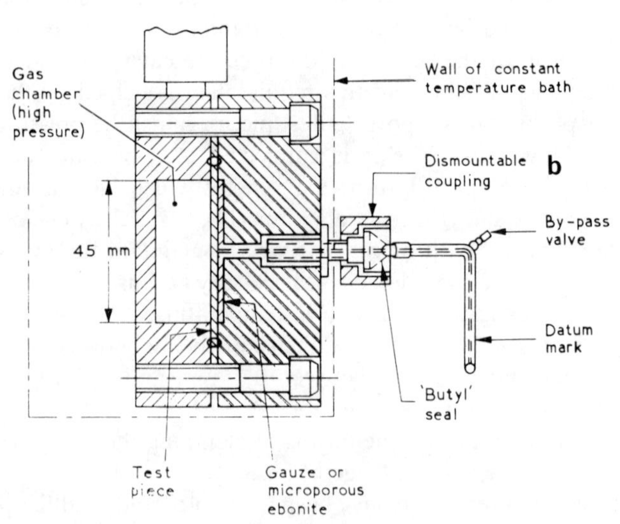

FIG. 17.1. Gas permeability apparatus, constant volume method. (a) Front elevation; (b) section.

conditioning period of at least 1 h to reach steady state conditions unless an approximate value of the diffusion coefficient is known, when the minimum conditioning time can be estimated from:

$$t = \frac{d^2}{2D}$$

where: d = test piece thickness, and
D = diffusion coefficient.

In the steady state a plot of pressure change against time should be linear. Any departure from linearity in the direction of increasing slope with time indicates that the steady state has not been reached. Leakage around the edges of the test piece will only result in an unexpectedly high rate of pressure rise. Any tendency for the slope to decrease with time is an indication of a leak from the low pressure side.

When a manometer system is used to measure pressure the reservoir height is adjusted to bring the liquid level above the datum mark with the by-pass valve open so that the pressure in the low pressure side is atmospheric. The by-pass valve is then closed. As the gas diffuses through the test piece the increase in pressure causes the liquid level to fall and as the meniscus passes the datum line a clock is started (i.e. at zero time). The reservoir is then raised to bring the meniscus above the datum line and both time and the manometer reading are noted when again the meniscus passes the datum line. This process is repeated to give a series of readings. In this way the pressure reading (manometer reading) is always taken at constant volume of the low pressure side of the cell.

The apparatus and procedure described require great care in setting up and in operation. The effort is eased considerably if an automatic pressure measuring device operating at effectively constant volume is used instead of the manometer.[9] Improvements as regards accuracy and sensitivity can be obtained by, for example, having a vacuum instead of atmospheric pressure on the low pressure side and using a Pirani gauge to monitor pressure change, but the apparatus then becomes more complicated and is probably of more interest for testing very low permeability plastic films than rubbers.

The permeability of the test piece is calculated from:

$$P = \frac{\mathrm{d}h}{\mathrm{d}t} \frac{V_0 \times d \times 273 \times 9.81 \times 10^3 \times \rho}{A \times p \times T \times 10^5}$$

where: dh/dt = rate of manometer rise (m/s),
V_0 = effective volume of low pressure side of the cell (m^3),
d = test piece thickness (m),
ρ = density of manometer liquid (Mg/m^3),
A = effective test piece area (m^2),
p = pressure difference across the test piece (Pa),
T = test temperature (K), and
10^5 = approximate atmosphere pressure (Pa).

17.1.2 Constant Pressure Method

The test cell is similar to that for the constant volume method, the test piece dividing the cell into high and low pressure cavities. The essential difference is that the low pressure side is connected to a device to measure the volume increase as gas diffuses to the low pressure side whilst maintaining constant pressure. In ISO 2782[6] and BS 903:Part A30[8] a graduated capillary tube is used to measure the volume change and the tube may be arranged either vertically or horizontally. In the vertical arrangement a U-tube capillary together with a reservoir is used so that the apparatus is effectively exactly the same as shown in Fig. 17.1. In the horizontal arrangement only a straight length of capillary is needed which contains a single drop of liquid which is pushed along as volume increases. The capillary cross section must be known to within 1% and areas between 0.7×10^{-6} to $2 \times 10^{-6}\ m^2$ are suggested.

The operation of the apparatus is very similar to the constant volume procedure. In the case of a horizontal capillary the movement of the liquid drop is monitored as a function of time. With a vertical capillary zero time is taken when the meniscus passes the datum line and readings of the level of the meniscus are taken as a function of time, the pressure being compensated for before each reading by adjusting the height of the reservoir to keep the height in the two legs of the tube equal.

As commented in Section 7.1.1, this type of apparatus requires considerable care to set up and operate. There is the added necessity in the constant pressure method to accurately graduate the capillary; however, when a horizontal capillary is used the apparatus is a little simpler than that for constant volume measurements.

The permeability of the test piece is calculated from:

$$P = \frac{dl}{dt} \frac{b273p_1a}{dt\, A(p_2 - p_1)T101\,300}$$

where dl/dt = rate of displacement of liquid in the capillary (m/s),
b = test piece thickness (m),
p_1 = pressure in low pressure side (Pa),
p_2 = pressure in high pressure side (Pa),
a = cross-sectional area of capillary (m^3),
A = effective test piece area (m^3),
T = absolute temperature (K), and
101 300 = standard atmospheric pressure (Pa).

17.1.3 Carrier Gas Methods

Carrier gas methods for measuring permeability are those where the quantity of gas passing through the test piece is estimated from the change in chemical composition of the gas mixture on the receiving side of the test piece. The test gas flows on one side of the test piece and a second gas, the carrier gas, flows on the other side and is quantitatively analysed to determine the quantity of test gas which has passed through the test piece. In such procedures there is no need for a pressure differential across the test piece.

This type of method offers several advantages over the pressure differential methods discussed previously. With little or no pressure differential there is less problem from leaks, no difficulty in supporting the test piece, and the situation is more like that in many packaging applications. Greater sensitivity is possible enabling very low permeability materials to be tested conveniently and the different transmission rates of the components of gas or vapour mixtures can be measured. The principal disadvantage is that the apparatus is relatively expensive.

Because of the greater sensitivity of carrier gas methods and their particular value in packaging applications, most developments have been for plastic films. However, an early standard method for hydrogen through rubber was given in BS 903 in 1950 (now discontinued) in which the carrier gas was air and the concentration of hydrogen was found by measuring the change in thermal conductivity of the gas mixture.

Apparatus using the measurement of thermal conductivity to estimate permeability has been described, for example, by Yasuda and Rosengren[10] and Pasternak *et al.*[11] The use of a gas chromatograph to measure the concentration of transmitted gas enables more sensitive detection devices to be used and several types of apparatus have been described.[12–14] Other detection systems have been used including absorption of light to detect sulphur dioxide[15] and an oxygen specific coulometric device.[16] A mass

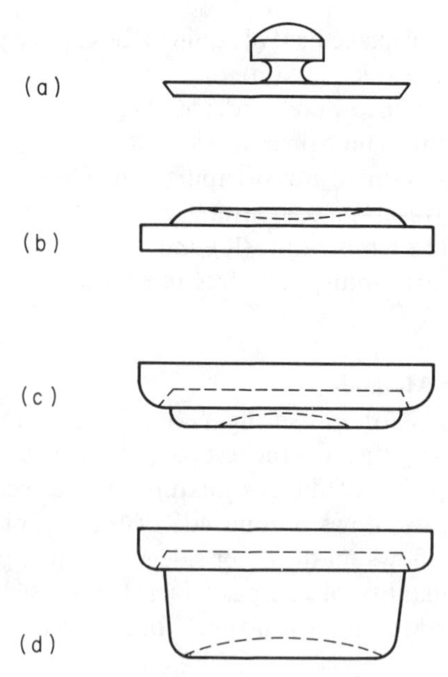

FIG. 17.2. Water vapour permeability dish. (a) Waxing template; (b) lid; (c) shallow dish for materials of normal permeability; (d) deep dish for materials of high permeability.

spectrometer offers the possibility of measuring several gases simultaneously.[17] All these procedures have been reviewed in detail by Lomax.[2,3]

17.2 VAPOUR PERMEABILITY

The commonest method of measuring vapour, especially water vapour, permeability is the dish method, detecting the quantity transmitted by change in weight. The apparatus illustrated in Fig. 17.2 is typical of that used for measuring water vapour permeability of sheet materials. A desiccant is placed in the dish and the test piece in the form of a thin disc is sealed with wax across the mouth of the dish, using a template to accurately define the effective test piece area. The dish assembly is then placed in a cabinet at a controlled humidity and weighed at intervals to measure the amount of water vapour transmitted and absorbed by the desiccant.

Such a method is detailed in ISO 2528[18] for sheet materials in general (ISO R1195 for plastics film has been withdrawn). To avoid leaks the wax seal must be applied very carefully using the templates specified, and the temperature and humidity during exposure of the sealed dishes must be controlled closely. The test must be continued until the increase in weight is substantially linear with time, i.e. equilibrium has been reached.

The result is always expressed as a transmission rate, not a permeability, and is hence dependent on test piece thickness. Generally, transmission rate is not a linear function of temperature or relative humidity and preferably test conditions are chosen to be as close as possible to those found in service.

Because of the existence of these international standard methods ISO TC 45 has decided not to publish an essentially identical method for rubbers.

The alternative to using a dish is to form the material into a bag and this so-called pouch or sachet method is often used for plastics films. The advantages are that a larger surface area is exposed, leaks through the wax seal are eliminated, and the conditions are more similar to packaging applications. It is less attractive for rubbers because they are not often used in that sort of packaging application and an alternative to heat sealing the pouch would be necessary.

Instead of putting the desiccant inside the dish, with a controlled humidity outside, the dish could contain water which is then transmitted out into a dry atmosphere and the amount transmitted measured by weight loss. By inverting the container the transmission rate when the water is in contact with the test piece can also be measured. The transmission rates measured by the various alternative procedures will be different because different vapour pressure gradients across the test piece are being used and, logically, the conditions most relevant to service would be chosen.

The procedure whereby the water is placed in the container is convenient for use with other volatile liquids and a standard method of this type has been published as ISO 6179[19] by TC 45. BS 903:Part A46[20] is identical. Despite the title of the international method it is intended for use with sheet rubber. A suitable apparatus is shown in Fig. 17.3, consisting of a lightweight aluminium container with a screw-on collar to retain the test piece. The rotating part of the collar applies pressure to the clamp ring through ball-bearings so that the test piece is not distorted when the collar is tightened. The two filling valves allow the liquid to be changed during test without disturbing the test piece and this is recommended

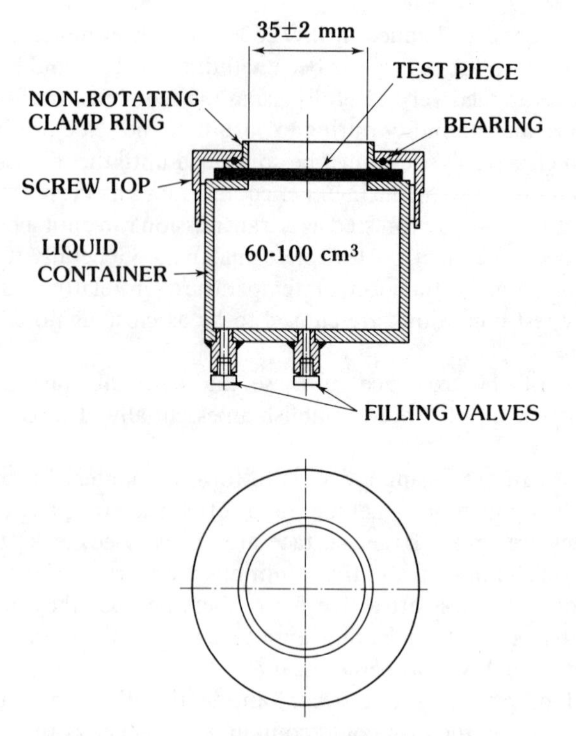

FIG. 17.3. Permeability cell for volatile liquids.

when a mixture of two or more liquids is used which are not transmitted at the same rate, so changing the properties of the liquid left in the cell.

In the standard two procedures are defined. In both cases the cells, after assembly, are inverted so that the liquid is in contact with the test piece and a conditioning period totally 96 h given. For procedure A the container is then emptied and re-filled at 24 h intervals until the weight loss per 24 h is effectively constant. In procedure B the weight loss is simply determined after 72 h. The standard does not warn that with procedure B equilibrium conditions have not been reached.

The various methods for vapour permeability discussed above are all essentially simple but require great care to achieve good reproducibility, are time consuming, and are not generally sensitive enough to measure very low transmission rates. Automation of the weighing process can be achieved by attaching the permeability cell to a balance arm.[21-23]

A considerable number of alternative techniques have been suggested for measuring vapour permeability of plastics and the following examples are by no means exhaustive. For water vapour, various moisture sensitive transducers can be used to detect the vapour transmitted[24,25] and such methods may reduce the time for testing, give increased sensitivity or be suitable for use with thick test pieces.[26] A rapid infra-red method has been compared with the gravimetric method[27] and the use of radioactive tracers is suggested for water[28] and has also been used for organic vapours. Attention is also drawn to the carrier gas methods using a gas chromatograph or a mass spectrometer as the detector described in Section 17.1.3 and to the review by Lomax.[2,3]

REFERENCES

1. Hennessy, B.J., Mead, J.A. and Stening, T.C. (1966). *The Permeability of Plastics Films*, Plastics Institute.
2. Lomax, M. (1980). *Polym. Test.*, **1**(2).
3. Lomax, M. (1980). *Polym. Test.*, **1**(3).
4. Yasuda, H. and Stannett, V. (1975). *Polymer Handbook*, 2nd Edn (Ed. J. Brandrup and E.H. Immergut), John Wiley and Sons.
5. ISO 1399, 1982. Determination of Permeability to Gases—Constant Volume Method.
6. ISO 2782, 1977. Determination of Permeability to Gases—Constant Pressure Method.
7. BS 903:Part A17, 1973. Determination of the Permeability of Rubber to Gases (Constant Volume Method).
8. BS 903:Part A30, 1973. Determination of the Permeability of Rubber to Gases (Constant Pressure Method).
9. Pye, D.G., Hoehn, H.H. and Panar, M. (1976). *J. Appl. Polym. Sci.*, **20**, 1921.
10. Yasuda, H. and Rosengren, K. (1970). *J. Appl. Polym. Sci.*, **14**, 2839.
11. Pasternak, R.A., Schimecheimer, J.F. and Heller, J. (1970). *J. Polym. Sci.*, **A–2**(8), 467.
12. Duskova, D. (1974). *Int. Polym. Sci. Tech.*, **1**(9), T/108.
13. Pye, D.G., Hoehn, H.H. and Panar, M. (1976). *J. Appl. Polym. Sci.*, **20**(2), 287.
14. Caskey, T.L. (Dec. 1967). *Modern Plastics*, 148.
15. Davies, E.G., Rooney, M.L. and Larkins, P.L. (July 1975). *J. Polym. Sci.*, **19**(7), 1829.
16. Gracie, J.W., SPE 33rd Annual Tech. Conf. Antec 75, Atlanta, GA., May 1975, p.592.
17. Ivashchenko, D.A., Krotov, V.A., Talakin, O.G. and Fuks, E.V. (Sept. 1972). *Vys. Soed.*, **14A**(9), 2109.
18. ISO 2528, 1974. Sheet Materials—Determination of Water Vapour Transmission Rate—Dish Method.

19. ISO 6179, 1981. Fabrics Coated with Vulcanised Rubber in Determination of Vapour Transmission of Volatile Liquids (Gravimetric Technique).
20. BS 903:Part A46, 1984. Determination of Vapour Transmission Rate of Volatile Liquids Through Rubber Sheet and Rubber Coated Fabrics.
21. Wang, Tsen Chen (Apr. 1975). *ACS Div. of Polym. Chem. Polym.*, Preprints 16, No. 1, 758.
22. Smit, F. (Feb. 1962). *Plastica*, **15**, 94.
23. Leslie, H.J. (Nov. 1963). *Packaging*, 82.
24. Rybicky, J. and Marton, J.P. (June 1975). *J. Paint Technol.*, **4**(605), 57.
25. Rudorfer, D., Pesa, O. and Tschamler, H. (Dec. 1973). *Mitt. Chem. Fiw. Ost.*, **27**(6), 299.
26. Hadge, R.G., Riddell, M.N. and O'Toole, J.L. (Dec. 1972). *SPE J.*, **28**(12), 58.
27. Bornstein, N.D. and Pike, L. ASTM STP548, Symposium New Brunswick, NJ, Apr. 1972, p.20.
28. Mayet, R., Margrita, R. and Mornas, A. (Oct. 1976). *Caout. Plast.*, No. 564, 85.

Chapter 18

Adhesion, Corrosion and Staining

Rubber is frequently used as a composite with other solids, for example in tyres, belting and coated fabrics, or may be in contact with other solids during use. The testing of composite materials or products containing rubber is, in general, outside the scope of this book but certain tests which are usually considered to be 'rubber tests' are included here. These are adhesion to metals, adhesion to fabrics, adhesion to cord, corrosion of metals and paint staining.

18.1 ADHESION TO METALS

Rubber is bonded to metal during processing to form a variety of products and in most cases a very strong bond is necessary for the product to perform satisfactorily. It is usually desirable to measure bond strength by testing the actual product but this is not always possible or convenient and, particularly for evaluating bonding systems, there is a need for tests using standard laboratory prepared test pieces. Whether the product or a test piece is used the bond should be strained in essentially the same manner as would occur in service, although this may be complex rather than, for example, in simple tension or shear.

Possible modes of straining for laboratory test pieces are illustrated in Fig. 18.1. With peel and direct tension tests, failure tends to occur in the rubber if the bond strength is high. It can be argued that if the bond is stronger than the rubber it is strong enough, but this attitude assumes that failure would be similar with another mode of straining and may not allow discrimination between a good and a very good bond. The tension test with cone-shaped metal end pieces was developed to encourage

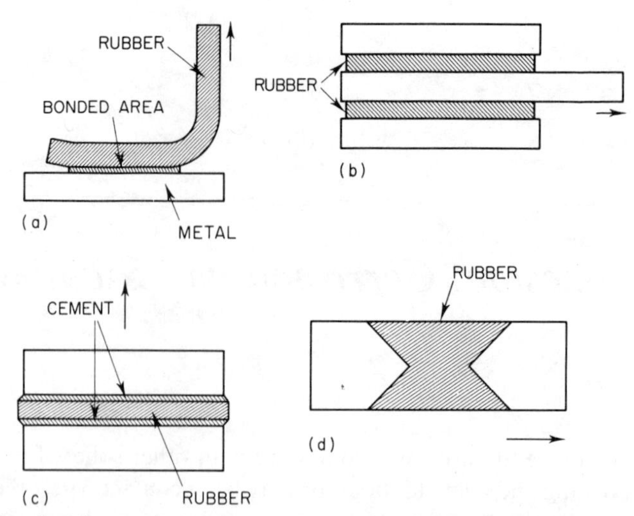

FIG. 18.1. Rubber to metal bond strength, modes of straining. (a) 90° peel; (b) shear; (c) direct tension; (d) tension with conical end pieces.

failure at the interface between rubber and metal because of a stress concentration at the tips of the cones. It is usual to report the type of failure as well as the numerical value of the bond strength. Symbols are commonly used as follows:

R = failure in the rubber.
RC = failure at the interface between the rubber and the cover cement.
CP = failure at the interface between the cover cement and the primer cement.
M = failure at the interface between the primer cement and the metal.

In practice it is not always possible to distinguish between RC and CP and in any case a single coat bonding system might have been used.

18.1.1 Standard Methods

Test methods have been standardised internationally for peel, direct tension, shear and 'tension' with conical ends.

PEEL TESTS

The method given in ISO 813[1] is a peel test using a test piece, 6 ± 0.1 mm thick and 25 ± 0.1 mm wide, bonded to a 1.5 mm thick metal strip along

25 mm of its length. The rubber is peeled at 50 ± 5 mm/min using (preferably) a low inertia tensile machine, having first started to strip the rubber from the metal using a sharp knife. This rather dubious procedure of cutting at the bond line is intended to lessen the probability of failure in the rubber and the standard states that if the rubber starts to tear during the test it shall be cut back to the metal.

The maximum force during stripping over 25 mm length expressed per mm of width is taken as the bond strength. It is suggested that an autographic recording of the strength over the 25 mm length is taken but no account taken of the possibility of the trace exhibiting several peaks and troughs. The measured adhesion strengths from peel tests are lower than those from tension tests and the theoretical aspects of this have been discussed by Kendall.[2]

The equivalent British standard BS 903:Part A21,[3] method B is essentially identical to ISO 813. The ASTM equivalent D429,[4] method B is very similar but uses slightly different test piece dimensions, these being direct metric equivalents of imperial sizes, and test speeds. Both the British and ASTM methods hint at extending the scope of the test by using a range of test temperatures or ageing treatments. The basic method is a closely specified test intended for quality control or the comparison of bonding systems but could readily be extended to investigate the effects of test piece dimensions, peeling angle, test speed, etc.

TENSION TESTS

The international standard method given in ISO 814[5] uses a disc test piece 3 ± 0.1 mm thick and between 35 and 40 mm in diameter, bonded to metal plates which are approximately 0.08 mm less in diameter than the rubber. The slightly smaller size of the metals is intended to prevent the rubber tearing from the edges of the metals during test.

The assembly is separated at 25 mm/min in a tensile machine and great care must be taken to ensure that the test piece is accurately aligned so that the tension is uniformly distributed over its cross section during test. Any misalignment will tend to introduce a peeling action. In practice, the stress at the rubber/metal interface does not remain even because shear forces are introduced as the rubber deforms under tension. Because of this the measured bond strength depends on the shape factor (see Section 8.3) of the rubber disc, the strength increasing with decreasing thickness.[6] The result is expressed as the maximum force divided by the cross-sectional area of the test piece.

The equivalent British standard, BS 903:Part A21, method A,[3] is essentially identical to ISO 814. The ASTM equivalent, D429, Method A,[4] is very similar but uses slightly different test pieces dimensions as noted for the peel tests described above.

A 'tension' method using conical metal end pieces is standardised in ISO 5600[7]. BS 903:Part A40[8] is identical and ASTM D429, Method C[4] is very similar. The test piece diameter is 25 mm and the cone angle 45° but the distance between the tips of the cones is 12 ± 1 mm in ISO and 11.5 ± 1.2 mm in ASTM. An earlier draft of ISO 5600 had the tolerance as ± 0.1 mm which perhaps implies that this dimension is critical. The grip separation rate is 50 mm/min which is double the speed used for the disc tension method. The result is simply expressed as the maximum force recorded.

The test was investigated by Painter[9] who showed that the stress is concentrated at the tips of the cones. The stress distribution is not even and the action is not pure tension but involves peel and shear forces. Painter's results showed that failure occurred at the interface rather than in the rubber and the measured strengths were lower than with a plain disc test piece of similar diameter, more in line with the results of peel tests.

SHEAR TESTS

A shear test using a quadruple element test piece is given in ISO 1747.[10] The double sandwich construction is intended to provide a very stiff test piece which will remain in alignment under high stresses. A double shear test piece (see Section 8.4.1) is being suggested for all shear tests on rubber and probably this would be satisfactory for adhesion measurements but it would seem sensible to conduct interlaboratory tests before its introduction. The present standard quadruple test piece uses rubber elements with the dimensions closely specified as 4 ± 0.1 mm thick and 20 ± 0.1 mm long, but the measured adhesion strength in shear is less affected by the test piece shape factor then tension tests[6] and these tolerances seem unnecessarily tight. The test piece is strained at a rate of 50 mm/min, in line with the speed for peel and conical end test pieces, and the result expressed as the maximum force divided by the total bonded area of one of the double sandwiches.

It is presently proposed in TC 45 to withdraw ISO 1747 but to modify the method for shear modulus (see Section 8.4.1) to allow the test to be taken to the failure point, i.e. effectively combining the methods. The

British method, which was virtually identical to ISO 1747 has in fact already been withdrawn simply because it was rarely used.

18.1.2 Other Methods

It is worth remembering, and this is applicable to adhesion to fabrics or any other substrate, that adhesion is only a tiny part of testing rubbers whereas it is the prime consideration for adhesives. Hence for a fuller understanding of the relevance and limitations of adhesion tests as well as for alternative test methods it is suggested that reference should be made to the great volume of literature from that industry.

With very good bonding systems it is often difficult to discriminate between the systems because of failure in the rubber and yet in service differences in performance may be evident. This situation was recognised by Buist *et al.*[6,11] who made comparisons of various methods and observed that in service, bonds may be subjected to impacts (i.e. high strain rates) or to repeated dynamic cycling (fatigue). Neither of these factors are considered in the standard methods discussed above.

Buist and Naunton[6] suggested impact methods based on Izod and falling weight apparatus, preferring the falling weight apparatus, with each test piece receiving a single blow. With the particular case of automobile bumpers in mind, Given and Downey[12] developed a high speed test using a double element shear test piece and a sophisticated servo-hydraulic universal test machine.

Impact methods can be used to test fatigue resistance of bonds by making repeated blows but this is not very convenient. Buist and Naunton[6] used the Goodrich Flexometer (see Section 12.1) with a modified test piece holder to fatigue bonds in tension and were able to discriminate between bonding systems which appeared equal in the standard tension test. Buist *et al.*[11] used the Goodrich Flexometer in compression prior to making the standard tension test and also developed a slow speed cycling test in shear. Beatty[13] used a modified 'Rotoflex' machine which fatigues the bond by bending. Modern universal tensile machines would seem very convenient for dynamically testing bonds in various modes of straining and at different strain rates but it is costly to utilise such machines in this way. Many *ad hoc* rigs have been constructed to test bonded components and it would seem reasonable to develop a fairly simple dynamic apparatus to fatigue standard test pieces. A variation of the shear test in which a metal rod is pulled out of a block of rubber has been described by Khromov and Yakovleva[14] and this test piece could presumably be tested dynamically. A BRMA publication,[15] which gives recommendations for testing rubber to metal bonded components in general, suggests conditions for carrying out dynamic tests.

18.1.3 Non-Destructive Tests

To be able to estimate bond strength by a non-destructive method is extremely attractive especially for quality control purposes. The possibility of using ultrasonics for this application has been recognised for a long time[16] and efforts have been made to standardise a procedure for the inspection of such components as engine mountings.

Ultrasonic flaw detection operates on the principle that the amount of ultrasonic energy transferred from one material to another is related to the difference between their acoustic impedances. For example, at a rubber/air interface there is a large difference in acoustic impedance and less ultrasonic energy will be transmitted than at a well-bonded rubber to metal interface. Hence, if there is an area of debonding at the rubber/metal interface and there is a thin layer of air or a vacuum between the two this can be detected by loss in the transmitted, or increase in the reflected, ultrasound. The basic technique has been described by Preston[17] who monitored the decay of internal reflections within the metal to increase the observed difference between a good and a bad bond.

Attractive and simple as the technique is in theory, in practice there are a number of difficulties which severely limit its value. Only areas of disbond, not a weak bond, can be detected although very weak areas can be made to part by pre-stressing, which is in any case necessary to separate the debonded areas. Even the detection of complete disbonds is limited; a disbond with a liquid at the interface or a break between the rubber and the bonding agent is often very difficult to find. In all cases great care has to be taken to eliminate variations in the coupling of the ultrasonic probe to the component and the angle at which the ultrasound strikes the bond to avoid masking the observed difference between a bond and no bond. Notwithstanding these remarks, there have been considerable developments in ultrasonic flaw detection in recent times although there has not been any widespread adoption of the technique in the rubber industry generally.

Other non-destructive tests have been suggested to estimate bond quality but such techniques as holography and radiography, and also ultrasonics, have mostly been used in the rubber industry for detection of flaws in tyres. It is not considered appropriate to cover non-destructive flaw detection in general here but reviews of applications in the polymer industry have been published[18–20] which contain in total 281 references.

18.2 ADHESION TO FABRICS

Rubber is used as a composite with textile fabrics in such products as belting and hose and also as a coating on the fabric to form 'proofed' materials.

Tests for adhesion are carried out in peel or direct tension, peel being the most common although tension tests are particularly useful for thin coatings where the rubber is too thin or too weak to successfully carry out a peel test.

18.2.1 Peel Tests

The international method, ISO R36,[21] for adhesion strength of rubber to fabrics uses a 25 mm wide strip test piece, long enough to permit separation over at least 100 mm. The fabric and rubber are separated by hand over a length of about 50 mm and the two ends placed in the grips of a tensile testing machine. The grips are separated at a rate of either 50 ± 5 mm/min or 100 ± 10 mm/min so as to give a rate of ply separation of 25 mm/min or 50 mm/min.

The angle between the two gripped 'legs' of the test piece is 180° (see Fig. 18.2). The plies should separate at a sharp angle but this will depend on the thickness and stiffness of the plies. The standard suggests that the thickness should be reduced if necessary so that the line of separation of the plies lies as closely as possible to the plane of the axis of the 'legs' of the test piece held in the grips. The unstripped portion of the test piece is left to find its own level during the test but variation in the angle (Fig. 18.2) will affect the measured result. The angle α depends on the relative stiffness of the plies B and C, the greater the stiffness ratio B/C the nearer the angle α approaches 180°. It would seem better to restrain the unpeeled portion A so that α is either 90° or 180°. Other possible forms of peel test are also shown in Fig. 18.2.

The stripping force is recorded continuously so that a trace as shown in Fig. 18.3 is obtained. How to obtain the adhesion value from this trace has been the subject of much debate and difficulty in reaching an agreement is the main reason why a revision of R36 has been delayed. However, consensus was reached and at the time of writing the new standard is in the course of publication. This revision refers to ISO 6133[22] which has three procedures, for traces having less than five peaks, 5–20 peaks and more than 20 peaks, respectively, which has already been discussed in Section 8.6.4. In the present (old) edition a rather complicated and by no means explicit method is used based on the mean of the lowest

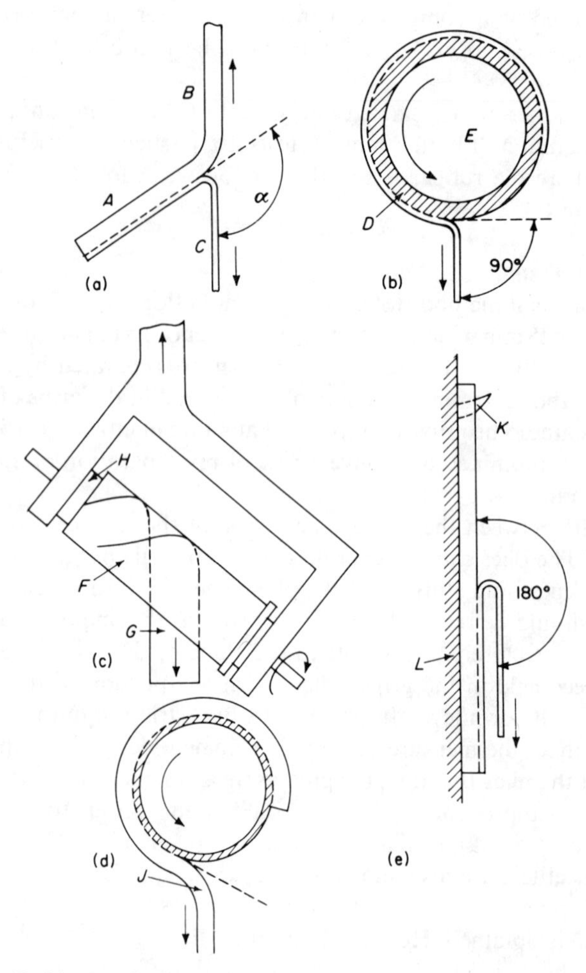

FIG. 18.2. Rubber to fabric peel test. (a) Strip test piece. A is the unstripped part, B is the remaining part, C is the separated ply; (b) ring test piece (D) mounted on mandrel (E); (c) test on hose with helical cord reinforcement. F is the test piece, G is the separated ply, H is the mandrel; (d) ring test piece with thick separated ply (J), giving acute angle of separation; (e) strip test piece supported at K against backboard (L) to ensure 180° separation.

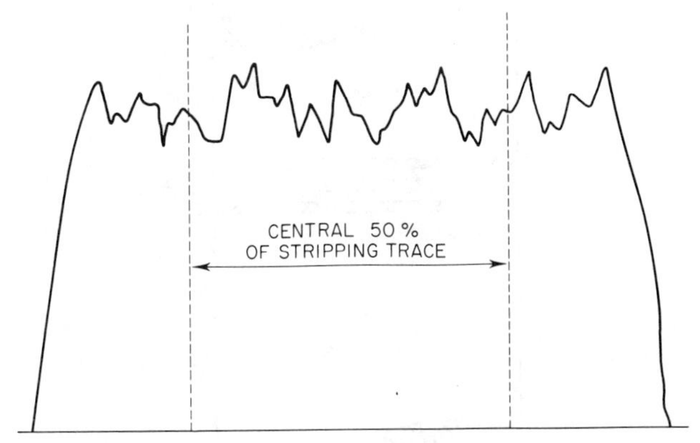

FIG. 18.3. Example of stripping force trace.

50% of peak values taken from the central 50% of the trace. The simplest, and a reasonable procedure which has not been standardised is to take the mean of the force trace and in practice many workers do this by drawing a line through the trace using a transparent rule. It should be noted that the peaks and especially the troughs are affected by the dynamic response for the testing machine.[23,24] For this reason only measuring systems having very low inertia should be used for this test.

The equivalent British standard BS 903:Part A12[25] is technically almost identical with ISO R36. It allows the use of ring test pieces of diameter smaller than 50 mm, which are peeled with a nominal angle of separation of 90°. The description of failure type is omitted but an alternative method of presentation of results is given for control test purposes which is simply the mean force level. It will doubtless be replaced by the new international method.

ASTM D413[26] contains a method similar to ISO R36 but has provision for holding the unpeeled portion such that there is 180° separation. The mean force level is taken as the adhesion strength. ASTM D413 also details methods for peeling the strip with the unpeeled portion held such as to give 90° separation, and for rings of up to 100 mm diameter.

ASTM D413 also gives a simple dead load method for adhesion strength whereby a mass, large enough to cause peeling, is hung from one leg of the test piece and the rate of separation noted. The problems of interpreting the results are discussed but tensile machines are common enough that there would seem to be little use for this type of procedure.

In addition to ASTM D413, a 'strap peel test' is specified in ASTM D2630.[27] Most of the standard is concerned with the details of moulding

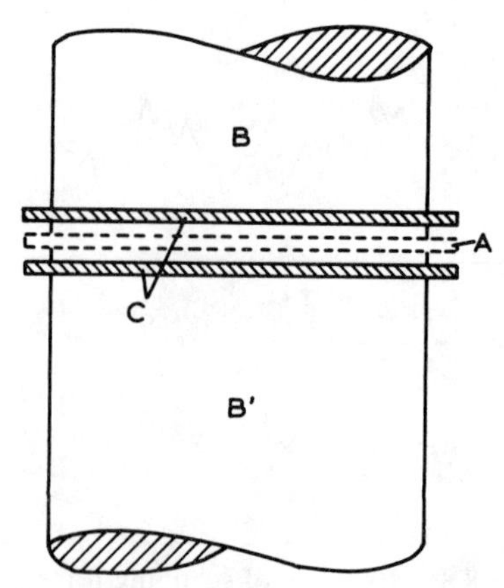

FIG. 18.4. Direct tension adhesion test: A is the rubber–fabric test piece; B and B′ are metal cylinders; C is cement.

a six-layer test piece. This test piece is optionally flexed using the Scott machine detailed in ASTM D430[28] before being peeled at 2.5 mm/min.

Because coated fabrics are generally dealt with in separate standards committees and because the thinner coatings are not strong enough to allow the use of the peel methods described above, separate standards have been developed for these products. The problem of failure in the coating is overcome by using reinforcements of fabric or cement. These methods are really product tests and outside the scope of this book but the appropriate references can be noted. The international method is ISO 2411,[29] the British methods are in BS 3424[30] and the ASTM method is D751.[31]

18.2.2 Direct Tension Tests

Borroff and Wake,[32] and later Meardon,[33] developed a direct tension method which more nearly measures the 'true' adhesion between fabric and rubber. It is particularly useful for discriminating between adhesive systems, when the peel tests can be misleading. The main objection to the method is practical in that the preparation of test pieces is rather difficult.

The method is covered in ISO 4637.[34] The test piece consists of two metal cylinders, 25 mm diameter, between which the composite to be tested is cemented (see Fig. 18.4). The metals are gripped in a tensile machine and separated at a rate of 50 mm/min and the maximum force recorded. The most important part of the test is the preparation of the metal/rubber/fabric test piece and international interlaboratory tests showed that unless very careful preparation of the metals was carried out failure occurred at the metal surface. ISO 4637 gives considerable detail on surface preparation, after machining the ends are lapped and degreased with trichloroethylene whilst the test piece is wiped with a solution of ammonium hydroxide in acetone. The assembly is cemented together with a cyanoacrylate adhesive using a special jig and it should be noted that the piece of fabric/rubber under test is a square of side approximately 32 mm and hence larger than the metal cylinders.

ISO 4637 was developed from the British standard BS 903:Part A27[35] which has itself now been revised to be identical to the international method. It is one of those regrettable lapses in standardisation that this revision had to wait 17 years and seven years after the international method was published.

18.2.3 Dynamic Tests

A dynamic ply separation test is really a fatigue test on the rubber/fabric composite to weaken the bond or to determine the number of cycles until the bond fails. In principle any flexing test (see Section 12.1) could be used but there is little evidence of particular methods being standardised. A noted exception is ASTM D2630, mentioned above, which refers to the Scott Flexer detailed in ASTM D430. In addition, ASTM D430 suggests that the Dupont Flexer can be used for testing composites. Tests for fatigue of coated fabrics are outside the scope of this book and are normally intended to induce cracking rather than coating separation.

The relative bond strengths of different rubber/textile systems may be different in dynamic and static tests. It would hence seem important to assess any composite that will be subjected to fatigue in service by a dynamic method and it is unfortunate that so few standard methods exist.

18.3 ADHESION TO CORD

The adhesion of cord, textile or metal to rubber is a specialised measurement in that virtually all interest centres on tyres and to some extent belting. Most static tests consist essentially of measuring the force to pull a cord out of a block of rubber into which it has been vulcanised and it is apparent that the result is critically dependent on the efficiency with which the test piece was moulded. The measured force is also dependent on the amount that the rubber deforms during the test.

A great deal of effort has been put into improving the procedure for test piece production and to finding the best test piece and supporting jig geometry and, largely because of various experts favouring different variations on the basic theme, progress to international standardisation has been slow.

The 'original' method is the 'H-pull' or 'H-block' test standardised in ASTM D2138[36] together with a variant, the 'U' test. In the former, two blocks of rubber are gripped in the tensile machine and in the latter a loop of cord is hooked onto one of the grips. ASTM D2229[37] gives a method for steel cord whereby the block of rubber is held in one grip and the cord held in the other grip. ASTM D1871[38] contains two methods of the same type for single strand wire together with a procedure whereby the rubber and wire are peeled apart.

An international version of the H-pull test for textile cord was published as ISO 4647[39] in 1982 and the method for steel cord, ISO DIS5603[40] is, at the time of writing, in the process of publication. BS 903:Part A48[41] is identical to ISO 4647. Factors affecting the measured adhesive strength and improvements to the standard methods have been discussed by, for example, Hicks *et al.*,[42] Skolnik[43] and Campion and Redmond.[44] DIS 5603 contains two methods of moulding, the second being equivalent to ASTM D2229 and the first results from the work of Campion in particular.

One of the main points of debate with the above methods is the stress distribution due to gripping the rubber block. Nicholson *et al.*[45,46] used a test with two cords embedded in the block of rubber and avoided holding the block in one grip of the testing machine. Further analysis was made by Brodsky[47] who used three cords. Ridha *et al.*[48] have calculated the stress fields in tyre cord adhesion test pieces and Mollet[49] has compared the various methods. Tyres are very definitely fatigued during use and, as mentioned for fabric/rubber adhesion above, it is very important to carry out dynamic tests to assess bond efficiency. Methods have not apparently been standardised but a variety of procedures have been reported.[50-56] Some workers have used the same or a similar test piece as in static tests and applied a cyclic tensile stress or strain whilst others have used some form of fatigue tester operating in compression/shear to repeatedly stress the cord/rubber composite or even to flex samples in the form of a belt.

18.4 CORROSION OF, AND ADHESION TO, METALS

Some rubber compounds can cause corrosion of, and tend to stick to, metal surfaces with which they are in contact and corrosion can even be caused to a metal in close proximity but not touching the rubber. Although not a very widespread problem, there has been sufficient concern, particularly for some military applications, for tests to be devised to assess the relative degree of corrosion and adhesion caused by different compounds.

Most tests are based simply on placing the rubber in contact with metal under load, ageing for a period under specified conditions and assessing corrosion and adhesion by visual inspection. It has proved rather difficult to obtain good reproducibility and it is essential that great care is given to the preparation of samples, in particular as regards cleanliness. An international standard method has been published as ISO 6505[57] in which rubber strips are sandwiched between the metals of interest (usually copper, brass, aluminium and mild steel) under a load of 10 kg and clamped. The sandwich is normally aged under relatively dry conditions, for example 7 days at 70 °C, and then visually examined for signs of adhesion or corrosion.

The same method is given in BS 903:Part A37[58] which also contains a second method for assessing the degree of corrosion when the rubber is not in contact with the metal. Zinc is used as the standard metal as this is fairly readily corroded. A strip of zinc and the rubber test piece are both suspended over distilled water in a stoppered container maintained at 50 °C. After a period of three weeks the corrosion products are removed from the zinc by immersion in chromium trioxide solution and the loss in weight used as the measure of degree of corrosion. Even more care has to be taken than in the contact method to avoid contamination and to obtain reproducible results.

18.5 STAINING

Paint, or other organic materials can become stained by rubber in contact with them or by water which contains leached out constituents of the rubber. Heat and/or light may intensify the degree of staining. The staining is caused particularly by such ingredients as anti-oxidants and the problem of the compounder is to achieve adequate environmental resistance without an unacceptable degree of staining. Staining is important in such consumer products as cars and kitchen appliances and, although no staining would be ideal, in practice some staining may have to be tolerated. Hence, tests to produce and measure staining are often included in specifications for products such as door seals.

A variety of test procedures have been in fairly widespread use for many years, virtually all of them being subjective in that visual assessment of the degree of staining is used. There has also been considerable confusion over the use of such terms as migration staining, which has been used differently in different commercial standards so that it was particularly helpful when an international standard was published.

As defined in ISO 3865:[59]

Contact stain is the stain which occurs on the surface directly in contact with the rubber.

Migration stain is the stain which occurs on the surface surrounding the contact area.

Extraction stain is the stain caused by contact with water containing leached-out constituents of the rubber.

Penetration stain is the staining of a veneer layer of an organic material bonded to the rubber surface.

It must be noted that the stain on the surface directly in contact with the rubber is always contact stain even if the stain has to be intensified by exposure to light after removal of the rubber.

To distinguish between colour changes caused by ageing of the paint rather than by the rubber a blank test assembly may be used. A blank is an assembly prepared and tested in the same manner as the samples under test but the rubber is replaced by an inert material such as aluminium.

By contrast, a reference sample is a test assembly using the rubber under test but which is protected from light during any irradiation exposure period and is hence used to distinguish between the effects of light and heat.

CONTACT AND MIGRATION STAIN

Contact and migration stain are generally measured at the same time. In ISO 3865 the rubber test piece in the form of a rectangle cut from sheet is sandwiched between painted metal panels (or other test material) and stored at 70°C for 24 h. One panel is then examined for both contact and migration stain and the other panel is exposed, without the rubber test piece, to artificial light before examination. If required, only the heat exposure may be carried out or, alternatively, the panel and test piece can be exposed to artificial light having omitted the heat ageing stage and the panel examined for migration staining. It is usual to carry out the full procedure.

EXTRACTION STAIN

Distilled water is dripped on to the test piece held at 30° to the horizontal so that the water runs down the test piece on to the painted panel. A segment of filter paper and a cotton cord conduct the water along the panel into a drip container. One litre of water is dripped over a period of 24 h after which the panel is examined for staining. Optionally, the panel can be exposed to artificial light before examination.

PENETRATION STAIN

A 0.5 mm thick veneer of white, non-discolouring rubber is applied under pressure to a sheet of the test rubber and the composite vulcanised. A test piece cut from the composite sheet is exposed to artificial light and examined for staining. The composition of the white veneer is left for agreement between the interested parties. Alternatively, part of a finished product incorporating a veneer may be used or the rubber test piece without the veneer coated with a paint.

18.5.1 Artificial Light Exposure

The preferred light source is a xenon lamp as used for artificial weathering tests and 24, 48 or 150 h is suggested as the exposure time. A mercury lamp is also allowed when exposure times of 2, 4 or 8 h are suggested. It is extremely doubtful that the two sources will give equivalent effects and it is rather unfortunate that the standard could not have been more precise. Variations in the light dosage and in surface temperature during exposure (given as 60 ± 15°C) are probably unimportant in cases where there is either no staining or heavy staining but in cases of dispute involving slight staining, care should be taken that only the effects of very similar light dosages are compared. Although the xenon arc is a more realistic source when compared to daylight, in practice the mercury lamp is often used because it is relatively inexpensive.

18.5.2 Assessment of Staining

ISO 3865 allows the degree of staining to be assessed by eye, by eye with the help of a grey scale, or using a reflectance spectrophotometer but in practice the purely visual method is almost exclusively used. Many interlaboratory disputes result due to lack of objectivity on the part of the operator, particularly if it is his material which has stained badly. In principle the instrumental method is much superior and interlaboratory tests have shown that relatively simple spectrophotometers can give satisfactory results. Unfortunately, due to the small area of staining, simple spectrophotometers are not suitable for assessing migration and extraction stains and for many applications these are the most important types of staining.

18.5.3 Other Standards

BS 903:Part A33[60] is identical to ISO 3865. ASTM D925[61] has methods for contact, migration and diffusion (equivalent to penetration) staining. The contact stain method is similar to ISO 3865 except that there is no provision for developing the stain by exposure to light. Migration stain is not measured at the same time.

The migration stain method omits any exposure to heat so that the paint panel and rubber are exposed only to artificial light. This separation of procedures for contact and migration stain tends to give the impression that contact stain is produced by heat and migration stain by light, which is not true. Light will intensify both contact and migration stain. Only the mercury type light sources are specified in ASTM and the requirements are given in considerable detail. This is obviously desirable from a standardisation point of view but it was found difficult to reach satisfactory agreement at the international level.

The ASTM diffusion (penetration) stain method is similar to that in ISO 3865 but more detail is given on preparation of the veneer, including a formulation.

There have been suggestions that further method is needed to measure staining caused by the close proximity of the rubber without actual contact. This type of staining can occur, for example, in the boot of a car due to airborne migration of constituents of the spare tyre and tests have been devised by individual motor companies.

REFERENCES

1. ISO 813, 1974. Determination of Adhesion to Metal—One Plate Method.
2. Kendall, K. (Jan. 1973). *J. Adhesion*, **5**(1), 77.
3. BS 903:Part A21, 1974. Determination of Rubber to Metal Bond Strength.
4. ASTM D429-82. Adhesion to Rigid Substrates.
5. ISO 814, 1974. Determination of Adhesion to Metal—Two Plate Method.
6. Buist, J.M. and Naunton, W.J.S. (1950). *Trans. IRI*, **25**, 378.
7. ISO 5600, 1979. Determination of Adhesion to Rigid Materials Using Conical Shaped Parts.
8. BS 903:Part A40, 1980. Determination of Adhesion to Rigid Materials using Conical Shaped Parts.
9. Painter, G.W. (1959). *Rubber Age NY*, **86**, 262.

10. ISO 1747, 1976. Determination of Adhesion to Rigid Plates in Shear—Quadruple Shear Method.
11. Buist, J.M., Meyrick, T.J. and Stafford, R.L. (1956). *Trans. IRI*, **32**, 149.
12. Given, D.A. and Downey, R.E., Polym. Sci. Tech., 9A, ACS Symposium Philadelphia, Pa., Apr. 1975, p.315.
13. Beatty, J.R. ACS Division. Rubber Chemistry Fall Meeting, 1973, Preprint 42.
14. Khromov, M.K. and Yakovleva, T.L. (1976) *Kauch. i Rezina*, No. 12, 48.
15. *Testing of Rubber to Metal Bonded Components*, BRMA Publication, 1982.
16. Heughan, D.M. and Sproule, D.O. (1953). *Trans. IRI*, **29**(5), 255.
17. Preston, T.E. (Mar. 1970). *Br. J. NDT.*, 17.
18. Trivisonno, N.M. (1985). *Rubb. Chem. Tech.*, **58**(3).
19. Bergen, H., ACS Rubber Division Meeting, Minneapolis, 1981, paper 44.
20. Scott, I.G. and Scala, C.M. (1982). *UDT Int.*, **15**(2).
21. ISO R36, 1969. Determination of the Adhesion Strength of Vulcanised Rubbers to textile Fabrics.
22. ISO 6133, 1981. Analysis of Multi-Peak Traces Obtained on Determinations of Tear Strength and Adhesion Strength.
23. Pickup, B., Proc. 3rd Rubb. Tech. Conf., Cambridge, 1954, p.439.
24. Eagles, A.E. and Norman, R.H. (1958). *Rubb. J. Int. Plastics*, **135**, 189.
25. BS 903:Part A12, 1975. Determination of the Adhesion Strength of Vulcanised Rubbers to Fabrics (Ply Adhesion).
26. ASTM D413–82. Adhesion to Flexible Substrate.
27. ASTM D2630–83. Adhesion to Fabrics (Strap Peel Test).
28. ASTM D430–73. Dynamic Fatigue.
29. ISO 2411, 1973. Fabrics Coated with Rubber or Plastics—Determination of the Coating Adhesion.
30. BS 3424:Part 7, 1982. Coating Adhesion Strength.
31. ASTM D751–79. Methods of Testing Coated Fabrics.
32. Borroff, E.M. and Wake, W.C. (1949). *Trans. IRI*, **25**, 199.
33. Meardon, J.I. (1962). *Rubb. and Plast. Weekly*, **142,** 107.
34. ISO 4637, 1979. Determination of Rubber to Fabric Adhesion—Direct Tension Method.
35. BS 903:Part A27, 1986. Determination of Rubber to Fabric Adhesion—Direct Tension Method.
36. ASTM D2138–83. Adhesion to Textile Cord.
37. ASTM D2229–80. Adhesion to Steel Cord.
38. ASTM D1871–84. Adhesion to Single Strand Wire.
39. ISO 4647, 1982. Determination of Static Adhesion to Textile Cord—H-pull Test.
40. ISO DIS5603. Determination of Adhesion to Wire Cord.
41. BS 903:Part A48, 1984. Determination of Static Adhesion to Textile Cord—H-pull Test.
42. Hicks, A.E., Chirico, V.E. and Ulmer, J.D. (1972). *Rubb. Chem. Tech.*, **45**(1), 26.
43. Skolnik, L., ACS Division of Rubber Chemistry Fall Meeting, Denver, Oct. 1973, Paper 13.
44. Campion, R.P. and Redmond, G.B. (Feb. 1975). *Rubb. Ind.*, **9**(1), 19.

45. Nicholson, D.W. Livingston, D.I. and Fielding-Russell, G.S. (1978). *Tire Sci. Tech.*, **6**(1).
46. Fielding-Russell, G.S., Livingston, D.I. and Nicholson, D.W. (1980). *Rubb. Chem. Tech.*, **53**(4).
47. Brodsky, G.I. (1984). *Rubb. World*, **190**(5).
48. Ridha, R.A., Roach, J.F., Erickson, D.E. and Reed, T.F. (1981). *Rubb. Chem. Tech.*, **54**(4).
49. Mollet, J.R. Int. Rubb. Conf., Paris, June 2–4, 1982, Paper 012.
50. Lunn, A.C., Evans, R.E. and Ong, C.J., ACS Meeting, Cleveland, Oct. 13–16 1981, Paper 20.
51. Bourgois, L., Davidts, J. and Schittescatte, M. (1973). *Text. Inst. Ind.*, **11**(1).
52. Campion, R.P. and Corish, P.J. (1977). *Plast. Rubb. Mat. Appln.*, **2**(2).
53. Kachur, V. and Weaver, E.J., ACS Meeting, Minneapolis, Apr. 1976, Paper 2.
54. Hewitt, N.L. (1985). *Rubb. Plast. News*, **15**(5).
55. Vorachek, J.J., Causa, A.G. and Fleming, R.A., ACS Meeting, New Orleans, Oct. 7–10, Paper 60.
56. Bourgois, L., ASTM Special Publication 694, 1978.
57. ISO 6505, 1984. Determination of Adhesion to, and Corrosion of, Metals.
58. BS 903:Part A37, 1979. Assessment of Adhesion to and Corrosion of Metals.
59. ISO 3865, 1983. Methods of Test for Staining in Contact with Organic Material.
60. BS 903:Part A33, 1985. Methods of Test for Staining in Contact with Organic Material.
61. ASTM D925–73. Staining of Surfaces (Contact, Migration and Diffusion).

Index